Viral Hepati

A Handbook for Clinicians and Scientists

Viral Hepatitis

A Handbook for Clinicians and Scientists

Elizabeth Ann Fagan MSc, MD, MRCP, FRCPath

*Departments of Internal Medicine & Pediatrics, Rush–Presbyterian–St Luke's
Medical Center, Chicago, IL 60612-3824, USA*

and

Tim J. Harrison PhD, FRCPath

*Royal Free and University College Medical School, University College London,
Department of Medicine, Royal Free Campus, London, NW3 2PF UK*

First published 2000

A CIP catalogue record for this book is available from the British Library.

ISBN 1 85996 0251

BIOS Scientific Publishers Ltd
9 Newtec Place, Magdalen Road, Oxford OX4 1RE, UK
Tel. +44 (0)1865 726286. Fax +44 (0)1865 246823
World Wide Web home page: http://www.bios.co.uk/

Published in the United States of America, its dependent territories and Canada by Springer-Verlag New York Inc., 175 Fifth Avenue, New York, NY 10010-7858, in association with BIOS Scientific Publishers Ltd.

Contents

Preface

We cast our minds back to recall the difficulties we faced as novices in viral hepatitis when trying to understand the field from our different viewpoints. A physician/scientist could benefit from applying scientific rationale to underscore clinical decision-making. A scientist interested in answering clinical questions requires a minimum working knowledge of medicine in order to prioritize scientific research and interpret the field alongside clinical colleagues. Although there are several excellent texts that address viral hepatitis, most take a conventional approach to the subject—as viewed primarily by the hepatologist or virologist.

Instead, this book aims to blend clinical and scientific aspects of viral hepatitis to assist non-specialized physicians and scientists to share across their disciplines and relate their knowledge to this vast and rapidly advancing field.

As the book evolved, we added another dimension—to address transatlantic differences in common terminologies and ways of approaching the subject. In the USA, preceptorship courses in viral hepatitis are custom-designed to address specific topics, such as hepatitis C, for specific interest groups. Representatives from the pharmaceutical companies typically seem to enjoy the clinical aspects of the course—meeting a patient and talking about their liver tests, diagnosis and treatment options—whereas, clinical gastroenterologists seem to appreciate especially an explanation of the virology when placed in a clinical context.

This book should be a useful addition to preceptorship courses aimed at pharmaceutical representatives and scientists that interact with clinically oriented personnel working in viral hepatitis. Equally, junior medical faculty and house-staff and non-specialist clinicians/internists should find this book useful, especially as an introduction to scientific aspects of the field. From the scientific perspective, graduate students and post-doctoral scientists working on viral hepatitis should appreciate the opportunity to get to grips with the myriad of clinical signs, symptoms and tests associated with these infections. We hope that this book will provide 'something for everyone' interested in viral hepatitis and complement the many excellent larger texts devoted to virology and hepatology.

We thank Dr Jonathan Ray at Bios for his patience as he watched the book evolve.

Abbreviations

3TC	the – enantiomer of 2′ deoxy – 3′ thiacytidine (lamivudine)
aa	amino acid
AFP	alpha fetoprotein
AIDS	acquired immune deficiency syndrome
AIH	autoimmune hepatitis
ALF	acute liver failure
ALT	alanine aminotransferase (=SGPT)
AMA	anti-mitochondrial antibodies
ANA	anti-nuclear antibodies
Anti HBc	antibody to the HBV core (nucleocapsid) protein, HBcAg
Anti HBe	antibody to the HBV e antigen, HBeAg
Anti HBs	antibody to the HBV surface protein, HBsAg
ARA-AMP	adenosine arabinoside monophosphate
AST	aspartate aminotransferase (=SGOT)
bp	base pair
BCP	basal core promotor
bDNA	branched DNA
bHCG	beta-human choriogonadotrophin
CAH	chronic active hepatitis
CBC	complete blood count (synonym, WBC: whole blood count)
ccc	covalently closed circular
CDC	Centers for Disease Control and Prevention
cDNA	complementary (copy) DNA
CEA	carcinoembryonic antigen
CIFN	consensus interferon
CLH	chronic lobular hepatitis
CTL	cytotoxic T-lymphocyte
CMV	cytomegalovirus (human herpesvirus 4)
CPH	chronic persistent hepatitis
DEPC	diethyl pyrocarbonate
DHBV	(Pekin) duck hepatitis B virus
DHF	dengue hemorrhagic fever

DNA	deoxyribonucleic acid
DR	direct repeat
DSS	dengue systemic shock
EBV	Epstein–Barr virus (human herpesvirus 5)
EGS	external guide sequence
EIA	enzyme immunoassay
ELISA	enzyme-linked immunosorbent assay
EPI	expanded program of immunization
FCH	fibrosing cholestatic hepatitis
G6PD	glucose-6-phosphate dehydrogenase
GBV	GB virus (named after patient GB with acute hepatitis)
geq	genome equivalents (copies)
GGT	gammaglutamyl transpeptidase
GM-CSF	granulocyte-macrophage colony-stimulating factor
GSHV	ground squirrel hepatitis virus
HAAg	hepatitis A antigen (determines single serotype)
HAI	histological activity index
HAV	hepatitis A virus
HBcAg	hepatitis B core antigen
HBeAg	hepatitis B e antigen
HBsAg	hepatitis B surface antigen
HBIG	hepatitis B immune globulin
HBV	hepatitis B virus
HCC	hepatocellular carcinoma (syn. hepatoma, primary liver cancer; PLC)
HCV	hepatitis C virus
HDAg	hepatitis D (delta) antigen
HDV	hepatitis D virus (the delta agent)
HEV	hepatitis E virus
HGV	hepatitis G virus (also known as GBV-C)
HHV-6	human herpesvirus 6
HIV	human immunodeficiency virus
HLA	the human major histocompatibility complex
HSV	herpes simplex virus
IFN	interferon
Ig	immunoglobulin
IG	immune globulin
IL	interleukin
INR	International Normalized Ratio (this is not the prothrombin ratio)
IRES	internal ribosome entry site
ITU	intensive therapy (care) unit (synonymous with MICU)
IVDU	intravenous drug use(r)
kb	kilobase (pairs), 1000 nucleotides
LFT	liver function tests
LKM	liver, kidney, microsomal (autoantibodies)
LOHF	late onset hepatic failure

mAb	monoclonal antibody
M_r	relative molecular mass
mRNA	messenger RNA
NAC	N-acetyl cysteine
NASBA	nucleic acid sequence-based amplification
non-A–E	non-A, non-B, non-C, non-D, non-E
NSAID	non-steroidal anti-inflammatory drugs
nt	nucleotide (pair)
NS	non-structural
NTP	nucleoside triphosphate
ORF	open reading frame
PAN	polyarteritis nodosa
PBL	peripheral blood lymphocyte
PBMCs	peripheral blood mononuclear cells
PCR	polymerase chain reaction
PEG-IFN	polyethylene glycol interferon alpha-2a
PHLS	Public Health Laboratory Service
PLC	primary liver cancer (syn. HCC)
PT	prothrombin time
PTH	post-transfusion hepatitis
Rf	rheumatoid factor
RIBA	recombinant immunoblot assay
RNA	ribonucleic acid
r.p.m.	revolutions per minute
RT–PCR	reverse transcription–polymerase chain reaction
SBP	spontaneous bacterial peritonitis
s.c.	subcutaneous
SGOT	serum glutamic oxaloacetic transaminase (=AST)
SGPT	serum glutamic-pyruvic transaminase (=ALT)
SLE	systemic lupus erythematosus
SMA	smooth muscle antibodies
SOD	superoxide dismutase
t.i.w.	three times each week (caution: this abbreviation could indicate twice each week)
TSH	thyroid stimulating hormone
TTV	TT virus (note: not an acronym for transfusion transmitted)
USSD	ultrasound scanning determination
UTR	untranslated region
VZV	varicella-zoster virus (human herpesvirus 3)
WBC	white blood cell
WHO	World Health Organization
WHV	woodchuck hepatitis virus
WMHBV	Woolly monkey hepatitis B virus
wt	wild type
YFV	yellow fever virus

To our families and other friends on either
side of the Atlantic Ocean

Chapter 1

General Introduction

1.0 Issues

Viral hepatitis in humans is caused by at least five viruses (*Table 1.1*) for which the liver is the major site of replication. A variety of other viruses (*Table 1.2*) occasionally show marked hepatotropism. The symptoms of acute and persistent virus replication in the liver are remarkably consistent despite the diversity of viruses involved.

Hepatitis A virus (HAV) and hepatitis E virus (HEV) are considered together in Chapter 2. Both are small, non-enveloped viruses with positive-sense RNA genomes of around 7.5 kb. These viruses are spread by the fecal–oral route and probably replicate in the intestinal tract prior to infecting the liver. The usual outcome is an acute hepatitis that resolves without evidence of chronic liver disease. Occasionally, viral replication may be protracted. Rarely, the acute hepatitis may pursue a fulminant course (acute liver failure, Chapter 5).

Hepatitis B virus (HBV) is considered in Chapter 3 with hepatitis D virus (HDV), the delta agent, which relies on HBV for transmission. HBV is a small (3.2 kb DNA genome), enveloped virus with a replication strategy similar to that of the retroviruses. HBV is spread predominantly by blood (parenteral) and sexual contact. HBV appears to replicate almost exclusively in the liver, although viral nucleic acids and antigens may be detected in peripheral white blood cells. HBV may cause persistent infections, especially following infection of babies and young children, and virus replication may persist for life. HBV infections persist in only around 1–5% of healthy adults. Liver damage relates mostly to immune-mediated mechanisms of the host, through lysis of infected hepatocytes by cytotoxic T lymphocytes. Chronic hepatitis may ensue, leading to cirrhosis and hepatocellular carcinoma.

HDV is an unusual agent, with similarities to plant viroids, and the smallest virus (1.7 kb circular RNA genome) known to infect man. HDV is found in association with HBV, its helper virus, which provides the viral envelope (HBsAg) required for transmission. HDV may infect simultaneously with HBV (co-infection) or superinfect individuals already persistently infected with HBV.

1

Hepatitis C virus, identified in the 1980s, is the major cause of parenterally-transmitted non-A, non-B hepatitis and is discussed in Chapter 4. GB virus C (GBV-C), also known as hepatitis G virus (HGV), was cloned in 1995 and resembles HCV in being transmitted predominantly by parenteral routes. These small, enveloped viruses are members of the *Flaviviridae* with positive-sense RNA genomes of around 9.4 kb. HCV causes persistent infections in the majority of cases and these may progress to chronic hepatitis, cirrhosis and primary liver cancer. The pathogenic potentials of HGV (GBV-C) are unclear (Chapter 8).

Many of the human herpesviruses, especially Epstein–Barr virus (EBV) and cytomegalovirus (CMV), are common viral pathogens. They cause overt hepatitis only under specific conditions as in the immunosuppressed host, including liver graft recipients (Chapter 6) and individuals infected with human immunodeficiency virus (HIV-1, Chapter 8) and in pregnancy and young children (Chapter 7). HIV-1 does not cause hepatitis *per se* but viral hepatitis is common in HIV-infected individuals, reflecting shared risk factors. Unusual viral infections of the liver occur in HIV-infected individuals as part of the spectrum of opportunistic infection.

Other 'exotic' viruses that may infect the liver are considered in Chapter 8. Yellow fever virus (YFV) and Rift Valley Fever virus, among others, can cause severe hepatitis. As transmission is by mosquitoes, most cases are restricted to tropical areas. The advent of international travel and encroachment of tropical forests by urbanization have encouraged the spread of these agents, presenting as hepatitis in the returning traveller.

Hepatitis in pregnancy and pediatric practice requires special consideration (Chapter 7). Ubiquitous viruses, especially the herpesviruses, occasionally cause significant hepatitis. Also, the propensity for chronic carriage of HBV is high following transmission to new-born babies and infants. Universal immunization against hepatitis B would reduce significantly the high social cost of chronic liver disease and primary liver cancer within the next one or two generations, despite lack of a vaccine against HCV.

Hepatotropic viruses occasionally cause acute liver failure (ALF, also referred to as fulminant hepatitis, Chapter 5). Survival without transplantation depends on viral etiology. Prediction of patients with ALF most likely to recover without grafting is crucial to rationalizing the very limited supply of donor livers. Liver transplantation (Chapter 6) is the final management option also for end-stage cirrhosis from chronic viral hepatitis. Selection of transplant candidates depends on viral etiology because survival of the graft and recipient depends on prevention of recurrence in the graft.

1.1 Major viruses (A–E)

Failure to grow viruses such as HBV and HCV in cell culture precludes classification using traditional virological methods such as classical serotyping and characterization of cytopathic properties *in vitro* (Table 1.1).

Table1.1. The major hepatotropic viruses

	HAV	HBV	HCV	HDV	HEV
Family Genus	*Picornaviridae* Hepatovirus	*Hepadnaviridae*	*Flaviviridae* (Hepacivirus)	Unclassified (Viroid-like)	*Caliciviridae* (Hepevirus)
Size (nm)	27	42	36	36–40	27–38
Poly A⁺	Yes	–	(poly U)ª	No	Yes
Nucleic acid	ssRNA	dsDNAᵇ	ssRNA	ssRNA	ssRNA
Genome size (kDa)	7.5	3.2	9.4	1.7	7.5
Genome structure	Linear	Circular	Linear	Circular	Linear
Open reading frames (ORFs)	One	Four	One	One	Three
Structural proteins	VP1–4	HBcAg HBsAg	Nucleocapsid E1 (gp35) E2 (gp70)	HDAg (HBsAg)	ORF2 (ORF3ᶜ) products

ª The untranslated region at the 3′ end of the HCV genome contains a polyU tract upstream of 98 nt of novel sequence (known as the 3′ X region) which is highly conserved among subtypes. ᵇ The HBV genome contains a single stranded region. ᶜ A structural role of the ORF3 product remains to be proven.

HAV, family *Picornaviridae* (genus hepatovirus), are 27 nm naked icosahedral particles containing positive-sense, 7.5 kb single-stranded RNA. They are spread by the fecal–oral route and cause acute hepatitis.

HBV, family *Hepadnaviridae* [hepadnaviridae are divided into orthohepadnaviridae (mammals) and avihepadnaviridae (birds)], are 42 nm enveloped particles containing 3.2 kb, partially single-stranded, circular DNA. They are spread by the parenteral route and cause acute and chronic hepatitis.

HCV, family *Flaviviridae* (candidate genus hepacivirus) are 35 nm enveloped particles containing positive-sense, 9.4 kb single-stranded RNA. They are spread by the parenteral route and cause mainly chronic hepatitis.

HDV, (the delta agent), is an unclassified, viroid-like entity with 36 nm enveloped particles containing 1.7 kb circular RNA base-paired into a rod-like structure. HDV is defective: it requires the HBV helper function—HBsAg envelope—for transmission. It is spread by the parenteral route and causes acute and chronic hepatitis in co-infection or superinfection with HBV.

HEV, family *Caliciviridae* (candidate genus hepevirus, but possesses characteristics of alphavirus supergroup and may be reclassified) are 28–34 nm naked particles containing positive-sense, 7.5 kb single-stranded RNA. They are spread by the fecal–oral route and cause acute hepatitis.

Table 1.2. Viruses and candidates causing hepatitis in humans

Major hepatotropic viruses
  Hepatitis A, B, C, D, E viruses

Hepatitis non-A through non-E
    GBV-C/HGV (controversial)
    TT virus (very controversial)
    SEN virus (limited information available)
    'Hepatitis F' (candidate viruses)

Herpesviruses
  Herpes simplex virus (HSV-1)
  Varicella zoster virus (VZV)
  Human herpesvirus 6 (HHV-6)
  Cytomegalovirus (CMV)
  Epstein–Barr virus (EBV)

Exotic viruses
  Flaviviridae
    Yellow fever virus
    Dengue virus
  Arenaviridae
    Lassa fever virus
    Junin virus
    Machupo virus
  Filoviridae
    Ebola virus
    Marburg virus
  Bunyaviridae
    Rift Valley Fever virus

} These viruses cause hemorrhagic disease hepatic involvement varies

Occasional causes of viral hepatitis
  Measles virus
  Adenoviruses
  Echoviruses
  Coxsackieviruses
  Influenza virus
  Parvoviruses
  Reoviruses
  Mumps virus

1.2 Epidemiology—the paradox of viral hepatitis

Viral hepatitis A, B and non-A–E are diseases reportable (notifiable) to the Public Health authorities in most countries. Unfortunately, serious under-reporting has undermined efforts to raise awareness of the massive global burden and adverse impact throughout the world (*Table 1.3*). Overwhelmingly, the burden of viral hepatitis is carried by the developing world. The clinical outcomes and durations of immunity following exposure to the major hepatotropic agents (A–E) in childhood determine the burden of infection in adults.

HAV and HEV are highly endemic in regions of the world where standards of sanitation are poor, supplies of clean drinking water limited and overcrowding common. Consequently, most children will have had exposure to these enterically-transmitted viruses before 5 years of age. Although most cases of viral hepatitis in children are

Table 1.3. Burden of acute and chronic viral hepatitis, 1991–1994, Sentinel Counties USA (Centers for Disease Control)[a]

	HAV	HBV	HCV	HDV
Acute hepatitis (annual × 1000)	125–200	140–320	35–180	6–13
ALF (total deaths)	100	150	NK	35
Chronic hepatitis (annual × 10⁶)	0	1–1.25	3.9	0.7
Chronic liver disease (total deaths, annual)	0	5000	8–10 000	1000

[a] Almost all of the few cases of symptomatic hepatitis E in the USA are in travelers returning from regions of high endemicity, such as Mexico and India. Rare cases without travel have been reported with distinct US genotypes of HEV detected in man, as well as domestic swine. NK, not known.

Key Notes Acute and chronic viral hepatitis

Acute and chronic viral hepatitis are diseases notifiable to the public health authorities regulated by government and state, such as the Public Health Laboratory Service (PHLS) in the UK and Centers for Disease Control (CDC) in the USA

---

symptomless, epidemics and outbreaks occur causing significant morbidity and mortality. Hepatitis E seems to be more common than hepatitis A in adults in certain developing countries, as HAV, but not HEV, infection in childhood results in life-long immunity. Conversely, hepatitis A is more common than hepatitis E among adults residing in the developed countries. A decline in herd immunity with improvements in hygiene is coupled with an increase in risk of exposure, especially through international travel. Also, symptoms of hepatitis A infection can be more severe with increasing age. Large outbreaks of hepatitis A and E are favored following contamination of a common source, such as drinking water or shellfish, with a high concentration of virus. In endemic countries, HEV is also a major cause of sporadic hepatitis and these isolated cases may help to maintain the reservoir of virus between epidemics.

Outbreaks of hepatitis A that occur in developed countries tend to be smaller, despite the lower level of herd immunity. Control of sanitation prevents large-scale spread, but infection can be maintained by secondary spread. Presumably humans and possibly also livestock and rats (for HEV) maintain the reservoir of these viruses. The socioeconomic burdens of hepatitis A and E are underestimated in the shadow of hepatitis B and C that cause chronic liver disease.

The majority of the estimated 350 million carriers of the hepatitis B virus live in Asia and sub-Saharan Africa (*Figure 1.1*). Within these continents, around 10% of the

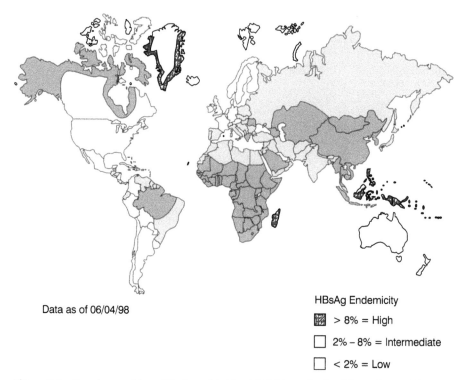

Data as of 06/04/98

HBsAg Endemicity

    > 8% = High

    2% – 8% = Intermediate

    < 2% = Low

Figure 1.1. Prevalence of hepatitis B worldwide, 1997 (data obtained from the World Health Organization at URL: www.who.int/gpv-surv/graphics/htmls/hepbprev.htm).

population are persistently infected with HBV (HBsAg carriers) and the majority of the remainder have evidence of past exposure (anti-HBs and/or anti-HBc). Viral hepatitis is a leading cause of cirrhosis worldwide. End-stage cirrhosis due to chronic hepatitis C ranks with alcoholic liver disease as the most common indication for liver transplantation in the USA and UK.

The prevalence of hepatitis C is reported to be falling in the West especially with the widespread screening of blood donors introduced in the 1990s. However, the true prevalence remains unknown as most epidemiological data derive from blood donors who self-select against high parenteral risk; groups with high-risk behavior are difficult to reach. Worldwide, the prevalence of HCV infection may be around 1–2%. The high prevalence rates reported from some countries (such as Japan) may be attributable to widespread parenteral practices with inadequately sterilized instruments in the past, and such practices continue in many developing countries today.

HBV and HCV are the main etiological agents implicated in chronic liver disease and its consequences, including cirrhosis and primary liver cancer, which ranks among the 10 most common cancers worldwide. In turn, the pattern of global distribution of these infections is determined by the predilection for vertical transmission of HBV, but not HCV, in the region. HBV accounts for most cases of chronic liver disease and its consequences in the East (excepting Japan) and Africa. HCV is the major agent implicated in the West and Japan, as well as in HBsAg seronegative individuals.

The epidemiology (and potential for eradication) of HDV follows the distribution of chronic hepatitis B. Areas of high prevalence, such as some Mediterranean countries and regions of South America, have been identified but rates of infection are declining in the former. Foci of high and intermediate endemicity continue to be identified. Clusters and outbreaks with significant morbidity and mortality are confined mostly to intravenous drug users who share contaminated needles and other high-risk behavior.

In the West (*Figure 1.2*), hepatitis B is uncommon and prevalence rates of HBsAg are well below 1%. Acute cases in adolescents and adults most commonly reflect sexual transmission or may be attributable to intravenous drug use. Rates of HBsAg carriage are rather higher in groups of individuals who have migrated from high prevalence areas. For example, the general prevalence rates for HBsAg seropositivity among adults of Asian origin range between 5 and 15%; up to 50–70% of the remainder will have detectable anti-HBc and anti-HBs, indicating previous exposure. Serological markers of current or previous infection are more common in males than in females of most ages. Prevalences for HBsAg in males peak at around 10–30 years of age and in females usually at around 40–50 years.

Seroprevalence data for immigrants and refugees from regions of high endemicity reflect their country of origin. Surveys from the Centers for Disease Control (CDC) detected HBsAg in around 14–15% of Vietnamese and Cambodian refugees living in the USA. HBsAg was detected in 6–14% of Koreans living in the USA, as well as in Korea. HBsAg was detected in around 8% of immigrants from the Philippines living in Alaska as well as in their native country.

Universal immunization of infants against hepatitis B has been adopted already by the majority of the wealthier countries in South-East Asia. Reduction of the carrier rate to below 2% can be achieved by this single strategy, without concerns over non-responsiveness to immunization and breakthrough infections (see Sections 3.9 and 7.12.4). Importantly, a significant reduction in the incidence of hepatocellular carcinoma

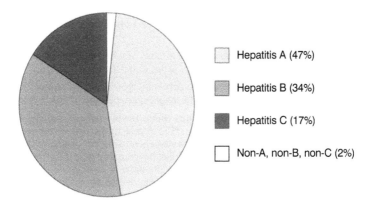

Hepatitis A (47%)

Hepatitis B (34%)

Hepatitis C (17%)

Non-A, non-B, non-C (2%)

Figure 1.2. Sporadic (community-related) acute viral hepatitis 1982–1995, Sentinel Counties USA (Centers for Disease Control).

> **Key Notes** Sentinal counties study by the CDC
>
> The Acute Viral Hepatitis Study Group from the Centers for Disease Control and Prevention collects data on the incidence and prevalence of viral hepatitis from four counties in the USA, Jefferson (Birmingham, Alabama), Denver (Denver, Colorado), Pinellas (St Petersburg, Florida) and Pierce (Tacoma, Washington), as communities representative of the USA

(HCC) in children, has been achieved within 15 years of introducing hepatitis B vaccine in countries such as Taiwan.

Caution is required before attributing hepatotropic potential to any newly discovered virus that is transmitted parenterally. Delineation of epidemiology becomes especially difficult when depending on data from individuals who share high-risk behaviors for viral hepatitis such as transfusion, intravenous drug use and multiple sexual contacts. Preliminary data implicate GBV-C/HGV, and also TTV, in a minority of cases of acute, non-A–E hepatitis and most infections with these viruses may be symptomless. Their propensity for chronic infection and prevalence in the general population remain to be determined.

1.3 Genetic variation

All of the major hepatitis viruses have high rates of mutation because of a low fidelity of genome replication. These evolve rapidly with a short generation time (*Table 1.4*).

Table 1.4. Virus variants

Wild-type (wt)	Classically, the non-mutant phenotype: generally used to denote the major (predominant) form of virus occurring naturally in a community or geographic region
Mutant	Virus with a genetic change leading to a phenotype distinct from the wild-type
Variant	Often used synonymously with mutant or where the underlying genetic cause of an altered phenotype is not understood
Serotype	Group of viruses that share a common, type-specific antigenic determinant. For example, the single serotype of hepatitis A virus
Genotype	Phylogenetically distinct groups within the diversity of sequences reported for a particular virus. (The degree of divergence within and between genotypes may vary between viruses.) For example, the single serotype HAV has been divided into seven genotypes (four in humans, three in non-human primates) on the basis of variation in one region of the genome (see section 2.2.4). At least six major genotypes of HCV are recognized and these are further divided into subtypes (see section 4.2.4)

As is typical for RNA viruses, HAV, HCV and HEV encode RNA-dependent, RNA polymerases for genome replication. These enzymes lack a proofreading function and error rates can be high (in the order of 10^{-4} for HCV). A similar rate of error may be expected for HDV which is replicated by the host RNA polymerase II.

The genomes of most DNA viruses, and host DNA, are replicated by DNA-dependent DNA polymerases. These enzymes have proofreading function to correct any miscorporation of nucleotides and limit the error rate (in the order of 10^{-8}). However, replication of the HBV genome resembles that of the retroviruses (see section 3.2.1). The viral DNA template is transcribed by the host RNA polymerase and the viral polymerase converts the pregenomic RNA to DNA; both processes lack proofreading.

1.3.1 Quasispecies

Individuals infected with RNA viruses such as HIV and HCV harbor viruses comprising a heterogeneous population of closely related sequences, known as quasispecies (*Table 1.5*). Quasispecies arise from the accumulation of errors during virus replication. In fact, infections may be established by an inoculum containing a mixture of virus sequences. On average, an error rate of 10^{-4} implies that replication of the 9.4 kb HCV genome will lead to misincorporation of around one base in each progeny genome.

Direct sequencing of polymerase chain reaction (PCR) products may overlook minority species but provides a rapid means of determining the consensus sequence. Heterogeneous virus populations in any infected individual are best defined by amplifying, subcloning and sequencing large numbers of individual genomes to detect minority species. Quasispecies may be detected by sequencing hypervariable regions, such as HVR 1 at the amino terminus of E2 of HCV.

1.3.2 Escape variants and persistent infection

Variants that arise during replication have an evolutionary advantage if they can escape the immune responses of the host or therapeutic interventions aimed at clearing the infection. A balance must exist between numbers and locations of mutations and viability of the virus. Large numbers of mutations are likely to be selected against

Table 1.5. Quasispecies

Definition	The genetic heterogeneity of a closely related virus population infecting an individual and usually centered around a dominant (master) sequence
Quasispecies effect	The cumulative genetic heterogeneity that is more common for RNA, than DNA, viruses. Any virus population is dynamic and prone to change through mutations during replication and selection by the host immune response and other factors
Master sequence	The most common sequence within the quasispecies in the infected individual
Consensus sequence	A sequence derived from the most common base found at each nucleotide position. This may be determined by direct sequencing and also may differ from the master sequence

if they are detrimental to the viability of the virus especially for compact genomes, such as HBV. All nucleotides are within open reading frames (ORFs) and encode protein. Around half are in regions where ORFs overlap and are translated in two phases. Consequently, small changes, even point mutations, may have a profound effect on gene expression.

What constitutes a 'mutant' rather than variation of the 'wild-type', is open to interpretation. For HBV, new sequences are usually compared to the variety in the database as well as a consensus derived from them, to determine whether changes in nucleotide, or predicted amino acid, sequences occur in other 'wild-type' sequences. For example, almost all isolates of HBV encode phenylalanine at amino acid 158 in the major surface protein. The few that encode leucine (a rare, *adw*4q$^-$ subtype, see Chapter 3) may be considered an uncommon variation of the wild-type, whereas a sequence encoding lysine is unprecedented and would be considered mutant. Some isolates of HBV are unable to synthesize HBeAg due to a point mutation (G to A at nt1896) in the pre-core region (see section 3.2.4.1). These 'precore mutants', although common, are not regarded as wild-type because they result in a phenotypic change. Similarly, hepatitis B viruses with mutations in the S ORF, resulting in changes in highly conserved residues of HBsAg, have been found in some anti-HBs seropositive individuals, including vaccine recipients (see section 3.2.5.4). These have been described as 'surface variants', 'vaccine escape mutants' and 'anti-HBs escape mutants' —such terms are used loosely and interchangeably.

1.3.3 Limitations on studying the clinical significance of virus variants

Limited data are available linking individual variants with pathogenicity, as most studies have used direct sequencing rather than the cumbersome analysis of numerous clones to detect mixed sequences. Further, practical limitations on sequencing complete genomes has led to denoting virus variants according to the relatively small region of the genome that was amplified and sequenced. Virus variants of HBV containing the pre-core stop mutation (G1896A) have been linked to more severe disease, including fulminant hepatitis B. However, such a mutation may be a 'surrogate' marker for virus heterogeneity. Consequently, mutations in other parts of the genome that co-segregate with the pre-core stop mutation, such as within the core promoter (nt1762/1764), may be responsible for the pathogenic phenotype. This area of research requires reappraisal now that improvements in PCR (including long-range PCR) and sequencing technologies make analysis of entire genomes feasible. Similarly, there is no clear consensus as to whether mutations within a region of NS5a of HCV genotype 1b (so called 'interferon sensitive region') render the virus resistant to interferon therapy.

Furthermore, there is a lack of *in vivo* (animal) models for assessing pathogenic effects of various strains of hepatitis viruses. Results of *in vitro* assays that assess replication efficiency do not necessarily correlate with pathogenic potential. Although all HAV isolates so far are the same serotype, individual case reports document variants of HAV with mutations and deletions in the 5'UTR, as well the 2B/2C region, that are associated with severe hepatitis A. Comparable data for HEV are likely to emerge as we learn more about its heterogeneity. So far, virus variants have not been implicated in the predilection for severe outcome for HAV infection in older adults and for HEV in pregnant women, respectively.

1.4 Serological diagnosis of acute and chronic viral hepatitis

The individual serological markers of infection accompanying each virus are discussed in the relevant chapters and only general principles are discussed here. Panels of tests (*Table 1.6*) are required for the accurate diagnosis of viral hepatitis A to E (and beyond), as clinical features do not distinguish between causes. Only rarely are tests requested individually. Also, as risk factors overlap, multiple agents should be sought; HBV and HCV (as well as HDV) may co-exist especially in the intravenous drug user and alcoholic patient.

Serial testing is important with negative or indeterminate initial results, especially for acute viral hepatitis (*Tables 1.7* and *1.8*). Seroconversion may be delayed. IgM antibodies to HBV (IgM anti-HBc) may be depressed with HDV co-infection. IgM anti-HDV and HEV may be detectable only transiently for acute hepatitis D and E, respectively. IgM antibody assays for HCV are not commercially available; seroconversion to anti-HCV positivity can be delayed many weeks, especially in the immunosuppressed with HIV and following liver transplantation (see Chapters 6 and 8) and in ALF (Chapter 5).

Additional or alternative etiologies, especially alcohol, drugs and autoimmune diseases must be considered in all patients presenting with clinical and biochemical features of viral hepatitis.

Key Notes Definitions of multiple infections

1) Superinfection: Primary virus infection on a background of persistent viral hepatitis. Serological markers may show transient reduction: for example, in HBV carriers levels of HBsAg, IgM anti-HBc and HBV DNA may fall during superinfection with other hepatitis viruses. Serial testing is recommended;

2) Co-infection: Simultaneous primary infection with more than one virus; the classical example is HBV and HDV; serological markers are: IgM anti-HBc and IgM anti-HDV. Markers of chronicity (IgG antibodies, anti-HBe) are not detected.

Recent advances have improved sensitivity and specificity of diagnosis using enzyme immunoassays to detect serological markers alongside sensitive molecular techniques, such as the PCR, to detect viral nucleic acids in serum. However, detection of a virus *per*

Table 1.6. Serological diagnosis of viral hepatitis

	HAV	HBV	HCV	HDV	HEV[a]
Acute infection	IgM anti-HAV	IgM anti-HBc	Assays for IgM not available	IgM anti-HDV	IgM anti-HEV
Virus replication	HAV RNA	HBV DNA	HCV RNA	HDV RNA	HEV RNA
Chronic infection	NA	HBV DNA HBs Ag	HCV RNA	HDV RNA[b]	NA
Past infection	IgG anti-HAV	anti-HBs/IgG anti-HBc	ND	IgG anti-D	IgG anti-HEV[a]

[a]Assays for anti-HEV antibodies are not standardized and may lack sensitivity and specificity, concentrations of IgG anti-HEV may wane rapidly. [b]usually with HBV DNA in serum. NA: not applicable; ND: not defined.

Table 1.7. Acute viral hepatitis—false negatives and pitfalls in monitoring

Hepatitis B

HBsAg, HBeAg and HBV DNA may be undetectable in fulminant hepatitis B

Levels of viraemia may be very low with replication of HBeAg-negative variants

Hepatitis D

Acute hepatitis may be overlooked: IgM responses may be transient

IgG antibodies become detectable early following acute infection

Hepatitis C

Anti-HCV antibodies may be delayed—seroconversion can take months

HCV RNA does not discriminate acute from chronic infection

Hepatitis E

Acute hepatitis may be overlooked: IgM antibodies may be transient

IgG antibodies become detectable early following acute infection

se does not implicate the agent in the etiology of the disease and interpretation of disease severity remains limited. Similarly, monitoring of conventional serological markers as responses to antiviral therapies does not predict accurately clearance of virus.

1.4.1 Enzyme immunoassays

These use a wide variety of formats to detect antibodies in serum. Usually, antibodies from the sample of interest are captured by antigen bound to a solid substrate. This reaction is detected by a second antibody (anti-antibody), conjugated to an enzyme such as horseradish peroxidase. The bound antigen may be protein derived from the native virus, a recombinant protein or a synthetic peptide.

Table 1.8. Serological diagnosis of acute viral hepatitis

	Acute infection	Pitfalls of antibody markers	Associated viremia
HAV	IgM anti-HAV	Reliable, may persist for 1 year	HAV RNA: transient
HBV	IgM anti-HBc	Reliable, may be suppressed with HDV coinfection	HBV DNA: rapid decline with clearance
	HBsAg	Insufficient: may indicate chronic infection	
HCV	Seroconversion to anti-HCV	May be delayed up to 1 year, or lacking with immunosuppression	HCV RNA may indicate acute-on-chronic infection
HDV	Seroconversion to IgM anti-HDV	Transient, not widely used	HDV RNA: transient
	Seroconversion to IgG anti-HDV		
HEV	Seroconversion to IgM anti-HEV	Transient, not widely available	HEV RNA: transient
HHV	Rise in paired titers	Rise may be delayed/completed	
EBV	Progression from early antigens	Rise may be delayed/completed	

The antibody response may recognize epitopes that are linear and/or conformational. Use of linear peptides may result in increased specificity but decreased sensitivity compared with more complex antigens with intact secondary (conformational) structure. Conversely, tests based on native or recombinant antigens will tend to be more sensitive (fewer false-negatives) but less specific than assays targeting linear epitopes. Conformational epitopes may involve amino acid residues that are not co-linear in the primary sequence and cannot be mimicked by synthetic peptides. The use of monoclonal antibodies for antigen capture or detection also may lead to false negatives if proteins with amino acid substitutions (or deletions) are not recognized.

Enzyme immunoassays (EIAs) also may be formatted (designed) to detect antigen. For example, anti-HBs bound to a solid substrate may be used to capture HBsAg, which is then detected by a second, labeled antibody.

Genetic variation of the viruses may lead to false-negative results. For example, the first generation test for anti-HCV antibodies relied on a protein encoded by a relatively variable region of the genome (NS4, see section 4.5.1). Antibodies raised by certain individuals against divergent HCV genotypes did not react with this test antigen. Second generation assays overcame this variability by including the relatively conserved nucleocapsid protein. A similar concern remains with testing for HBV. Assays relying on monoclonal antibodies that detect only specific epitopes within the *a* determinant may fail to detect viruses variant for HBsAg.

1.4.2 Recombinant immunoblot assays

These are used as the most common supplemental tests for specificity to evaluate repeatedly reactive results in the screening assays (EIAs) for anti-HCV antibodies, especially in the setting of testing low-risk individuals such as volunteer blood donors (see *Figure 4.4*). Other supplemental tests for antibody specificity to HCV include dot blot immunoassays and synthetic peptide assays used to quantitate antibody levels. The recombinant immunoblot assay (RIBA) is a strip immunoblot assay that tests for antibodies to virus (e.g. HCV) using viral antigens immobilized on a strip of membrane (e.g. nitrocellulose) and including recombinant proteins expressed in *Escherichia coli* as well as yeast. Various controls also are immobilized on the membrane including IgG in low and high concentrations.

As the RIBA and EIA share the same antigens (such as c100-3) these assays cannot be considered as scientifically independent. They do not improve detection in early infection. In the first generation assay (RIBA 100), cross-reacting controls as well as the recombinant c100-3 antigens are separately immobilized. This RIBA included a related antigen (5-1-1) expressed in *E. coli* and superoxide dismutase expressed in yeast to rule out cross-reactivity with yeast antigens or superoxide dismutase. Less than half (45%) of initially reactive blood donations were considered specific for anti-c100-3 antibody using the RIBA 100 test.

The widely used second generation assay, RIBA HCV 2.0, uses four antigens: 5-1-1 (*E. coli*), c100-3, c22-3 and c33c (yeast). An unequivocally positive RIBA is defined as when all four bands are strongly positive (4+, maximum intensity) with respective negative and positive controls and correlates with HCV RNA positivity in > 90% of sera. In the immunocompetent individual, low intensity bands (< 2+) and reactions only against 5-1-1 and/or c100-3 correlate well with absence of viremia—undetectable HCV RNA levels using RT–PCR.

The third generation assay, RIBA HCV 3.0, uses two recombinant antigens; a modified c33c and NS5 and three synthetic peptides, which replace 5-1-1, c100-3 and c22-3. A positive band (> 1+ intensity) for two to five of the HCV antigens is considered reactive. The increased sensitivity of the latest generation assays reflects improvements in the quality of the antigens such as c33c and c22, rather than the additional antigens such as E2 and NS5.

Problems with sensitivity and specificity of assays may be revealed, depending on the prevalence of infection in the population and the illnesses involved. In high-risk groups for HCV infection, such as intravenous drug users and multiply transfused individuals, an unequivocal (all four antigens positive) result in the latest generation of RIBA correlates well (>90%) with the presence of viremia (HCV RNA). In contrast, in low prevalence populations such as volunteer blood donors, who already have been screened for parenteral risk factors, less than 50% of samples repeatedly reactive for anti-HCV antibody (by EIA) are judged as true positives (HCV RNA detectable). In such low-risk groups, the repeated reactivity tends to be against individual antigens such as c33c or c22-3 and HCV RNA remains undetectable in most cases. Interpretation of these indeterminate results remains difficult in clinical practice. Conversely, these antigens may be the only ones detected in immunosuppressed individuals, such as recipients of organ grafts in whom HCV RNA can be detected in some cases.

1.4.3 Nucleic acid hybridization

Base pairing in double-stranded nucleic acid depends upon the non-covalent, hydrogen bonding of cytosine with guanosine and of adenosine with thymidine (DNA) or uridine (RNA). Under appropriate conditions, two single-stranded nucleic acid molecules of complementary sequence will anneal or hybridize. This reaction is sequence-dependent and highly specific.

A simple hybridization assay is illustrated by the 'dot blot' used to detect HBV DNA. Serum is treated to disrupt any virus particles present and any DNA is denatured with alkali. Single-stranded DNA is bound to a nitrocellulose or nylon membrane by direct spotting or filtration through a manifold. Following a prehybridization step to saturate sites on the filter which bind nucleic acid non-specifically, an HBV probe is allowed to hybridize to DNA on the filter. Typically, the probe comprises denatured or single-stranded, cloned DNA labeled *in vitro* with radioactivity or a non-radioactive compound such as biotin or digoxygenin. Synthetic oligonucleotides also may be used as probes. Finally, the filter is washed to remove non-hybridized probe and processed through the appropriate detection system.

Hybridization assays for HBV DNA:

Dot blot hybridization:
- Acute vs. chronic infection
- Monitoring antiviral therapies
- Inverse correlation with inflammation

bDNA assay:
- Less sensitive than PCR
- Easy for bulk (routine) testing
- Semiquantitiative

In clinical practice, problems of quantitation and standardization of results have limited the validity of comparing results between centers and patients. Limits of sensitivity are typically denoted at around 0.5–1.0 pg. This measurement usually refers to quantitation of the 'spots' using densitometric analysis of autoradiographs and scintillation counting of radiolabel. Results are calibrated by comparison to a serially diluted external standard, such as a recombinant plasmid. Sensitivities are less using internal standards, which are processed in parallel with the test samples.

A more detailed investigation may be carried out using Southern blotting. For example, total DNA may be purified from liver and serum, digested with a variety of restriction enzymes and the resulting fragments separated by size using agarose gel electrophoresis. DNA may then be transferred from the gel to a membrane and hybridized with an HBV DNA probe as described above. Use of appropriate restriction enzymes allows, for example, replicating and integrated HBV DNA (see *Figure 3.7*) to be distinguished.

1.4.4 Branched chain DNA assay

This is an alternative to dot hybridization and PCR for detection of HBV and HCV nucleic acids. The general limit of detection is around $2.5–3.5 \times 10^5$ genome equivalents (copies) per ml. The basis of the assay is specific hybridization of synthetic oligonucleotides, such as to the 5'UTR and nucleocapsid region of the HCV genome, which allow the target nucleic acid to be captured onto a microtiter well. Synthetic, branched DNA molecules and multiple copies of an alkaline phosphatase-linked probe are hybridized to the immobilized complex, amplifying the signal. The bound alkaline phosphatase is detected by incubating the complex with a chemiluminescent substrate and measuring the resultant light emission.

The branched chain DNA (bDNA) assays show several advantages over PCR-based techniques. These include less labor intensive preparation allowing handling of larger numbers of clinical samples. Also, standard curves are linear over a wider range of virus concentrations, allowing easier and more reliable quantitation of viremia. The bDNA assay relies on boosting signal rather than amplifying the target nucleic acid, this improves reproducibility and is less susceptible to contamination than PCR-based assays. The major disadvantage of the bDNA assay is its poor sensitivity compared with the PCR, leading to false negativity with low-level ($< 2.5 \times 10^5$ genome equivalents (geq)/ml) viremia (*Table 1.9*). Up to 15% of RIBA-reactive samples are negative in the bDNA assay but positive using RT–PCR. Typically these record below 1000 geq/ml by quantitative RT–PCR. Monitoring using nested PCR assays may be necessary to confirm clearance of virus especially following antiviral therapy.

1.4.5 Polymerase chain reaction

The PCR amplifies geometrically a segment of DNA between a pair of primer binding sites (*Figure 1.3*). A DNA virus can be amplified directly following extraction of this nucleic acid. Detection of viral RNA genomes and transcripts first requires conversion of the RNA to DNA using reverse transcriptase (RT–PCR). The reverse transcriptase step may be primed by random, hexameric oligonucleotides or the antisense PCR primer—the former may be more sensitive but use of the latter permits a 'one-tube' reaction and minimizes the chances of contamination.

Oligonucleotides are synthesized which are complementary to the primer binding sites on opposite strands of the DNA and usually between 200 and 2000 nucleotides apart. The reaction comprises repeated rounds of heat denaturation, annealing of primers and extension using a heat-stable DNA polymerase (e.g. *Taq*, from *Thermus aquaticus*). Reaction products from each extension act as templates in the following cycle producing the geometric 'chain reaction'. The final product may be detected as a band of specific size following agarose gel electrophoresis.

Table 1.9. Limitations in sensitivities of assays to detect viral nucleic acids

Assay	Lower limit of detection (geq/ml)
Nested PCR	1
Dot-blot hybridization	10^5 (0.1–1.0 pg HBV DNA)
bDNA	$2.5–3.5 \times 10^5$

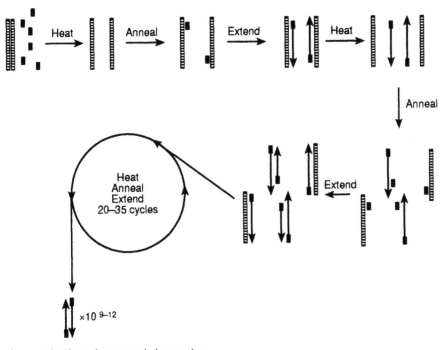

Figure 1.3. The polymerase chain reaction.

Unfortunately, reproducible results can be difficult to achieve and many clinical samples fail to amplify probably through inappropriate handling of specimens. Sensitivity can be improved by increasing the volume of serum extracted. Typical in-house sensitivities were down to around 3000 genome equivalents (copies) per ml when equated with the bDNA assay (2–5×10^5 genome equivalents/ml). Sensitivity was improved significantly (to around 100 genome equivalents/ml) by increasing the volume of extraction (50–100 μl serum) and using proteinase K and phenol–chloroform for extraction of nucleic acids. Further sensitivity and specificity may be gained by hybridization with an oligonucleotide probe which does not overlap the primers or a further PCR with an internal primer pair (nested PCR).

Quantitation using PCR also is desirable, particularly for monitoring responses to antiviral therapies. Quantitation of HCV RNA can be accomplished by dilution assays but this consumes large quantities of reagents. Competitive methods use, for example, samples spiked with a known quantity of a modified template, such as a synthetically mutated target RNA, which is co-amplified. Such templates may have an additional restriction enzyme site, which can be distinguished on gel electrophoresis, while otherwise being identical to the natural template. Spiking with DNA does not monitor the efficiency of reverse transcription in RT–PCR. Spiking with RNA transcribed *in vitro* is preferable for monitoring RT–PCR because the reverse transcription step is controlled also. An inherent problem with quantitative PCR assays is the intra-assay variation (and intra-sample variation) of amplifying the target, especially in the presence of contaminants.

Commercial assays are being modified to include internal standards for calibration. Typical linear ranges are between 10^3 and 10^6 geq/ml. These limitations have clinical significance. Levels below 1000 geq/ml may be reported with variable results (including undetectable) whereas results may not be comparable between patients and assays for very high levels of viremia (>1 000 000 geq/ml). Also, risks of cross-contamination are reduced by including an incubation step with specific enzymes such as uracil-N-glycosylase, which cleaves uracil-containing contaminating DNA (from previous PCR reactions).

Strand-specific PCR typically aims to detect the negative strand of positive-sense RNA viruses, such as HCV. In diagnostic PCR, the positive strand is amplified (and sometimes quantitated) and results are taken to reflect virus replication, or more accurately, viremia. Specific amplification of the negative strand correlates directly with virus replication, for example, in particular tissues or cell types. However, it is difficult to eliminate priming of the positive strand by contaminating fragments of nucleic acid.

Contamination problems. PCR-based techniques remain research intensive and difficult to adapt for large-scale routine use. Their exquisite sensitivity also leads to a major drawback of false-positive results from cross-contamination. Stringent precautions must be taken to avoid carry-over with the products of previous reactions. This entails dedicated workspace, reagents and equipment for sample preparation and for setting up reactions that are separate from the laboratory where downstream analysis is carried out. Segregation of equipment into 'dirty' and 'clean' areas is mandatory. Stocks of reagents (primers, buffers, water etc.) should be prepared frequently and dispensed into single-use aliquots. False negatives may result with poor extraction of nucleic acid. RNA is destroyed rapidly by repeated freezing and thawing and by RNases, which are thermostable and ubiquitous. The specificity and sensitivity of the PCR are greatly improved with a second round of amplification, 'nested' PCR, and the use of several sets of primers to different but conserved regions of the viral genome, such as NS3 (helicase) as well as 5'UTR regions for detecting HCV RNA.

Limitations of a positive PCR result are:

- Etiology—not proven
- Disease activity—little correlation
- Monitoring limited with HBV DNA
- Contamination (false positives)

1.4.6 Nucleic acid sequence-based amplification

Nucleic acid sequence-based amplification (NASBA) has been used successfully for the detection of RNA of various viruses, including HCV. This methodology relies on the direct detection of RNA using specific primer pairs, one of which contains a promoter sequence within the 5' end that is recognized by T7 RNA polymerase. This enzyme transcribes as many as 100 copies from each template molecule. Reverse transcriptase in the reaction converts the target to a cDNA/RNA hybrid and, following RNaseH digestion of the RNA strand, to double-stranded DNA. Second strand synthesis is primed by an oligonucleotide containing the T7 promoter sequence and the double-stranded DNA product acts as a template for transcription. The single-stranded RNA transcripts may enter the cyclic phase, being reverse transcribed and increasing the pool of templates. Over time, the single stranded RNA product accumulates. No specific equipment is necessary, such as thermal cyclers. The amplified product, which is single-stranded RNA, is detected by Northern hybridization using labeled, internal oligonucleotide probes. The risks of contamination are reduced compared to conventional two-step RT–PCR, by containing all reagents in one tube. Also, amplification of around 10^9-fold can be achieved at a single temperature in around 2 hours, because each transcription step generates around 10–100 copies of RNA. Only a few cycles are necessary to achieve high amplification and NASBA can be modified to allow quantitation of RNA by spiking several competitor RNAs as internal standards.

1.5 Collection and storage of clinical samples

Falsely negative or falsely positive results arise too frequently from inadequate handling and storage of clinical samples. All samples should be collected if possible to standards suitable for analysis by PCR-based techniques.

1.5.1 Principles

The principles are to work quickly to avoid degradation of nucleic acids and cleanly to avoid contamination. Storage of samples in multiple small aliquots avoids repeated freeze/thaw cycles which accelerate degradation of nucleic acids, especially RNA. Containers for freezing samples should be clean and labeled with indelible ink beforehand to minimize handling and cross-contamination between samples. Padded gloves, clean goggles and clean long-handled forceps should be available to retrieve tubes and tissue from the bottom of liquid nitrogen containers.

Containers for handling and freezing serum and tissue samples should be washed in DEPC (diethyl pyrocarbonate)-treated water and autoclaved to remove RNases and DNases which destroy nucleic acids. Pestles and mortars for triturating tissue under liquid nitrogen should be baked at 80°C for in excess of 8 hours. Many plastics (Universal tubes, collection bags, autoclave bags etc.) do not withstand baking or freezing at temperatures below −20°C. Many plastic bags in common use become stiff and crack on freezing. Many 'permanent' ink markers 'run' on thawing of labeled plastic containers and bottles. Glass containers should be avoided to minimize cracking, especially during freeze/thaw cycles.

Hands should be washed thoroughly between patients. Gloves and collection tubes should be dispensed from unopened packages; fresh for each patient and labeled clearly before collection. A minimum of two separate aliquots should be prepared for each specimen and patient to minimize repeated cycles of freezing and thawing.

Degradation of RNA and DNA occurs with time regardless of the storage temperature. Loss is increased with multiple freeze/thaw cycles. Avoid refrigerators and freezers that 'automatically' defrost! Refrigeration should be as low as possible (i.e. −75°C is preferable to − 40°C and −20°C). Samples, especially lumps of tissue, should be snap-frozen prior to refrigeration to ensure rapid freezing throughout the specimen; avoid large lumps, which freeze slowly in their centers and are difficult to sub-divide once frozen. Snap-freezing is essential when handling necrotic tissue, such as liver in fulminant hepatitis, to minimize hydrolysis of nucleic acids by the many enzymes in liver. Aliquots of valuable samples should be stored in different freezers and locations.

Decontamination. Flasks for transporting liquid nitrogen should be prepared in advance by washing thoroughly with detergent which destroys lipid (enveloped viruses). Mortars and pestles should be treated with acid to prevent carry-through of nucleic acids prior to preparation for re-use. Containers should be soaked in commercially available fluids which specify capability of destroying HBV. Sodium hypochlorite (bleach) can be used instead. Recommended dilutions should be followed; higher concentrations do not necessarily improve the efficiency of decontamination. All containers should be rinsed thoroughly afterwards in clean water and dried with disposable towels. Tubes, pestles and mortars and glassware should be soaked overnight in DEPC-treated water at 37°C and autoclaved to remove nucleases. Note: DEPC is potentially carcinogenic but becomes harmless on autoclaving.

1.5.2 Serum

This should be separated by gentle centrifugation (3000 r.p.m. for 5–10 min, 4°C if possible) soon after formation of the clot. Serum should be aliquoted, using disposable plastic pipettes (pastettes) into sterile, RNase-free and DNase-free (autoclaved) plastic tubes which withstand snap-freezing in liquid nitrogen. Attention should be paid to avoiding contamination of samples and tubes by handling in a clean area and using a decontaminated centrifuge and rotor buckets. Containers for transporting liquid nitrogen (flasks) should be decontaminated before use and samples from individual patients kept separate during collection and separation.

1.5.3 Liver

Tissue samples should be cut rapidly into small pieces using disposable scalpels—a fresh one for each patient and each analysis—on a clean surface. (Large pieces of tissue are difficult to handle when frozen.) Tissue can be placed in pre-labeled tubes, which withstand snap-freezing in liquid nitrogen, in a decontaminated container.

1.5.4 Fresh frozen tissue

Immunohistochemical and molecular techniques (*in situ* PCR, *in situ* hybridization) are being applied to liver tissue to correlate the localization of viral antigens and nucleic acids with liver injury (hepatocellular necrosis, apoptosis etc.) and with inflammatory cell infiltrates. Studies are optimal on cryostat tissue sections from tissue snap-frozen using protocols which protect tissue architecture.

Other diseases may coexist with viral hepatitis. Samples should be stored in anticipation of assisting with additional diagnoses such as malignancy, Wilson's disease, metabolic disorders etc. Various embedding mediums, such as OCT, can be used to help preserve tissue architecture and should be considered for storage of samples of potential diagnostic value—fresh-frozen biopsy for cancer etc.

Frozen tissue wrapped in aluminum foil is adequate for estimations of copper, as in suspected Wilson's disease, provided collection of tissue including the biopsy technique uses copper-free instruments (biopsy needles) and containers. Fresh frozen tissue should be reserved in cases of suspected fatty liver (as in acute fatty liver of pregnancy). Fat is best detected on fresh liver and stained with oil-red 'O'.

An aliquot should be considered for electron microscopy, especially in cases of suspected herpesviruses and unusual infections. Tiny samples only should be cut without crushing using a very sharp scalpel and placed into cacodylate-buffered gluteraldehyde (not formaldehyde). Five to six crumb-sized fragments are adequate for most analyses and do not exhaust the fixative. These samples can be stored at 4°C for several months.

1.6 Clinical practice

1.6.1 Issues

The application of molecular techniques to clinical hepatology has revolutionized the diagnosis and management of viral hepatitis. Discoveries of hepatitis C and E viruses have led to reappraisal of the classification of viral hepatitis, making obsolete the term non-A, non-B hepatitis. The relegation of unknown, indeterminate causes of presumed viral hepatitis into non-A through non-E (non-A–E) hepatitis masks an heterogeneous group of conditions, some undoubtedly of non-viral origin.

Clinical features overlap sufficiently to exclude their usefulness in diagnosis in the individual patient. Furthermore, multiple infections commonly co-exist with viral hepatitis. Risk factors for their transmission are shared, especially use of intravenous drugs, multiple transfusions, sexual promiscuity, crowding and poor sanitation.

Specific management strategies and outcome, especially of ALF (Chapter 5) and selection for liver transplantation (Chapter 6) rely heavily on the accurate diagnosis of viral etiology. Biochemical liver 'function' tests (LFTs) offer limited guidance in diagnosis, assessment of inflammatory disease activity in the liver and potential clinical outcome. Serological detection of viral nucleic acids, such as HBV DNA and HCV RNA, does not equate with pathogenesis of disease or its severity.

Chronic viral hepatitis B and C are likely to progress to cirrhosis with repeated bouts of inflammation and necrosis in the liver. Prognosis and clinical course depend on severity and persistence of necroinflammatory activity as well as viral etiology. In contrast, little correlation exists between outcome and serological indicators of disease activity, such as liver enzyme activities. These aspects have been incorporated into the reclassification of chronic viral hepatitis, especially the histopathological features. Qualitative descriptive terms such as chronic persistent hepatitis (CPH) and chronic active hepatitis (CAH) were borrowed historically from the field of autoimmune hepatitis. These terms are likely to become obsolete but require definition as they have been used widely in the literature (see section 1.8.3).

1.6.2 Liver 'function' tests

The standard tests do not measure liver function. Instead, these measure dysfunction qualitatively and in a limited way. Caution is required over their interpretation, especially with single tests and time-points. Liver tests do not discriminate between or within etiologies of disease, whether of viral or non-viral origin. Also, overall they are unreliable as predictors of severity of liver damage, survival and outcome. Although derangement of any liver test usually indicates significant disease, mild perturbations, or values within the normal range do not exclude equally significant liver disease.

Liver enzymes [aspartate aminotransferase (AST), alanine aminotransferase (ALT)], serum bilirubin and albumin typically are less deranged in ALF than acute uncomplicated hepatitis. An elevated prothrombin time, the best available indicator of recent impairment of synthetic function (of clotting factors), depends on sufficiency of vitamin K and variation in test reagents (thromboplastins). The prothrombin time only becomes a sensitive predictor of adverse prognosis in ALF when grossly elevated (>100 s UK, > 30 s USA).

Liver function tests (LFTs) can be within, or near, the normal range despite chronic hepatitis and cirrhosis. By the time the prothrombin time (uncorrected by vitamin K1) becomes elevated in chronic liver disease, liver damage is extensive, decompensated, and usually irreversible, leaving few options except liver transplantation.

1.6.2.1 Serum bilirubin levels and jaundice

Jaundice is rarely noticeable with serum bilirubin levels below around 1.5 mg/dl (20 μmol/l) in adults and children and around 5 mg/dl (85 μmol/l) in the new-born. The serum total bilirubin level is an insensitive test of early liver dysfunction because the healthy liver has a large functional reserve for conjugating and clearing around three times its capacity for synthesis. A rising serum total bilirubin reflects the inability of the diseased liver to clear bilirubin and usually indicates significant dysfunction (see Child's–Pugh classification, Table 1.23). The serum levels achieved also are limited by the capacity of the healthy kidney to excrete large amounts of urobilinogen. Accordingly, the highest levels of bilirubin in liver disease typically are found with concomitant renal dysfunction.

The majority of the bilirubin is conjugated ('direct bilirubin') in viral hepatitis. The presence of significant quantities (> 20% of total level) of the unconjugated ('indirect bilirubin') form should alert awareness to additional causes especially hemolysis—Wilson's disease, alcoholic hepatitis (Zieve's syndrome), disorders of red cells (G6PD deficiency, sickle cell disease, among others) and the congenital hyperbilirubinemias, including Gilbert's disease. The total (conjugated and unconjugated) serum bilirubin level may be higher (>340 μmol/l) than anticipated when significant hemolysis is superimposed on a diseased liver, such as sickle cell crisis in a multiply transfused patient who has cirrhosis from chronic hepatitis B and/or C.

Clinical awareness of jaundice depends on observer proficiency. Although obvious jaundice (high serum bilirubin level) indicates significant liver dysfunction, a similar degree of hepatocellular damage can occur without jaundice (anicteric hepatitis). Obvious jaundice is not invariable in ALF. Overall, there is no direct correlation between serum bilirubin level (and degree of jaundice) and severity of acute hepatitis. Serum bilirubin levels may rise slowly and remain only modestly elevated in very severe ALF, due to annihilation of metabolic pathways.

In ALF, the prognosis without grafting depends on the time interval from onset of jaundice (not serum bilirubin level) to the development of encephalopathy. A short time interval (<7 days) favors survival. This paradox presumably reflects retention of significant metabolic capability to handle bilirubin until near the time of development of cerebral edema. This delineation is used in one of the definitions of acute liver failure (Chapter 5) to indicate severity and, hence, adverse outcome. Most cases of ALF due to hepatitis A or B have a short time interval and over half recover without resorting to liver transplantation. In contrast, the majority of cases of ALF attributed to non-A–E hepatitis show protracted time intervals (>7 days) between onset of jaundice and encephalopathy. Most cases of late onset hepatic failure fall into this category and have a very poor prognosis without grafting.

Viral hepatitis typically is anicteric (without jaundice) in the majority of acute infections especially in hepatitis B and C, which progress to chronic liver disease. Anicteric acute hepatitis B is more likely to progress to chronic hepatitis B than acute infections which present with obvious jaundice. This inverse correlation between severity of acute infection and propensity for chronic carriage reflects the impact of the immune response aimed at clearing virus-infected hepatocytes.

1.6.2.2 Cholestatic hepatitis

Cholestasis arises when bilirubin, which is synthesized by the liver, fails to reach the intestinal tract. In the context of viral hepatitis, cholestasis is 'intrahepatic'—arising predominantly from impairment of transport of bilirubin from hepatocytes to intrahepatic bile ducts and bile canaliculi. Rarely, as in CMV infection, extrahepatic cholestasis ('obstructive jaundice') can arise from damage to the common bile duct and gallbladder.

Cholestasis tends to arise later in the course of acute viral hepatitis, often coinciding with improvement in clinical well-being. Patients with hepatitis A typically delay presentation until the cholestatic phase when they become aware of jaundice and pruritus (itching). Also, jaundice may recur in the few cases that relapse.

Serum levels show some correlation with severity of chronic hepatitis and, more so following development of cirrhosis. However, chronic liver disease may remain anicteric without marked elevations in serum bilirubin.

Serum bilirubin:

- Conjugated (direct) bilirubin usually comprises up to 30% of total bilirubin
- Unconjugated hyperbilirubinemia is likely when 'direct' is < 20%
- Renal threshold for conjugated bilirubin is low and urine detection is sensitive
- Conjugated bilirubin may be detected in urine despite a 'normal' serum bilirubin and no jaundice

1.6.2.3 Serum enzyme activities

Serum activities, 'levels', of ALT (*Table 1.10*) and AST (*Table 1.11*) reflect leakage from cells, mostly from hepatocytes. Accordingly, high values (> 40 μkat/l or > 2500 IU/l) reflect liver cell damage and are found early in massive hepatocellular necrosis, typical of acute viral hepatitis, including ALF. Similar ranges can occur with direct hepatotoxins, such as carbon tetrachloride, and do not allow discrimination between different viruses and other, non-viral causes of liver cell necrosis.

In clinical practice, absolute activities depend on the timing of sampling and peak usually before the serum bilirubin—when the patient is likely to present with jaundice. Serum levels may be lower than anticipated due to their short half-lives (18–48 h), clearance at various sites other than the kidney and limited regeneration of hepatocytes available for future destruction. Elevated levels, especially those below 8 μkat/l (<500 IU/l), are not specific for etiology or severity of liver cell damage and do not indicate prognosis in most cases (see ALF, section 1.7.1).

Serum levels of alkaline phosphatase (*Table 1.12*) rise following damage and obstruction to bile ducts. Elevations reflect synthesis of enzyme rather than damage to hepa-

Table 1.10. Alanine aminotransferase (ALT)

- Historical synonym: serum glutamic-pyruvic transaminase (SGPT)
- A cytosolic enzyme which catalyzes the transfer of alpha amino group of alanine to the keto group of ketoglutaric acid to form pyruvic acid
- Found predominantly in liver: more specific test for liver damage than AST

Table 1.11. Aspartate aminotransferase (AST)

- Historical synonym: serum glutamic oxalo-acetic transaminase (SGOT)
- An enzyme, found predominantly in mitochondria in the normal liver and heart muscle, which catalyzes the transfer of the alpha amino group of aspartic acid to the keto group of ketoglutaric acid, to form oxalo-acetic acid. In health, the low circulating levels reflect enzyme from cell cytosol
- Mitochondrial fraction elevated disproportionately to ALT in alcoholic hepatitis and myocardial infarction. AST: ALT greater than 2:1
- Found in liver, heart and skeletal muscle, kidneys, brain, pancreas, lungs and white and red blood cells

Table 1.12. Serum alkaline phosphatase

- A group of enzymes containing isoenzymes of unknown function
- Found in liver, bone, intestinal tract, kidneys, leukocytes and placenta
- Bound to membranes in bile canaliculi and liver sinusoids
- Isoenzymes differentiated by heat labilities and electrophoretic mobilities
- High value (> 4 ×) characteristic of cholestatic phase of hepatitis

tocytes (contrast AST and ALT) or failure to clear (contrast serum bilirubin) from a diseased liver. In acute and chronic viral hepatitis, elevated levels tend to parallel the degree of intrahepatic cholestasis. Levels tend to be higher in chronic hepatitis due to HCV rather than HBV, reflecting the predilection for damage to bile ducts. An elevated level also may occur in any space-occupying lesion of the liver such as HCC.

Serum levels of alkaline phosphatase are higher than the normal range for adults during bone growth, pregnancy and many other situations (*Table 1.13*). An elevation in serum level of gammaglutomyl transpeptidase (GGT) confers liver specificity on an elevated level of alkaline phosphatase (*Table 1.14*). The serum GGT level is elevated in alcoholic liver disease which often accompanies viral hepatitis. Elevations in GGT also can occur in pancreatitis, diabetes mellitus and following induction by various drugs including antiepileptic drugs [phenytoin (epanutin) and phenobarbitone] and warfarin.

1.6.2.4 Liver synthetic function

The degree of liver damage is reflected in impairment of synthesis by the liver of coagulation factors and albumin.

Prothrombin time (PT). The prothrombin time in plasma is a test of acute liver synthetic function because this measures activities of four coagulation factors synthesized in the liver: II (prothrombin), V (proaccelerin), VII (prothrombin conversion accelerator) and X (Stuart–Prower factor) as well as I (fibrinogen) and circulating inhibitors.

Table 1.13. Pitfalls with serum alkaline phosphatase

- No discrimination with non-viral causes of cholestasis
- Always assess with other tests for cholestasis (GGT etc.)

Misleading high values in:

- Obstructive jaundice: biliary tract disease with gallstones
- Infiltrations, masses[a] and granulomata in the liver

Higher serum levels than normal range for adults:

- Healthy babies and children (reflects osteoid bone growth)
- Healthy adolescent males (reflects bone growth)
- Third trimester of pregnancy (placenta; heat-stable alkaline phosphatase): > 2 × upper limit of normal
- Certain tumors (usually heat-stable component)

[a]Space-occupying lesions: tumors, abscesses, large cysts.

Table 1.14. Gamma glutamyl-transpeptidase (GGT)

- Involved in hydrolysis and transpeptidation of glutathione and transport of amino acids across cell membranes
- Found in cell membranes in liver, pancreas, many other tissues
- Characteristically high in cholestasis, biliary tract disease
- Characteristically elevated in alcoholic liver disease
- Confers liver specificity of elevated alkaline phosphatase
- Not elevated in bone disorders and pregnancy
- Elevated alkaline phosphatase and GGT: signifies liver disease
- Elevated alkaline phosphatase, normal GGT: signifies bone disease, pregnancy

The PT is sensitive to rapid changes in synthetic function because the half-life of each factor is a few hours. The test is especially sensitive to deficiencies of factors VII and X. Also, synthesis and activation of the four coagulation factors from the liver require vitamin K for their carboxylation. Vitamin K1 is utilized poorly in significant paren-chymal liver disease and malabsorbed in obstructive cholestatic liver disease.

A prolonged prothrombin time in clinical practice:

- Does not predict risk of obvious hemorrhage
- Does not predict reliably the risk of bleeding after liver biopsy
- Does not assess platelet count or function
- Occurs also in disseminated intravascular coagulation (detectable fibrin degrada-tion products), anticoagulant therapies, inhibitors (e.g. lupus anticoagulant) and antibodies to clotting factors
- Fresh frozen plasma masks the PT result and rarely prevents bleeding
- Liters of fresh frozen plasma are required to correct a PT > 100 s (> 50 s USA) in ALF
- Serial PT tests are essential for monitoring prognosis of ALF

Pitfalls with the prothrombin time (PT):

The clinical specimen
- Incorrect sample volumes and handling give spurious results
The patient
- Vitamin K1 dependent test: PT corrects with obstructive jaundice
- Fresh frozen plasma masks the result and rarely prevents bleeding
Laboratory processing
- Requires rigid quality control
- Normal ranges vary according to tests, handling and thromboplastins
- Thromboplastins: British comparative (12–14 s) versus USA rabbit brain (9–11 s)
- The same sample can yield different results with different thromboplastins
- Data comparisons difficult with significant interlaboratory variation
- The PT index (ratio of patient versus control) cannot be compared unless the same batch of thromboplastin is used. (Thromboplastins in Europe are more responsive than in the USA to the anticoagulant effects of warfarin)
- Conversions between results: inaccurate (non-linear); e.g. percentage of normal

Plasma (not serum) is separated from blood collected in specimen tubes containing a decalcifying agent (citrate). The time to clot is tested against control plasma following addition of calcium and tissue thromboplastin (human or rabbit brain, lung or placenta). Tissue thromboplastin contains phospholipid and excess glycoprotein called tissue factor. These serve to trigger the extrinsic pathway and common portion of the intrinsic pathway of the clotting cascade.

One-stage prothrombin time. This test was devised by Quick in 1935. He presumed this measured deficiency only of prothrombin but this predated discovery of other clotting factors. The prothrombin time test measures factors I (fibrinogen) and the vitamin K dependent factors; II, V, VII, IX and X against a standardized control substrate. Results are expressed as seconds versus control, e.g. 20 s versus 12 s, or expressed as a ratio, 20/12 (1.7). This prothrombin ratio will differ between assays, depending on the variation in control thromboplastins. The ratio must not be confused with the International Normalized Ratio (INR, see below), which serves to correct for variations in thromboplastins and allows for comparisons between assays in different laboratories.

Authorities in coagulation recommend restricting use of the INR to monitoring anti-coagulation therapy in patients without liver disease. However, some liver centers use the INR to monitor patients with severe liver disease.

Thrombocytopenia (< 75 000 × 10^6/l) in liver disease:

- Low platelet counts are common with splenomegaly
- Splenomegaly is common in chronic liver disease with portal hypertension
- Preparation for percutaneous liver biopsy
- Risk of hemorrhage increases with the PT prolonged > 2 s and reduced platelet numbers
- Percutaneous liver biopsy is contraindicated with low platelet counts (< 75 000 × 10^6/l)
- The bleeding time (normal range, 5–10 min) assesses function of platelets and clotting factors *in vivo*

Prothrombin time (PT) in acute liver damage:

- The most sensitive test of recent acute liver synthetic function
- Much more sensitive than serum albumin levels in acute liver disease
- Best indicator of prognosis in acute liver failure
- Less useful in chronic liver disease: normal in compensated cirrhosis
- Normal range (s) set per lab: e.g. 9–11 s (USA) and 10–14 s (UK)
- A change of >1 s is significant
- Changes occur within hours with deterioration in liver function
- Rising values correlate with necessity for liver transplantation

Key Notes International Normalized Ratio (INR)

This index was developed to allow comparisons of degrees of anticoagulation with warfarin therapy in patients at risk of thrombosis. The INR is the prothrombin time (PT) ratio that would have been obtained had the reference thromboplastin been used

INR = prothrombin time ratio (patient/mean of normals) (ISI)
N.B. This is on a log scale

ISI = International Sensitivity Index; value assigned by the manufacturer to each thromboplastin after calibration against an International reference thromboplastin (World Health Organization). The ISI for reference human brain is 1.0. In Europe, typical ISI values are 1.1–1.3 versus 2.2–2.6 (double) in the USA.

Prothrombin time in viral hepatitis:

Acute uncomplicated viral hepatitis
- Normal

Acute liver failure
- Elevated: best test of acute (recent) liver synthetic function
- Indicator of adverse prognosis
- >100 s (UK) or >50 s (USA); negligible survival without tranplantation
- Result masked with fresh frozen plasma

Chronic viral hepatitis
- Elevated in end-stage (decompensated) liver disease

Serum albumin. The half-life of albumin, also synthesized by the liver, is around 21 days. Serum levels remain within the normal range in acute viral hepatitis. A low serum albumin, in the absence of loss from the kidney (proteinuria), skin (burns) and malnutrition, indicates chronic liver dysfunction. Levels may be lower than anticipated in alcoholics due to their poor nutrition. A low level is unusual in a patient presenting with ALF and should trigger a search for underlying chronic liver disease, especially alcohol-related cirrhosis or Wilson's disease, or other causes of protein loss such as the nephrotic syndrome.

1.7 Acute hepatitis

This describes an acute inflammation of the liver. Viral and non-viral causes cannot be distinguished readily without additional tests, especially serological markers. The illness may resolve completely with clearance of the infection for example hepatitis A or E. Acute hepatitis may progress to chronic hepatitis for example hepatitis B and C. Rarely, liver failure may develop as a complication of acute hepatitis.

1.7.1 Acute liver failure

This term embraces all definitions and etiologies. Acute liver failure (ALF) replaces the popular term fulminant hepatitis (see section 5.1). There is no consensus definition, especially on exclusion criteria such as underlying chronic liver disease.

1.7.2 Biochemical profiles in acute hepatitis, including ALF

The biochemical profiles are similar for acute hepatitis A through E; the different etiologies cannot be readily distinguished because variations and fluctuations within tests and individuals are as great as for between the types of viral hepatitis (*Tables 1.15* and *1.16*).

Characteristically, in mild anicteric acute hepatitis B, serum levels of bilirubin and AST and ALT show modest elevations (<5 × upper limit of normal) whereas the viremia (HBV DNA levels) may be high. The most severe immune attack is seen in fulminant hepatitis B; serum AST and ALT and bilirubin levels are very high (>10 × upper limit of normal) early on reflecting massive destruction of hepatocytes. Serum HBV DNA levels are very low (typically undetectable by dot-blot assay) reflecting the exaggerated immune attack which leads to clearance of virus; subsequent chronic infection is rare.

Table 1.15. Biochemical profile in acute viral hepatitis: uncomplicated, cholestatic phase and ALF

	(Normal ranges)	Early	Cholestatic phase >2 weeks symptoms	Acute liver failure
Jaundice and cholestasis				
Total bilirubin	(2–15 µmol/l)	10–75	75–300	20–150
	(0.2–0.9 mg/dl)	1–5	4.5–18	2–9
	(0.8–2.2 µkat/l)	4–11	11–44	8.0–22.0
Alkaline phosphatase	(30–130 IU/l)	50–300	300–600	50–500
	(0.5–2.2 µkat/l)	0.8–5.0	5.0–10.0	0.8–8.0
GGT	(10–50 IU/l)	20–150	100–300	20–100
	(0.1–0.5 µkat/l)	0.2–1.5	1.0–3.0	0.2–1.0
Liver cell damage				
AST (SGOT)	(5–40 IU/l)	1000–8000	100–300	200–10 000
	(0.17–0.83 µkat/l)	34–133	3.4–15	7–340
ALT (SGPT)	(5–40 IU/l)	2000–10 000	100–500	200–10 000
	(0.08–0.67 µkat/l)	32–160	1.6–8.0	0.8–16.0
Liver synthetic function				
Albumin (35–50 g/l)		Normal	Normal	Normal
Prothrombin time (10.5–14.5 s)		Normal	10.5–16	20 to >100 s (UK) 20–50 s (USA)
After vitamin K1 administration		Normal	Corrects to normal	Minimal correction

Table 1.16. Biochemical profiles in acute hepatitis A (typical case)

Normal	Bilirubin < 17 µmol/l (< 2.2 µkat/l)	AST < 40 IU/l (< 0.83 µkat/l)	ALT < 40 IU/l (< 0.67 µkat/l)	Alkaline phosphatase < 130 IU/l (< 2.2 µkat/l)	GGT < 50 IU/l (< 0.5 µkat/l)	Prothrombin times (s) UK (12–15 s)[a]	USA (10–12 s)[a]
Early, uncomplicated	51 (6.6 µkat/l)	3760 (63 µkat/l)	4800 (80 µkat/l)	180 (3.0 µkat/l)	1.25 (75 IU/l)	14	11
Cholestatic phase	170 (22 µkat/l)	480 (8.2 µkat/l)	520 (8.6 µkat/l)	742 (12.4 µkat/l)	13.6 (820 IU/l)	18[b]	15[b]
Early ALF	139 (18 µkat/l)	4800 (182 µkat/l)	5563 (92 µkat/l)	240 (14.0 µkat/l)	6.5 (390 IU/l)	47[c]	26[c]
Late, severe ALF	160 (21 µkat/l)	376 (6.3 µkat/l)	448 (7.4 µkat/l)	210 (3.5 µkat/l)	4.7 (215 IU/l)	85[c]	48[c]

[a] Prothrombin times: different thromboplastins used in UK and USA; absolute seconds and PT ratios are not comparable. [b] Corrects to normal with vitamin K1. [c] Not fully correctable with vitamin K1.

Serum levels of ALT and AST in viral hepatitis:

Early acute hepatitis
- High (>8.3 μkat/l: >500 IU/l)
- Serum ALT and AST rise before bilirubin
- ALT and AST equally elevated in viral hepatitis (ALT > AST)
- Levels fall rapidly with recovery and normal liver synthetic function

Chronic viral hepatitis
- Typically below 8.3 ukat/l (< 500 IU/l)
- May fluctuate: intermittently within the normal range
- No correlation with severity of liver damage
- May be within normal ranges in cirrhosis

The pitfalls with serum ALT and AST levels are summarized in *Table 1.17*.

Alkaline phosphatase and GGT in viral hepatitis:
- Do not discriminate between etiologies, stages, severities
- GGT elevated with superimposed alcoholic hepatitis

Early acute viral hepatitis
- Low (< 5 ×) in uncomplicated infection
- High (> 5 ×) in cholestatic phase

Chronic viral hepatitis
- Levels reflect cholestasis and jaundice
- May be high in hepatitis C (bile duct damage)

1.7.3 Histopathology

Advances in serological assays for markers of viral infections have made liver histology secondary in the diagnosis of acute and chronic viral hepatitis in the immunocompetent host (*Table 1.18*). Histological features are non-specific for the diagnosis of viral hepatitis, except for ground-glass hepatocytes in hepatitis B and characteristic inclusion bodies in hepatitis due to CMV.

Table 1.17. Pitfalls with serum AST and ALT levels

Poor correlation
- Severity of clinical illness and prognosis
- Severity of liver damage (necrosis of hepatocytes) on histology

Misleading low levels (<8.3 μkat/l)
- Acute liver failure (ALF): the PT remains elevated
- Normal range values can occur in cirrhosis

Misleading high levels (>8.3 μkat/l)
- In acute myocardial infarction
- Skeletal muscle injury
- Early acute obstruction of the common bile duct
- AST > ALT in viral hepatitis superimposed on alcoholic liver disease

> **Case History – 33-year-old woman presents with a three week history of malaise**
> ALT 1599 IU/l; AST 428 IU/l; total bilirubin 22 mg/dl (374 μmol/l); PT 13 s prolonged.
> Four days later, she is admitted for confusion:
> ALT 487 IU/l; AST 179 IU/l; total bilirubin 29 mg/dl (494 μmol/l); PT 22 s prolonged;
> HBsAg negative, anti-HBs positive, total anti-HAV negative.
>
> **Comment**
> Consistent with acute liver failure. The patient was IgM anti-HBc positive.
> The falling liver enzymes reflected the failing liver in the presence of a rising pro-
> thrombin time.

Acute uncomplicated (non-fulminant) hepatitis. General features include damage to hepatocytes, infiltration of the liver by inflammatory cells (mostly mononuclear cells, lymphocytes and plasma cells) and disarray of liver cell plates. Damaged hepatocytes become swollen (ballooned) and undergo acidophilic degeneration (shrinkage) with apoptosis. Surviving hepatocytes undergo regenerative hyperplasia leading to disarray of the architecture of the liver plate (*Table 1.19*).

Table 1.18. Indications for liver biopsy in acute hepatitis

- Serological markers are the mainstay for diagnosis
- Histology usually is unnecessary for diagnosis of hepatitis A, B, B + D or E in the immunocompetent host
- Histology is helpful for diagnosis in:

Acute hepatitis
- Acute infection in immunosuppressed host (including graft recipients) – serological markers may be misleading
- Hepatitis C – serological markers do not discriminate acute from chronic infection
- Other viruses: herpes, unusual infections
- Cholangiolytic hepatitis

Unresolved disease
- Acute-on-chronic hepatitis including reactivation of chronic hepatitis (hepatitis B, B and D, C)
- Unsuspected underlying chronic liver disease from other causes e.g. alcoholic hepatitis, autoimmune hepatitis, alpha-1-antitrypsin deficiency, Wilson's disease etc.

Histological evolution of acute viral hepatitis:

- Spotty (focal) necrosis: variation in severity of damage to hepatocytes within a zone
- Periportal necrosis: typical of severe hepatitis A (reversible)
- Bridging necrosis: necrotic hepatocytes linking portal tracts to terminal hepatic venules following collapse of confluent necrosis in zone 3. A precedent to cirrhosis in hepatitis B

Acute liver failure:

- Panacinar necrosis: confluent necrosis of entire acinus
- Multiacinar necrosis: confluent necrosis of adjacent acini

Table 1.19. Histological features of cell degeneration and death

Acidophilic degeneration	Necrosis of hepatocytes typically seen in acute viral hepatitis. The hepatocytes are characterized by dense staining and are irregular in outline
	Degenerating hepatocytes may undergo apoptosis (syn. acidophilic bodies, Councilman bodies)
Apoptosis	Programmed cell death (physiological suicide). A natural process which removes cells that have served their function
Councilman bodies	Acidophilic degeneration of hepatocytes (apoptotic bodies). Described in Yellow Fever but may be abundant in other hemorrhagic fevers and present in other causes of acute liver injury
Erythrophagocytosis	Engulfment of red blood cells (erythrocytes) by macrophages such as Kupffer cells. Seen in some etiologies of viral hepatitis, especially those due to EBV, HSV and VZV

Damage occurs throughout the acinus but is maximal around hepatic venules (zone 3 of Rappaport). Damage also may be prominent around portal tracts (zone 1) in hepatitis A. Kupffer cells are prominent. Bile ducts and bile canaliculi usually are not affected except in hepatitis C. Lipofuscin granules may be seen in Kupffer cells in resolving acute hepatitis. Fibrosis is absent; its presence indicates chronic liver disease.

Histological features do not necessarily point towards a specific viral etiology. Features suggestive of herpesvirus infections (HSV, VZV but not CMV) include confluent necrosis with hemorrhage and viral inclusion bodies. The inflammatory cell infiltrate may predominate in sinusoids (similar in hepatitis C) rather than alongside hepatocytes. Necrosis of hepatocytes is less prominent than in hepatitis B. Mitoses may be abundant in regenerating hepatocytes and epithelioid granulomata may be present. Erythrocytes (red blood cells) may undergo phagocytosis by Kupffer cells (erythrophagocytosis).

A patient transplanted for long-standing chronic hepatitis B may develop severe acute hepatitis B which progresses to acute liver failure and loss of the graft. The histological features of fibrosing cholestatic hepatitis (FCH, see section 6.3.3) differ from acute hepatitis B and ALF in the native liver.

1.8 Chronic hepatitis and cirrhosis

Chronic hepatitis defines a chronic inflammation of the liver with necrosis of hepatocytes. This is often accompanied by fibrosis, and is without improvement for 6 months or more. This definition is used regardless of etiology (viral or non-viral) (*Table 1.20*).

Table 1.20. Causes of chronic hepatitis

- HBV infection
- HDV superinfection with HBV
- HCV infection
- Autoimmune hepatitis
- Wilson's disease
- Alpha 1-anti-trypsin deficiency
- Drugs (methyl dopa, anti-TB[a] etc.)
- Cryptogenic (unknown)

[a] Anti-tuberculosis drugs such as isoniazid, many drugs cause a histological picture of CAH.

Chronic viral hepatitis usually is accompanied by abnormal liver enzyme tests for 6 months or more following acute hepatitis. The 6 month time limit to distinguish chronic, from acute, hepatitis is arbitrary but seems to predict most cases in which chronic inflammation will persist.

1.8.1 Biochemical profiles in chronic hepatitis

Significant liver dysfunction must be present to explain a persistently elevated level of total serum bilirubin. Serum levels of bilirubin, ALT and AST characteristically fluctuate in chronic hepatitis B and C and may return intermittently to within their normal ranges. In symptomless chronic hepatitis B, serum levels of ALT and AST typically are only mildly elevated ($< 3 \times$ upper normal limits) with high levels of viremia. This paradox reflects the relative lack of immune attack in chronic carriers of HBV; liver histology typically shows mild inflammatory activity (historically denoted as chronic persistent hepatitis). Conversely, in severe chronic hepatitis B (historical: CAH with marked necroinflammatory activity in the liver) liver enzymes show moderate elevations, albeit periodically. The HBV DNA level typically is low, reflecting the attempt of the host to eradicate HBV from infected hepatocytes (*Table 1.21*).

1.8.2 Child–Turcotte–Pugh classification: disease severity in chronic hepatitis

The Child–Turcotte classification, modified by Pugh, was formulated to assess disease severity in chronic liver disease, regardless of etiology (*Tables 1.22* and *1.23*). The aim was to help predict prognosis and operative risk of portocaval shunting for chronic liver disease, especially following hemorrhage from esophageal varices. These formulae were developed before the discovery of HCV but overall find usefulness in standardizing disease severity between patients and clinical trials (compare Knodell score, see section 1.9.2).

An elevated level of serum bilirubin and PT and a low serum albumin correlate with poor survival, especially in chronic liver disease due to non-viral causes such as alcoholic cirrhosis and primary biliary cirrhosis. An elevated serum bilirubin remains the most important predictor of adverse prognosis, especially in alcoholic cirrhosis when combined with an elevated PT (vitamin K resistant).

Table 1.21. Typical biochemical profiles in chronic hepatitis

Normal	Albumin (35–50 g/l)	Bilirubin <17 μmol/l (<2.2 μkat/l)	AST <40 IU/l (<0.83 μkat/l)	ALT <40 IU/l (<0.67 μkat/l)	Alkaline phosphatase <130 IU/l (<2.2 μkat/l)	GGT <50 IU/l (<0.5 μkat/l)	Prothrombin time UK (12–15s)[a]	Prothrombin time USA (10–12s)[a]
Well compensated	44 g/l	9 μmol/l (1.1 μkat/l)	65 IU/l (1.5 μkat/l)	45 IU/l (0.7 μkat/l)	100 IU/l (1.1 μkat/l)	43 IU/l (0.4 μkat/l)	12 s	11 s
Decompensated[b]	26 g/l	85 μmol/l (11 μkat/l)	480 IU/l (8.2 μkat/l)	520 IU/l (8.6 μkat/l)	742 IU/l (12.4 μkat/l)	820 IU/l (13.6 μkat/l)	22 s[e]	17 s[e]
Child–Pugh score[c]	3 points	3 points	NA	NA	[high: liver tumor][d]	[high: alcohol][d]	3 points	3 points

[a] Prothrombin times: different thromboplastins used in UK and USA: absolute seconds and PT ratios are not comparable. [b] Decompensated: acute-on-chronic exacerbation with ascites (the patient had developed HCC). [c] Child–Pugh score (see Table 1.22) assesses serum albumin level (3 points since below 2.8 g/dl), serum bilirubin level (3 points as 3.1 mg/dl) and prothrombin time (3 points as > 6.1 s prolonged) as well as presence of hepatic encephalopathy and ascites (see section 1.8.2). This patient had mild ascites (two points) but without hepatic encephalopathy and scored 11 points. [d] Note the AST was higher than the ALT and reflected the continued alcohol excess. The GGT also was high due to excess alcohol. Alkaline phosphatase was high due to a space-occupying lesion (liver tumor), alpha fetoprotein was elevated in this patient. [e] Not correctable with vitamin K1. NA, not applicable.

Table 1.22. Child–Turcotte–Pugh classification for severity of chronic liver disease: Minimum (best) score is 5, maximum (worst) score is 15

Points scored	1	2	3
Bilirubin (μmol/l)	<25	25–40	>40
Albumin (g/l)	>35	28–35	<28
Prothrombin time[a] (s)	1–4	4–6	>6
Ascites	None	Slight	Moderate
Grade encephalopathy[b]	0	I or II	III or IV

[a] PT: seconds prolonged from control. [b] Grades I-IV: degrees of confusion to coma.

Table 1.23. Interpretation of Child–Turcotte–Pugh score for severity of chronic liver disease

Grade	Points scored	Operative risk
A	5–6	Good
B	7–9	Moderate
C	10–15	Poor

1.8.3 Chronic viral hepatitis

Chronic carrier (hepatitis B). HBsAg seropositivity for a minimum of 6 months.

Historically this term had predictive value because most chronic hepatitis B carriers remained HBsAg seropositive with elevated liver enzymes (ALT, AST) for many years. Antiviral therapies were reserved for chronic carriers. Following acute hepatitis B, patients typically had to wait a minimum of 6 months to demonstrate lack of seroconversion from HBsAg to anti-HBs before being considered eligible for antiviral therapy. Today, this time delay is considered unnecessary. Chronic hepatitis B is likely following acute hepatitis if HBV DNA in serum remains detectable (by dot-blot hybridization) after the peak elevation of liver enzymes. Serial analyses of HBV DNA over 1–3 weeks should give an indication of the likelihood of clearing hepatitis B infection (see also section 3.6.2).

Historically, the word 'carrier' often was interpreted as being benign and applied particularly to potential blood donors who were symptomless when detected by routine screening. There was reluctance to biopsy seemingly well individuals. Subsequently, longitudinal follow-up and histological assessments showed that chronic carriers were at risk of progressive liver disease, including development of cirrhosis and HCC.

1.8.4 Predicting and monitoring responses to antiviral therapies

Elevated baseline pre-treatment enzyme activities (>3 × upper limit of normal) are favorable indicators of likely seroconversion from HBeAg to anti-HBe in HBsAg seropositive patients chronically infected with HBV. In contrast, elevated levels do not correlate with likelihood of clearance of HCV in chronic hepatitis C.

Caution must be exercised also when extrapolating therapeutic strategies and patho-genetic mechanisms of chronic infection between chronic hepatitis B and C. In chronic hepatitis B elevations in AST and ALT above 3 × upper normal limits during interferon therapy predict clearance of virus in most cases. Clearance is achieved via stimulation of the cellular immune response of the host (cytotoxic T cells) targeted to cause necro-sis of infected hepatocytes. In chronic hepatitis B, elevated baseline activities indicate some attempt, and residual capacity, by the host to clear HBV from infected hepato-cytes. The interferon therapy serves to boost this cytotoxic immune-mediated damage and favors clearance of virus from infected hepatocytes. Conversely, in the minority of patients with chronic hepatitis C who maintain clearance of virus (HCV RNA levels remain undetectable by RT–PCR), serum enzyme activities tend to fall rapidly without showing this preliminary peak. This discordance indicates differences in actions of interferon and responses of the liver in the two diseases.

Many clinical trials have interpreted a reduction in elevated serum activities of AST and ALT as indicative of a 'response' to antiviral therapies in chronic hepatitis B and C. The assumption has been that normalization of enzyme activities equates with cessa-tion of virus replication and, consequently, a lessening of necroinflammatory activity of the liver (see Knodell Score, below). By further extrapolation, these events are assumed to herald clearance of HBV or HCV from the liver.

Although normalization of AST and ALT levels seems to be a prerequisite for clearing HBV or HCV in the long-term, this 'complete response' in the short-term (< 6–12 months) does not guarantee eradication of virus from the liver. In most cases during or after cessation of therapy, viral nucleic acids remain detectable in serum using molecular techniques. Also, in the majority with chronic hepatitis B or C, liver enzyme activities increase following cessation of antiviral therapy indicating a tempo-rary suppression, rather than cessation, of virus replication. Elevations in enzyme activities do not correlate with disease severity. Similarly, reductions, including to within the normal range, do not necessarily reflect improvement in inflammatory activity in the liver or clearance of virus, especially in chronic hepatitis C and follow-ing antiviral therapy. In chronic hepatitis C, normalization of enzyme activities during antiviral therapy with ribavirin is achieved often independently of levels of HCV RNA (on RT–PCR). Significant hemolysis, especially in patients with glucose-6-phosphate dehydrogenase deficiency, can confound interpretation of AST levels.

1.8.5 Cirrhosis

Cirrhosis (Kirrhos, Greek for tawny) defines a disease affecting the whole liver char-acterized by fibrosis and regenerating nodules from surviving hepatocytes (liver parenchyma). Cirrhosis is irreversible and frequently precedes the development of HCC. Cirrhosis is believed to follow from chronic viral hepatitis when there is repeated necroinflammatory activity of the liver parenchyma, which is favored by continuing virus replication. Cirrhosis is common with chronic hepatitis C. Accordingly, large numbers of patients are at risk of developing HCC. Cirrhosis following infection with HBV or HCV probably takes many years to develop. Rapid development of cirrhosis has ensued within months in the grafts of some recipients transplanted for chronic viral hepatitis.

Cirrhosis may be symptomless for many years. Clinical and histological features do not distinguish viral from other causes of cirrhosis. Clinical presentations and complications are similar for cirrhosis regardless of etiology. Patients with cirrhosis related to chronic hepatitis B, B+D, and/or C are at increased risk of developing primary liver cancer.

Decompensated cirrhosis is a clinical term relating to the development of complications of advanced, end-stage liver disease when the liver begins to fail. Features such as encephalopathy and ascites reflect the critical loss of liver function.

Regeneration nodules formed in the cirrhotic liver restrict the flow of blood, almost two-thirds of which comes from the portal vein, leading to portal hypertension. The opening up of a collateral circulation can result in the formation of varices in the lower end of the esophagus. These are liable to rupture. Endoscopic sclerotherapy or banding may be used for emergent obliteration. Insertion of a TIPSS (transjugular intrahepatic portosystemic shunt) is aimed at providing a bypass to reduce the portal hypertension.

1.8.6 Mechanisms of carcinogenesis

Repeated episodes of inflammatory activity predispose to necrosis of hepatocytes and, ultimately, cirrhosis. The pathogenetic mechanisms predisposing to cirrhosis remain unclear. Cirrhosis is the common denominator in the majority of patients with HCC regardless of cause. Mechanisms of oncogenesis are likely to differ between HBV and HCV. HBV DNA integrates into the host genome whereas HCV RNA remains in the cytoplasm. HCC may develop without cirrhosis in some cases with chronic hepatitis B; cancer without cirrhosis is exceptionally rare in chronic hepatitis C.

1.9 Role of liver biopsy

Liver histology is invaluable in all cases of suspected chronic viral hepatitis, especially in interpreting objectively the response to antiviral therapies (*Table 1.24*). In chronic hepatitis, serological markers may be unaltered or change slowly and reduction in levels of liver enzymes may not reflect accurately the impact on the liver.

Table 1.24. Role of liver biopsy

• Confirm diagnosis	• Limited in viral hepatitis[a]
• Exclude other lesions	• Many diseases indistinguishable
• Grade the necroinflammatory activity	• Grading may increase or decrease
• Stage fibrosis and progression of disease	• Staging typically progresses
• Evaluate results of therapies	• Semiquantitative objective scores

[a] Absence of pathognomonic histological features in most cases except ground-glass cells containing HBsAg in hepatitis B.

Liver histology for chronic viral hepatitis:

- Essential for staging liver disease activity and severity
- Essential for monitoring antiviral therapies
- Serological markers and liver enzyme tests; unreliable indicators of disease severity
- Cirrhosis may be present with normal liver enzyme tests
- Other diseases may be present: alcoholic liver disease, HCC, alpha-1-anti-trypsin deficiency etc.

1.9.1 Historical terminology

Chronic hepatitis describes also the historical histological terms, chronic persistent hepatitis (CPH) and chronic active hepatitis (CAH). These were descriptive terms of liver morphology based on inflammatory activity and cell infiltrations and were applied alike to viral and non-viral (especially autoimmune) liver diseases.

This classification was based on prognosis; the outcome with CPH or chronic lobular hepatitis (CLH) is better than with CAH. These terms do not discriminate between viral and non-viral causes such as autoimmune hepatitis and drug-related hepatotoxicity. CPH, CLH, CAH form a spectrum of severity. Classification in the individual may be difficult given the significant sampling error of needle biopsy and lack of consensus over interpretation of the presence and importance of piecemeal necrosis in determining prognosis. These terms are widely used but are becoming obsolete with the appreciation that these categories form a spectrum with considerable histological and clinical overlap.

1.9.1.1 Chronic persistent hepatitis (CPH)

CPH is defined as chronic inflammatory infiltrate confined predominantly to the portal tracts, with little or no evidence of piecemeal necrosis. The architecture of the liver lobule and histological relationship between hepatic venules and portal tracts is preserved. The term CPH was used to define lesions considered more benign (non-progressive) than CAH but should be discarded because progressive liver disease and development of cirrhosis can occur.

1.9.1.2 Chronic lobular hepatitis (CLH)

This category was added to the general classifications later than CPH and CAH and describes features of acute hepatitis, with lymphocytic infiltrates within the lobules, which persist beyond 6 months.

1.9.1.3 Chronic active hepatitis (CAH)

A histological term, usually accompanying the clinical description of chronic aggressive hepatitis (synonym, aggressive chronic hepatitis). CAH is defined by a chronic inflammatory infiltrate in the portal tract which extends beyond the limiting plate into the parenchyma of the lobule, separating hepatocytes. This term was used to signify a more aggressive histological lesion than CPH with the propensity to develop cirrhosis. The architecture of the liver lobule is disturbed but there are no regeneration nodules (a hallmark of cirrhosis). CAH can coexist with cirrhosis: 'active cirrhosis' (historical term).

1.9.2 Knodell score (histological activity index)

Objective assessments of liver histology are essential in evaluating the influence of treatments (e.g. antiviral therapies) in controlled clinical trials. Semi-quantitative assessments have been developed, such as the Knodell histological activity index (HAI) (*Table 1.25*). Numbers are added but are non-linear (scores 1,3,4) and non-parametric statistical analyses are applied. The Knodell score indicates the severity of inflammation and is useful, for example, in monitoring improvement during anti-viral therapy. In contrast, the Child's–Pugh score (see section 1.8.2) measures overall severity of liver disease in terms of clinical outcome. Cases with an extremely poor prognosis and high Child's–Pugh score paradoxically may have low Knodell scores because the liver shows little inflammation, for example, in 'burnt-out cirrhosis'.

1.9.3 The METAVIR method

This scoring system was formulated from the consensus of a group of 10 pathologists in France and is designed for interpreting liver needle biopsies of hepatitis C. The METAVIR method scores (mild, moderate, severe from 0–3) for necroinflammatory activity (A0–A3), portal inflammation (P0–P3), polymorphonuclear cells, fat and bile duct injury as well as fibrosis from F0–F4. The METAVIR scoring method has been used mostly to assess rates of progression of fibrosis in chronic hepatitis C and alcoholic liver disease.

Table 1.25. The Knodell score

A semi-quantitative histological activity index (HAI)—usually modified

Inflammatory activity

    Periportal necrosis, interface hepatitis, bridging necrosis, multilobular necrosis (0–10)
    Intralobular necrosis (0–4)
    Portal inflammation (0–4)

Stage
    Fibrosis (0–4)

Many authorities also include
    Fat (0–4)
    Iron (0–4)

Knodell inflammatory activity scores are considered severe if above 12 (maximum 18).
Most patients with chronic HCV or HBV infection score 5–11.

1.9.4 Reclassification of chronic hepatitis

The term chronic hepatitis replaces CPH, CLH and CAH because these are considered a spectrum and CPH and CLH no longer are considered benign, non-progressive, lesions. Histopathological distinctions between CPH and CAH have become less clear especially with chronic hepatitis C; seemingly mild pathology (CPH, CLH) and absence of CAH do not necessarily predict a favorable outcome. Revised classifications include etiology because outcomes differ between chronic hepatitis B and C as well as for non-viral etiologies. Also, continuing virus replication is implicated in causing repeated liver injury—an important determinant in the progression to cirrhosis (*Table 1.26* and *Figure 1.4*).

1.9.5 Histopathology

Chronic hepatitis is a syndrome characterized by an inflammatory cell infiltrate, comprised mostly of mononuclear cells (lymphocytes and plasma cells) indicative of chronic inflammation, and by various forms of degeneration and necrosis of hepatocytes. This necroinflammatory activity is accompanied by varying degrees of bile duct damage (especially in chronic hepatitis C) and fibrosis of the portal tracts. The inflammatory infiltrate is located primarily in the portal tracts but may 'spill-over' into the lobule with invasion of the limiting plate (CAH). Lymphocytes may form aggregates, lymphoid follicles, characteristically seen in chronic hepatitis C (*Table 1.26*).

> **Key Notes** Histopathology
>
> Histopathological reports for signing out for the clinician should be improved with an internationally agreed standardized format. For example:
>
> Chronic hepatitis due to X (etiology, virus if known) with 1–3 grade of necroinflammatory activity and 1–3 stage of fibrosis. Cirrhosis is present or absent

Table 1.26. Suggestion for a simple scoring system for grading and staging chronic hepatitis (adapted from Hytiroglou *et al.*, 1995)

Necroinflammatory activity		Fibrosis/Cirrhosis	
0 None		0 None	
Interface hepatitis[a]	Lobular activity		Fibrous septa
1 Mild and lobular activity	Yes	1 Mild	None
2 Moderate and lobular activity	Yes	2 Moderate	Extend into lobules (limited[b])
3 Severe +/– bridging necrosis	Yes	3 Severe	Septa extend to vascular structures

[a] Synonymous with piecemeal necrosis. [b] Limited: not reaching vascular structures, terminal hepatic venules or portal tracts.

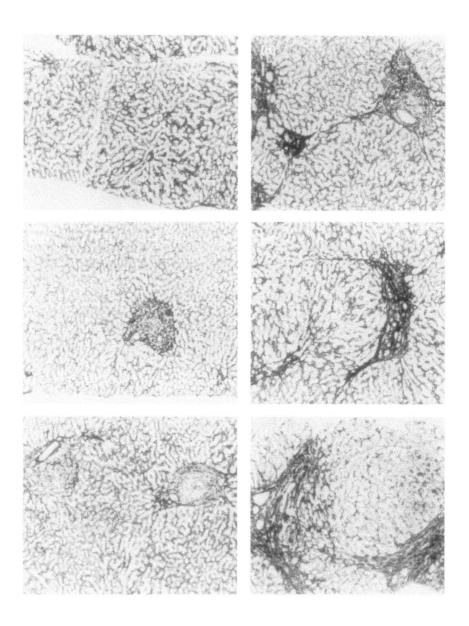

Figure 1.4. Progression of fibrosis to cirrhosis. (a) Normal liver. (b) Periportal inflammation. (c) Stage I: periportal fibrosis. (d) Stage II–III: bridging established between two portal tracts. (e) Stage III: bridging fibrosis. (f) Stage IV: cirrhosis, note nodule formation. (Reproduced with permission from Wong *et al.* (1997) *J. Virol. Hep.* **4**: 255–264. © 1997 Blackwell Science Ltd.)

Histological features do not necessarily point towards a specific viral etiology. Serological tests for viral markers remain the basis for diagnosis of viral hepatitis and complement histopathological assessment. Liver histology allows grading of the severity of necroinflammatory activity within lobules of the liver parenchyma, interface hepatitis and bridging necrosis, as these have predictive value for the likely development of cirrhosis.

Cirrhosis is believed to follow bridging necrosis of hepatocytes, defined as confluent necrosis linking vascular structures—portal tracts to terminal hepatic venules—within the hepatic parenchyma. The laying down of collagen—fibrosis—in areas of bridging necrosis in the portal tracts produces septa (strands) which extend into the lobules and link the vascular structures. Hepatocytes undergo regeneration to form nodules, a hallmark of cirrhosis.

1.9.6 Immunohistochemistry

Histochemical stains, such as Shikata's orcein or Victoria blue, will identify HBsAg within hepatocytes and complement the finding of ground glass hepatocytes in hepatitis B. Immunohistochemical methods using antibodies will demonstrate the distribution of HBcAg, HBeAg and HBsAg. Chronic hepatitis B of mild activity (CPH) typically is associated with HBcAg staining in nuclei, rather than in cytoplasm. Chronic hepatitis B of moderate or severe inflammatory activity (CAH) typically is associated with HBcAg in cytoplasm as well as nuclei. Immunohistochemical stains are being developed for detecting HCV antigens in liver tissue. Their potential applications are of interest because no serological counterparts are available.

These techniques remain predominantly research tools. They give some indirect indication of the degree of virus replication in the liver but have been superseded by molecular techniques which detect (and quantitate) viral nucleic acids in serum directly.

1.9.7 The immunocompromised host

The immunosuppressed individual, such as the liver transplant recipient (Chapter 6) or patient with HIV and acquired immune deficiency syndrome (AIDS) (Chapter 8), experiences an increased frequency of infections by the same range of viruses which infect the immunocompetent host. Clinical and histopathological features may differ. Liver histology is important for diagnosis in the immunocompromised host where antibody tests, such as anti-HCV, may be falsely negative. However, inflammatory activity can be misleadingly mild. Liver damage may be limited in patients with AIDS by the reduction in numbers of cytotoxic lymphocytes. Classical histological features may be absent, such as granulomata in tuberculosis. An immunoproliferative disease may occur with EBV infection. Lymphomas and Kaposi's sarcoma can affect the liver in patients with HIV and AIDS. A histological picture resembling sclerosing cholangitis with damage to bile ducts may be a prominent feature in HIV and AIDS.

Liver histology may be very helpful in the diagnosis of herpesvirus infections (CMV, EBV, HSV, VZV), especially in the immunocompromised host. Hepatitis with CMV may mimic sepsis: polymorphonuclear cells may predominate in the liver biopsy in clusters simulating microabscesses. Rapid diagnosis is essential; early treatment with antiviral agents, such as acyclovir or ganciclovir, improve prognosis.

1.9.8 Ultrastructural analysis

Electron microscopic analysis of tiny fragments of the liver biopsy, preserved in gluteraldehyde, may prove diagnostic in cases of hepatitis from unusual infections such as adenovirus, herpesviruses and exotic infections. Characteristic changes in ultrastructure occur in the mitochondria in Reye's syndrome within 24 h of hepatic encephalopathy which may assist the clinical diagnosis. Sampling must occur early on because these features become less prominent within days of the illness. Electron microscopy adds little to the diagnosis of hepatitis from the well-known hepatotropic viruses.

1.9.9 In situ *hybridization*

Detection (and quantitation) of viral nucleic acids directly in liver tissue should be the most accurate way of assessing virus replication. This approach can be applied to archives of embedded (formaldehyde-fixed, paraffin wax) tissue and could prove a valuable research tool when snap-frozen tissue is not available. So far, results add little to serological analyses except in the uncommon case without the typical serological profiles (cryptic infections). The clinical significance of the distributions of HBV RNA or HCV RNA within the hepatic lobule is unknown but may become more clear when interpreted in combination with cell subsets in inflammatory infiltrates.

Key notes Diagnosis of viral hepatitis

- Clinical features, epidemiology, not discriminative

- Diagnosis depends on specific serological markers

- Multiple agents may co-exist (e.g. B + C/+D)
        common routes of transmission (e.g. B/C/D)
        shared high-risk groups e.g. IVDU

- Serial/ repeat tests for suspected
        delayed seroconversion (HCV, Herpes)
        transient IgM (HDV, HEV)

- Acute-on-chronic liver disease may present as acute hepatitis

- Auto-antibodies (< 1 : 160) may be detectable in viral hepatitis

1.10 Clinical features

Discrimination between the many viruses infecting the liver based on incubation periods and clinical features can be misleading (*Table 1.27*). The patient with chronic viral hepatitis may be symptomless despite severe inflammation of the liver and coexisting cirrhosis. Presentation may be with non-specific symptoms such as fatigue and malaise. Multiple infections with hepatitis B, C and D viruses should be considered because of shared high-risk factors such as IVDU. Chronic liver disease may be suspected only on results of abnormal liver enzyme tests and serological markers of previous viral infections. Alternatively, the patient may present with complications of cirrhosis. Bleeding from esophageal varices signifies portal hypertension. Ascites, other microbial infections or a liver mass and loss of weight due to HCC are additional complications of cirrhosis. Clinical features and liver enzyme tests do not reflect accurately etiology or severity of liver disease. Liver enzymes occasionally may be within the normal range or show only minor elevations in cirrhosis.

1.10.1 Differential diagnosis

The diagnosis relies on interpretation of serological markers for viral infections, a detailed clinical history to exclude ingestion of drugs and exclusion of autoimmune (among other) liver diseases. Hepatitis, related to drugs, autoimmune disease, alcohol among other injuries, alpha-1-antitrypsin and Wilson's disease, may present acutely and be clinically and histopathologically indistinguishable from viral hepatitis. Features suggestive of drug-toxicity include fatty change (seen also in hepatitis C) and the presence of excess eosinophils and granulomata (excluding herpesvirus infections). Some patients may have concomitant viral and non-viral diseases.

Table 1.27. Clinical features of viral hepatitis

	A	B	C	D (with B)	E
Incubation (weeks)	2–6	4–26	4–?	4–26	2–10
Major routes	Fecal–oral	Parenteral, sexual	Parenteral	Parenteral	Fecal–oral
Sexual	Uncommon	Common	Uncommon	Uncommon	Rare
Vertical	Rare	Yes	Uncommon	Yes	Possible
Acute hepatitis	Yes	Yes	Uncommon	Yes	Yes
Chronic hepatitis	No	1–10% (adults) 90–95% (children)	>80%	Yes	No
ALF	$1 : > 10^5$	1 : 1000	Rare	Rare	Rare[a]
Vaccine	Yes	Yes	No	Yes[b]	No
Immune globulin	NHIG	HBIG	None	HBIG[b]	None

[a] Rare except for predilection in epidemics for women in third trimester of pregnancy.
[b] Immunoprophylaxis against HBV protects against HDV (except for superinfection of HBsAg carriers). NHIG, normal human immune globulin; HBIG, hepatitis B immune globulin.

In the liver graft recipient, confusion may arise in the presence of cellular rejection, ischemia and graft-versus-host disease. Immunohistochemical stains and *in situ* hybridization techniques to detect viral nucleic acids using RNA or cDNA probes on paraffin-embedded sections may be helpful in the immunocompromised patient.

1.11 Other disease associations

Viral hepatitis superimposed on other liver diseases, such as autoimmune liver disease, Wilson's disease and alcoholic hepatitis, may be sufficiently severe to force the patient to seek help. Consequently, viral hepatitis should not be diagnosed without consideration of why the patient chose to present at a particular time.

Extrahepatic manifestations are as likely to be due to additional diseases or complications as to be part of the clinical spectrum of hepatitis. Cardiac dysrhythmias arising during fulminant hepatitis are as likely due to metabolic disturbance (acid–base imbalance and low serum potassium) and raised intracranial pressure as due to myocardial disease. Purpura has been reported with hepatitis B and C and usually indicates disruption of the capillaries from immune complexes when accompanied by glomerulonephritis and arthralgias. Bruising, which can be mistaken for purpura, may occur when the low platelet count associated with hypersplenism is reduced further ($<30,000 \times 10^6/l$) as part of immune-mediated disease.

Elevated levels of serum amylase and lipase and overt pancreatitis have been reported in hepatitis B, including after liver transplantation. Biochemical and clinical pancreatitis are not uncommon companions of viral hepatitis especially on a background of excess alcohol and drugs such as azathioprine, corticosteroids and interferons, among many others, used in the treatment of chronic liver disease.

An unusual course, such as severe hepatitis A or B with renal failure in a child, should prompt extension of the search for additional precipitating factors—chronic liver disease such as shistosomiasis, or dengue and accompanying glomerulonephritis. Similarly, jaundice and a disproportionate elevation in AST (over ALT) in an otherwise well patient should lead to consideration of hemolysis from red cell abnormalities, alcoholic hepatitis and Wilson's disease, among many other conditions.

1.11.1 Extrahepatic manifestations

Reports of various extrahepatic manifestations and their diversity are increasing as more viruses are discovered (*Table 1.28*). Sufficient overlap exists between the different viruses and their unusual manifestations to prevent their clinical usefulness in diagnosis and management of the individual except for possibly polyarteritis nodosa associated with hepatitis B virus (see section 3.4.5.1) and porphyria cutanea tarda associated with hepatitis C virus (see section 4.3.6.4).

Cryoglobulinemia seems to occur most frequently (30–60%) in the context of chronic hepatitis C (see section 4.3.6.1) but is reported uncommonly in relapsing hepatitis A

Table 1.28. Extra-hepatic manifestations of viral hepatitis

Features	Comments
Skin rash, urticaria, purpura	Non-specific
Lichen planus, erythema multiforme	HBV: non-specific
Regional lymphadenopathy	Non-specific
Aplastic anemia	Non-A–E
Coombs' positive hemolytic anemia	Non-specific: rare
Sialadenitis	HCV (non-specific)
Acute pancreatitis	Non-specific, following liver transplantation
Acalculous cholecystitis and common bile duct obstruction	HAV, CMV
Myocarditis	Non-specific, coxsackieviruses (rare)
Landry–Guillain–Barre syndrome	Non-specific
Arthritis	HBV, HCV
Arthralgia	HAV, HBV, HCV
Cryoglobulinemia[a]	HAV, HBV, HCV, EBV
Mixed (types II and III)	HCV
Membranoproliferative glomerulonephritis	HBV and HCV
Vasculitis	HAV, HBV, HCV
Kawasaki-like disease	HBV
Porphyria cutanea tarda (sporadic)	HCV
Polyarteritis nodosa	HBV, ?HCV
Autoantibodies	
Anti-nuclear (ANA)	HCV, HDV
Liver Kidney Microsomal	
LKM-1	HCV
LKM-3	HDV
Anti-thyroid	HCV

[a] Polyclonal IgG and rheumatoid factor accompanied by monoclonal IgM in type II, and polyclonal IgM in type III, cryoglobulinemias.

(<5%), hepatitis B (10–15%), EBV and in patients with HIV infection. Cryoglobulins are detectable in other liver diseases, such as primary biliary cirrhosis, with comparable frequency (around 5–30%) to viral hepatitis, excepting hepatitis C. In hepatitis A, the immune complexes contain IgM anti-HAV antibodies and viral RNA. In hepatitis B, HBsAg and HBV DNA are detectable in most cryoprecipitates. In chronic hepatitis C, high concentrations of HCV RNA, viral antigens and anti-nucleocapsid antibodies have been detected. Cryoglobulins are associated with symptoms in a few patients with chronic hepatitis C. Symptoms are rare in such patients with chronic hepatitis B although their occurrence is more common with cirrhosis and, possibly, with various core and precore mutations associated with severe disease.

The overlap between autoimmune hepatitis and autoantibodies is also discussed in section 4.3.6.2 and porphyria cutanea tarda is discussed in section 4.3.6.4. The impact of iron overload in the liver in relation to diagnosis and treatment responses in chronic hepatitis C is discussed in section 4.3.6.6.

1.11.2 Aplastic anemia

This is a rare but well-recognized association with non-A, non-B hepatitis. Now, the majority of cases are recognized as non-A–E. Hepatitis accompanies aplastic anemia in around 2–5% of cases in the West. In contrast, this association seems especially common in the Far East; 4–23% of cases of aplastic anemia arising during childhood have concurrent viral hepatitis. There is a predominance of males (M : F = 2 : 1) with peak occurrences in childhood and adolescence. Aplastic anemia arises typically around 1–3 months following the onset of jaundice. The pathogenesis is obscure but immune mechanisms probably are involved. Reductions occur in CD4+ and CD8+ lymphocytes, which can be corrected with immunosuppressive therapy. Parvovirus B19 and GBV-C/HGV have been implicated in individual cases but in most instances a history of previous exposure to blood and blood products cannot be excluded.

1.11.3 Autoimmune features

Autoimmune disease should be excluded in all cases of suspected viral hepatitis. Testing for autoantibodies is essential when there is suspicion of autoimmune chronic hepatitis: abundant plasma cells (also seen in hepatitis A) and rosette formation of hepatocytes. The therapeutic options differ and have adverse consequences for the wrong disease. Corticosteroids increase virus replication and may lead to decompensation (ascites, spontaneous bacterial peritonitis, encephalopathy etc.) in chronic viral hepatitis. Interferons may aggravate autoimmune chronic hepatitis.

Separation of autoimmune hepatitis from chronic viral hepatitis may be difficult in clinical practice. First, the diseases may coexist. Chronic hepatitis B and/or C may follow an unscreened blood transfusion for bleeding varices, complicating cirrhosis from autoimmune hepatitis. Second, autoantibody markers are detectable in some cases of unequivocal viral hepatitis and do not necessarily indicate an autoimmune etiology. There are no specific 'cut-off' values for titers and variety of autoantibodies to segregate the different etiologies. High titers (>1 : 640) favor autoimmune hepatitis coexisting with viral hepatitis (detectable HBV DNA, HCV RNA and other viral markers). Third, autoantibodies may become detectable only during acute viral hepatitis, such as hepatitis A, or following instigation of antiviral therapy for chronic hepatitis B and C. Hyperthyroidism or hypothyroidism have been reported following interferon therapy. Rarely, antibodies to thyroglobulin and thyrotropin receptor have been detected in some of these treated patients.

Most difficulties come with attempting to interpret lower titers of autoantibodies in the presence of serological markers of viral hepatitis. Antibodies against smooth muscle and anti-nuclear antibodies (ANA), may be present in low titers (arbitrarily defined <1 : 160) in up to 10% of cases of chronic viral hepatitis, especially hepatitis C and D. Many patients with anti-LKM-1 antibodies (liver, kidney, microsomal) are infected with HCV. In most cases this reaction seems to represent non-specific cross-reactivity. In type 2 autoimmune hepatitis (AIH) antibodies to LKM-1 react against P450IID6, the cytochrome mono-oxygenase enzyme. Most patients with HCV RNA have undetectable anti-P450IID6 reactivity. Specificities of autoantibodies (for example anti-P450IID6, anti-cytokeratins, anti-F actins) and certain antigens such as the human leukocyte antigens (e.g. HLA-DR4) and changes in serial titers may help distinguish AIH from viral hepatitis (*Table 1.29*).

1.12 Management

1.12.1 Acute viral hepatitis

The vast majority of patients do not require any specific therapy and recover without hospitalization. No specific drug therapies are indicated for uncomplicated acute viral hepatitis. Data are insufficient regarding the potential benefits of using interferons especially in acute hepatitis B or C to hasten resolution of the acute illness and, importantly, to prevent persistent infection. Corticosteroids have no place in the management of acute viral hepatitis. Rarely and cautiously they should be used to alleviate intense, intolerable itching (pruritus) which can accompany the cholestatic phase of hepatitis A.

Table 1.29. Autoimmune hepatitis (AIH) and viral hepatitis

- Viral infections may trigger the presentation of AIH
- Treatment of AIH with immunosuppressive drugs (corticosteroids) may unmask viral hepatitis
- Interferons for chronic hepatitis B or C may unmask AIH
- HAV may trigger presentation of type 1 AIH

In hepatitis C

- False positive anti-HCV antibody tests (first generation) in hypergammaglobulinemia
- 1–4% of patients with anti-HCV antibodies (second, third generation) have features of AIH type 1
- In Italy most patients with anti-LKM-1 have HCV RNA

Case History – 43-year-old woman with a 1-week history of malaise and jaundice
ALT 7454 IU/l; AST 3654 IU/l; total bilirubin 4 mg/dl (68 μmol/l); PT 3 s prolonged; HBsAg positive, IgM anti-HAV negative.

Comment
Consistent with acute hepatitis B; the IgM anti-HBc tested positive.

1.2.1.1 Bed rest and diet

Rest and low-fat, high-protein, meals speed clinical well-being and clinical recovery but there is no evidence that these alter the outcome and prognosis for any given type of viral hepatitis. Bed rest is sensible but need not be enforced in the absence of fatigue and malaise and resolving serum enzyme activities (AST and ALT) accompanied by a normal PT. Restricted mobility usually is self-determined by fatigue. Rigorous exercise and sporting activities should be avoided until normalization of liver tests because many patients experience premature fatigue. Loose stools with occasional diarrhea may occur in hepatitis A or hepatitis E.

Patients may develop an early aversion to certain foods especially those of high-fat content, tea and coffee and cigarette smoking. Altered taste and anorexia may precede obvious manifestations of illness especially jaundice. Intake of an adequate nutritious diet is dictated more by taste and appetizing appeal than scientific rationale. Dietary intake of protein should be encouraged to limit weight loss and favor liver regeneration. Advice to restrict dietary protein and salt (sodium) is given in the mistaken belief that this will precipitate hepatic encephalopathy in acute hepatitis.

1.12.1.2 Abstinence from alcohol

Disagreement continues over the necessity of avoiding alcohol and recommended period of abstinence. Most patients without alcohol dependence seem to abstain spontaneously until liver enzyme activities begin to decline. Whether small amounts of alcohol (< 2 units/day) impair or delay recovery from acute uncomplicated hepatitis, remains unclear. Few studies have addressed intake of large quantities and do not control for the likelihood of having viral hepatitis superimposed on alcohol-related liver disease. Also, most predate adequate testing for HCV, and often also HBV, which are likely to confound any analysis.

Common sense would recommend abstinence during the acute phase, until liver enzyme activities return to near, or within, the normal range. Biochemical recovery is likely to take longer than anticipated with underlying chronic liver disease and should raise suspicion of additional etiologies such as alcoholic hepatitis. Scientific data are lacking to support recommendation that abstinence should extend for 6 months after normalization of all liver tests.

1.12.1.3 Incidental medications

Patients should continue without modification their regular medications such as estrogens (the oral contraceptive pill, hormone replacement therapy), anti-epileptic drugs, anti-hypertensive agents, anti-coagulation, among others. Any need to adjust anti-coagulation therapy, such as warfarin for previous deep vein thrombosis, should prompt critical reassessment of liver synthetic function, especially the PT.

1.12.1.4 Predicting complications

The key to managing any patient with acute viral hepatitis is to predict the minority who will pursue a complicated course with relapse and the rare patient who will develop ALF. The only clinical indicators of potential severity are age above 40 years (hepatitis A), pregnancy (hepatitis E and herpesviruses), and concurrent underlying chronic liver disease such as alcoholic hepatitis.

All patients should give a detailed clinical history, especially of potential risk factors in lifestyle (such as alcohol, recreational drugs, sexual behavior) for transmission. A thorough medical examination is necessary to exclude concomitant diseases; stigmata of chronic liver disease should be sought. A 'normal sized' liver (absence of hepatomegaly) may indicate viral hepatitis superimposed on cirrhosis which prevents enlargement of the liver. Splenomegaly may occur especially with EBV hepatitis but should alert the attending physician to other possibilities such as chronic liver disease with portal hypertension, Wilson's disease, hematological malignancy (such as lymphoma), among other conditions.

Ascites which is clinically detectable is uncommon in acute, uncomplicated viral hepatitis but is a feature of some patients with late onset hepatic failure. In clinical practice, obvious ascites more often indicates significant chronic liver disease (cirrhosis) and is accompanied by a low serum albumin—unusual in typical uncomplicated acute viral hepatitis. Other causes of ascites should be excluded such as veno-occlusive disease (HIV-1) related to Budd–Chiari syndrome, and non-hepatic causes, including abdominal malignancy and renal disease (nephrotic syndrome) which occasionally complicates hepatitis B and C.

Apart from serological markers for diagnosis, all patients should have liver tests carried out, particularly the prothrombin time. Tests should be repeated, if necessary after 24 h following correction of the PT for cholestasis with parenterally administered vitamin K1 (oral vitamin K may be absorbed slowly). Any elevation of PT should prompt early transfer to a referral center, especially in the face of falling enzyme activities (AST, ALT). These alone give no indication of severity and prognosis.

1.12.1.5 Anticipating persistent infection

The key for the other viruses, especially B, B+D and C, is to predict those infections that are most likely to persist because these patients will pursue a chronic course with the attendant adverse sequelae. Mild anicteric acute hepatitis B infection is more likely to persist than a severe case with obvious jaundice. Most presentations of hepatitis C are acute exacerbations of chronic hepatitis (acute-on-chronic) rather than *de novo* cases of acute hepatitis.

All patients should be closely followed with serial estimations of serological markers. Detection of HBV DNA in serum in serial samples over 1–4 weeks by dot-blot

Key Notes Management of acute viral hepatitis
- Viral hepatitis is a notifiable public health disease
- Viral hepatitis requires serological tests for diagnosis
- No specific recommendations and restrictions in most cases
- Regular diet, abstain from alcohol, rigorous exercise until normal ALT, AST
- Steroids should not be used except rarely for intolerable itching in hepatitis A

Key Notes Management strategies
Assessment of potential severity of acute illness
  Serological diagnosis B, C, D, E (pregnant) and herpes (pregnant) versus hepatitis A
  Underlying chronic liver disease e.g. alcoholic hepatitis
  Test for PT as well as liver enzyme activities
Discrimination for potential chronicity
- Acute resolving hepatitis (A, E) from potential chronic hepatitis (B, C, B+D)
- Unnecessary to wait 6 months to define chronic hepatitis B (HBsAg persists)
- Resolving acute hepatitis B or C: undetectable or falling levels of viremia
- Chronic hepatitis likely with persistence of HBV DNA (dot-blot) and HCV RNA (RT–PCR), especially high levels detected serially (e.g. > 2–4 weeks)
- Refer all babies and children early to special centers for hepatitis B, C, D (high chronicity)
- Antiviral therapies for chronic hepatitis B or C—liaise with special centers for protocols (clinical trials)
Care of contacts and family
- Notify public health officials
- Consider potential risk of outbreak/epidemics (closed institutions, food handlers etc.)
- Screen, immunize susceptible (seronegative) contacts, especially family members
  - IG for hepatitis A. Add vaccine for high risk individuals and containing potential outbreaks and epidemics e.g. among IVDU, homosexuals, within institutions
  - HBIG and vaccine for hepatitis B—especially babies and children
- Advise sanitary precautions for hepatitis A and E (washing hands, etc.)
- Advise against parenteral and sexual spread for hepatitis B (needles, safe sex, etc.)
- Pregnant women, children and the immunosuppressed—special considerations

hybridization (rather than by PCR) usually indicates persistent infection. These patients and all children with hepatitis B, B+D and C, without exception, should be referred to a center with expertise in managing viral hepatitis.

1.12.2 Chronic viral hepatitis

Assessment strategies are regardless of viral etiology. These should focus on evaluating objectively for monitoring and prognosis, the level of virus replication (HBV DNA, HCV RNA levels), histopathological assessment of the severity of the inflammatory activity of the underlying chronic hepatitis and presence of cirrhosis. Serum liver tests alone are insufficient for evaluating and monitoring liver inflammatory activity and the presence of cirrhosis. Serial analysis of liver histopathology is essential for evaluating the efficacy of therapeutic agents and should be conducted in the setting of a controlled clinical trial.

Assessment should include the potential risk for complications and decompensation, such as from spontaneous bacterial peritonitis, potential for bleeding if there are large red esophageal varices and high portal pressures, and operative risk (e.g. Child's–Pugh score, see section 1.8.2). All patients with cirrhosis, regardless of etiology, should be recruited into a program to search prospectively for early HCC. Regular estimations (3–6 monthly) of serum alpha-fetoprotein (AFP) and ultrasound examination (USSD) for liver masses are necessary because these tests are not very sensitive. The serum AFP may remain low in up to 50% of primary liver cancers, especially with hepatitis C. USSD examination is operator-dependent and unlikely to detect the very small (< 1–2 cm diameter) tumors which are most amenable to local resection and have the most favorable prognosis if removed before they spread (metastasize) beyond the liver.

Other chronic liver diseases, especially alcohol-related, may coexist and require evaluation. Multiple viruses may coexist such as hepatitis B, D, C viruses and GBV-C/HGV in the IVDU. Evaluation of their differing contributions to the chronic liver disease may be impossible.

The patient with chronic viral hepatitis (without cirrhosis) of moderate inflammatory activity on liver histology and with virus replication should be referred to a specialist center for consideration of antiviral therapies, preferably under clinical trial conditions.

Case History – 45-year-old man, with known history of alcohol excess presents with confusion and jaundice
AST 3800 IU/l; ALT 1750 IU/l; total bilirubin 12 mg/dl (204 μmol/l);
Total anti-HAV positive; HAV IgM negative;
HBsAg negative, anti-HBc seropositive, anti-HCV seropositive.

Comment
Consistent with alcoholic hepatitis (AST > ALT) with chronic hepatitis C and previous exposure to hepatitis A and B.

Key Notes Management of chronic viral hepatitis

- Define etiology for prognosis and management options
- Serological markers; multiple viruses may coexist
- Other chronic liver diseases common, especially alcohol-related
- Assess complications of chronic liver disease
- Ascites, spontaneous bacterial peritonitis, esophageal varices, portal hypertension, risk of variceal bleeding etc.
- Exclude malignancy
- Serum AFP, USSD of liver for HCC
- Determine level of virus replication for prognosis and infectivity status
- Assess suitability for antiviral therapies
- HBV DNA (dot-blot), HCV RNA (RT–PCR)
- Compensated liver disease (preferably Child's–Pugh grade A)
- Refer to specialist center for treatment and monitoring with antiviral therapies
- Assess suitability and urgency of liver transplantation
- ABO and Rhesus blood grouping, sex and body size
- Discuss with local transplant center (waiting lists, United National Organ Sharing (UNOS) status)
- Screen contacts and family members
- Immunize all babies and children in the household (hepatitis B)
- Immunize susceptible (seronegative) contacts of HBV with hepatitis B vaccine
- Counseling on safe sex, limiting parenteral spread

The patient with cirrhosis requires further evaluation to assess the feasibility of antiviral therapy with its inherent risks of precipitating decompensation and potential need for liver transplantation. Histopathological assessment of the degree of inflammatory activity within the liver is essential; serum liver enzyme activities are at their most misleading with accompanying cirrhosis. Antiviral therapy does not reverse the fibrosis or the risk of developing primary liver cancer once cirrhosis has developed. Antiviral therapy may be justified to improve clinical well-being, and possibly portal pressure which may accompany the reduction in moderate and severe inflammatory activity. Current antiviral therapies have little to achieve in 'burnt-out' cases—when only minimal inflammatory activity accompanies the cirrhosis. Whether these improvements reduce or delay the long-term development of malignancy, shall be answered only with follow up of large cohorts of patients beyond 20 years.

> **Case History – 52-year-old man presents with fatigue and skin itching (pruritus)**
> ALT 39 IU/l; AST 25 IU/L; Alkaline phosphatase 140 IU/l; GGT 204 IU/L; total bilirubin 1.3 mg/dl (22.1 μmol/l); total protein 6.7 g/dl; serum albumin 2.6 g/dl (26 g/l); PT 4 s prolonged; serum cholesterol 110 mg/dl; hemoglobin 9.9 g/dl; platelet count 104 000 cells/μl;
> HBsAg seronegative, anti-HBc seropositive, anti-HCV antibody positive, HCV RNA 4.7×10^6 geq/ml (bDNA assay).
>
> **Comment**
> Consistent with chronic hepatitis C and cirrhosis with previous exposure to hepatitis B: poor liver synthetic function (low serum albumin, low serum cholesterol), hypersplenism (causing thrombocytopenia). Note the normal range for AST and ALT despite severe disease.
>
> The elevated GGT and alkaline phosphatase were consistent with marked bile duct damage and fatty change on liver histology. The low hemoglobin level reflected occult bleeds from portal gastropathy and esophageal and gastric varices.

1.13 Emerging strategies in antiviral therapeutic research

Current strategies have targeted chronic hepatitis B and chronic hepatitis C rather than predicting and preventing evolution of acute viral hepatitis into chronic hepatitis with all the attendant adverse sequelae. This approach is not surprising given the vast numbers of symtomless acute infections which present only later as complications of chronic liver disease. More effort is required to predict, detect and follow patients and their contacts with acute hepatitis B with likely persistent infection. In clinical practice, hepatitis C rarely presents as an acute *de novo* infection; most acute illnesses represent flare-ups—exacerbations of acute-on-chronic hepatitis.

1.13.1 Suppressing virus replication

Antiviral strategies have focused on suppressing virus replication as the first step to eradicating the infection. The most commonly used antiviral agents such as interferons and ribavirin have multiple modes of action (and diverse side-effects) on the viruses and host. In clinical practice, an undetectable level of serum HBV DNA (by dot-blot hybridization) or HCV RNA (by nested PCR) has been deemed to indicate suppression of virus replication. In most instances, repeatedly undetectable levels are accompanied by persistent normalization of liver enzyme activities (serum ALT, AST). However, the extrapolation that an undetectable level of HBV DNA or HCV RNA, and/or normalization of liver enzymes, equates invariably with clearance of virus is naive. Many patients with undetectable HBV DNA levels by dot-blot hybridization are positive using the much more sensitive PCR. Levels of viremia fluctuate widely and serial estimations by sensitive (nested) PCR are required before 'clearance' of virus is assured.

Liver enzyme activities are used widely to monitor responses to antiviral therapies, especially with interferon and ribavirin. Evaluation of 'response' requires qualification especially in the context of end-points in clinical trials. Normalization of serum AST and ALT activities alone is insufficient evidence to assume clearance of virus, especially with chronic hepatitis C. Viral nucleic acids (HBV DNA, HCV RNA) should be sought with serial testing before declaring any antiviral therapy as successful.

1.13.1.1 Immunostimulation

Current antiviral drugs rely on the immune response of the host to effect eradication of the virus. Although the interferons show some immunostimulatory properties, their boost given to the cytotoxic T-cell response is inadequate to achieve eradication of virus in most instances. This is not surprising given that inadequacy of the immune response probably is a major determinant in the persistence of chronic hepatitis B and C. Accordingly, clearance is rarely achieved for chronic hepatitis B or C in the immuno-suppressed host with HIV and infected graft following transplantation.

The rebound in activation of the immune response, especially of T lymphocytes which follows withdrawal of immunosuppression, has been used to 'kick-start' the immune response of the host with chronic hepatitis B. Rapid withdrawal of a short course of corticosteroids has been followed by administration of interferon (IFN)-α in an attempt to sustain the attack by cytotoxic T cells responsible for clearing HBV from infected hepatocytes. This rebound is temporary and, at best, improves the chances of clearance by only 5–10% above those using IFN alone. These statistically marginal benefits must be balanced against the considerable risks and side-effects of giving immunosuppressive drugs, particularly corticosteroids, to the individual patient with chronic liver disease.

Other therapeutic agents with immunostimulatory properties, such as thymosin, are undergoing evaluation in chronic hepatitis B and C. Most studies have been uncontrolled with insufficient numbers of patients to prove unequivocally any benefit of combination therapy over IFN alone.

1.13.1.2 Reducing virus load

Antiviral prophylaxis in transplantation candidates aims to reduce the burden of replicating virus available to the graft. This assumes that levels of viremia before transplantation affect outcome afterwards. This has some rationale for chronic hepatitis B where low or undetectable HBV DNA levels (on dot-blot hybridization) prior to grafting correlate with better survival of the graft and recipient than high baseline levels of viremia. In contrast, for chronic hepatitis C, there is no evidence to correlate levels of HCV RNA (pre- or post-grafting) with their survival.

This discordance may reflect the different sensitivities in assays between HBV DNA (dot-blot hybridization, around 10^6 copies/ml) and HCV RNA (RT–PCR, theoretically down to 10 copies/ml).

1.13.1.3 Preventing graft infection

The alternative strategy is to delay instigation of antiviral therapy until infection of the graft is confirmed. This seems less ideal because there is no consensus on how to define recurrence. Until recently, detection of viral nucleic acid was considered insufficient evidence for commencing antiviral therapy following grafting. Current criteria include waiting for abnormal liver enzymes and histopathological features of graft infection. This area requires urgent reappraisal. Serum ALT levels may remain within the normal range in around 50% of recipients with grafts infected with HCV. Histopathological features of HCV infection in the graft and cellular rejection overlap sufficiently often to make impossible delineation between the two diseases. Also, eradication of virus is more difficult to achieve in the post-transplant setting of immunosuppression and higher than baseline viremias. The risks of treatment and side-effects of antiviral therapy must be balanced against likely success in achieving eradication of the virus.

1.13.2 Side-effects of IFNs

Many of the systemic symptoms that accompany influenza and other viral infections result from IFNs produced endogenously. Similarly, they respond to acetaminophen (paracetamol) and non-steroidal anti-inflammatory drugs (NSAIDs) when given prophylactically around 30–60 min before injecting the IFN. Acetaminophen is safe in these circumstances whereas NSAIDs should be used with caution (see *Table 4.6*).

Most side-effects are dose-dependent, occur with doses above 3 MU thrice weekly and resolve promptly following discontinuation of the IFNs (*Table 1.30*). A reduction in

Table 1.30. Side-effects of interferons (IFNs)

Common (and early on)	
Flu-like symptoms—malaise, fatigue, fever, chills, headaches, sinusitis	
Myalgia (muscle aches), irritability and anxiety (mood changes)	
Depression, anorexia, weight loss, hair loss (alopecia), thrombocytopenia, leukopenia	
Rare	
Skin	Rashes, itching (pruritus) and dry skin, exacerbation of herpes labialis (cold-sores), psoriasis
Nervous system	Depression which may be severe with suicidal thoughts
	Altered sleep patterns, forgetfulness, drowsiness, seizures, psychosis, altered vision (ischemic retinopathy), ototoxicity, sensory and motor neuropathies
Gastrointestinal	Abdominal discomfort, reactivation of peptic ulcer
	Gastrointestinal bleeding, pancreatitis
Cardiovascular/pulmonary	Alterations in blood pressure, dysrhythmias, myocardial infarction, myocardiopathy, congestive cardiac failure, pulmonary edema, interstitial lung disease
Renal	Impairment of renal function, proteinuria, electrolyte disturbances
Metabolic	Elevated uric acid
Hematological	Leukopenia, thrombocytopenia, pancytopenia, hemolytic anemia
Autoimmune	Various autoantibodies, thyroid dysfunction, vasculitis, arthritis, systemic lupus erythematosus (SLE)

dose for non-specific side-effects usually is not necessary because these tend to lessen over time.

Autoantibodies become detectable in over half of the patients given IFNs for more than 3 months. Features of an autoimmune condition, other than autoantibodies, develop in around 1–5% given this therapy. Hypothyroidism or hyperthyroidism may become manifest and be mistaken for side-effects of IFNs. Serial screening for autoimmune features, especially anti-nuclear, anti-thyroid and anti-smooth muscle antibodies, is recommended in all patients prior to instigation of IFN therapy. Risks of developing overt autoimmune disease do not correlate directly with the presence, variety or titer of autoantibodies, except, possibly for antimicrosomal antibodies and subsequent thyroid dysfunction. Symptoms and titers of autoantibodies usually resolve on discontinuation of IFN therapy. Anti-IFN antibodies have been detected with recombinant preparations but their clinical significance remains unclear (*Table 1.31*).

Table 1.31. Types and functions of naturally occurring interferons

Types	Source	Genes	Amino acids	Glycosylation	Functions
Type I					
Alpha	Leukocyte	>20	16 AA	No	As monomer
Beta	Fibroblast	One	16 AA	Yes	As dimer
Omega	Trophoblast	One	172 AA	Yes	
Type II					
Gamma	Immune	One gene (three introns)	146 AA	Yes	As dimer

Alpha and beta receptors are encoded on chromosome 21; gamma receptors on the long arm of chromosome 12, chromosome 6 but require accessory factors (chromosome 21)

Key Notes Autoimmune manifestations of IFN therapy

- Autoantibodies (anti-thyroid, anti-nuclear, anti-smooth muscle, anti-microsomal)
- Hyperthyroidism or hypothyroidism
- Thrombocytopenic purpura
- Vasculitis
- SLE-like syndromes
- Hemolytic anemia
- Rheumatoid arthritis
- Diabetes mellitus

1.13.2.1 Absolute contraindications to IFN therapy

(i) Psychiatric. IFNs should be not used in patients with present or past psychosis, severe depression and uncontrolled psychiatric problems. Depression may be severe with suicidal thoughts and require psychiatric care. Altered mood and depression can arise suddenly in patients without a previous history of psychiatric illness. New onset of depression can be sudden and severe, especially in patients predisposed with an antecedent history. Close follow-up and antidepressant therapy should be anticipated to minimize the possibility of successful suicide. Antiviral therapy should be discontinued promptly for uncontrolled depression.

(ii) Neurological. Present or past history of seizures.

(iii) Cardiovascular. Symptomatic heart disease.

(iv) Hematological. Neutropenia, leukopenia or thrombocytopenia: reductions in numbers of white blood cells (WBCs) and platelets are common with IFN therapy but usually do not cause clinical problems. Antiviral therapy should be discontinued for WBC numbers below $1000 \times 10^6/l$ because the risk rises for severe microbial infections. Antiviral therapy should be discontinued for platelet counts below $70\,000 \times 10^6/l$ because the risk of bleeding rises. All patients receiving IFN should be followed closely and monitored frequently (full blood counts; platelet counts, white cell counts) for impairment of renal and liver function and myelosuppression.

(v) Hepatic. Decompensated cirrhosis with ascites, encephalopathy, elevated PT (INR). Organ transplant recipients, except for liver transplantation (with caution see sections 6.3 and 6.4) as IFN therapy increases the risk of rejection.

(vi) Pregnancy, unreliable contraception.

(vii) Relative contraindications to IFN therapy include uncontrolled diabetes mellitus and autoimmune disorders. Vigilance is necessary to detect pancreatitis (elevated serum amylases and lipases) hypothyroidism [elevated thyroid stimulating hormone (TSH) levels], hyperthyroidism (elevated thyroid hormones T3, T4 and undetectable TSH) and autoimmune disorders (autoantibodies, vasculitis, arthritis, SLE) among many other problems.

1.13.3 Approaches under development

Future approaches are likely to use combinations of antiviral agents especially those which target specific steps in the replication cycle of each virus:

(i) **Inhibition of virus replication by targeted delivery to hepatocytes of:**
  - Antisense oligonucleotides, which bind the viral RNA and prevent translation, including RNA-based external guide sequences;
  - RNA molecules which contain ribozyme sequences may bind to and cleave viral RNAs;
  - Genes expressing antisense molecules, α-IFNs and other antiviral molecules.

(ii) Virus-vector mediated delivery of antiviral RNA sequences
- HDV as a decoy vector in chronic HBV infection to infect the liver but carry antiviral molecules.

(iii) Targeting enzymes involved in virus replication
- Approaches being used to develop antiviral drugs against HIV-1 are being adapted for viral hepatitis. Viral proteases, such as NS3, are essential for the replication of HCV. Recombinant NS3 and co-factor NS4 are being used in *in vitro* systems in a search for protease inhibitors;
- Herpesviruses encode the enzyme assemblin which is a serine proteinase essential in virus replication and assembly. Antiviral agents are being developed which target assemblin and its precursor.

1.14 Emerging strategies in immunoprophylaxis

(i) DNA-based vaccines
- DNA vectors containing the HBV surface gene encoding HBsAg are immunogenic (T- and B-cell responses) in animal models and may provide a relatively cheap alternative to the licensed recombinant subunit vaccines.
- DNA vectors containing HCV genes encoding core and envelope glycoproteins induce T-cell responses in animal models. Immunogens are being designed to take into account the HLA haplotype which affects the level of immune responses.

(ii) Cytotoxic T lymphocyte responses
- Specific epitopes may be targeted in therapeutic vaccines against hepatitis B and C.

Further reading

Alter, M.J., Kruszon-Moran, D., Nainan, O.V., *et al.* (1999) The prevalence of hepatitis C virus infection in the United States, 1988 through 1994. *New Engl. J. Med.* **341**: 556–562. *NHANES serological study of approximately 40 000 people aged 2 months to 89 years. Anti-HCV prevalence was 1.8% (95% CI: 1.5–2.3%) viz 3.1–4.8 million people in the USA are infected with HCV. Of all positives, 65% were in the age range 30–49 years. Black men aged 40–49 years had the highest rate of seropositivity for HCV (9.8%).*

Anonymous. (1993) Nomenclature of the human interferon genes. *J. Interferon Res.* **13**(1): 61–62.

Borden, E.C. (1992) Interferons—expanding therapeutic roles [editorial comment]. *New Engl. J. Med.* **326**(22): 1491–1493.

Czaja, A.J. (1994) Autoimmune hepatitis and viral infection. *Gastroenterol. Clin. North Am.* **23**: 547–566.
A detailed review of serological markers of autoimmune disease and their interpretation when detected in viral hepatitis.

Desmet, V.J., Gerber, M., Hoofnagle, J.H., Manns, M., Scheuer, P.J. (1994) Classification of chronic hepatitis: diagnosis, grading and staging. *Hepatology* **19**: 1513–1520.

Hutin, Y.J.F., Pool, V., Cramer, E.H., *et al.* (1999) A multistate, foodborne outbreak of hepatitis A. *New Engl. J. Med.* **340**: 595–602.
Outbreak in USA during 1997 involving 258 cases reported from Schools in Michigan and Maine States and linked to consumption of frozen strawberries. Linked by genetic sequence analysis.

Hytiroglou, P., Thung, S.N., Gerber, M.A. (1995) Histological classification and quantitation of chronic hepatitis: keep it simple. *Sem. Liver Dis.* **15**: 414–421.

Knodell, R.G., Ishak, K.G., Black, W.C., *et al.* (1981) Formulation and application of a numerical scoring system for assessing histological activity in asymptomatic chronic active hepatitis. *Hepatology* **1**: 431–435.
Semiquantitative histological disease activity index for assessing progression of liver disease.

Kwok, S., Higuchi, R. (1989) Avoiding false positives with PCR. *Nature* **339**: 237–238.
Seminal paper on precautions necessary with the PCR.

Martin, P., Friedman, L.S. (Eds) (1994) Viral hepatitis. *Gastroenterol. Clin. North Am.* **23**: 429–619.
Update Series on acute and chronic viral hepatitis, including changing epidemiology, molecular biology techniques and diagnosis.

Samuel, C.E. (1991) Antiviral actions of interferon. Interferon-regulated cellular proteins and their surprisingly selective antiviral activities [Review]. *Virology* **183**(1): 1–11.

Scheuer, P.J. (1995) The nomenclature of chronic hepatitis: time for a change. *J. Hepatol.* **22**: 112–114.

Sen, G.C., Ransohoff, R.M. (1993) Interferon-induced antiviral actions and their regulation [Review]. *Adv. Virus Res.* **42**: 57–102.

Chapter 2

The Enteric Hepatitis Viruses: HAV and HEV

2.0 Issues

Hepatitis A and E viruses (HAV and HEV) share a major route of transmission (fecal–oral) and cause self-limiting, acute disease. Their limited duration of viremia restricts parenteral transmission. Noticeable differences occur in epidemiological and clinical aspects. Other environmental and host factors must impact on their spread and outcomes of infection. Information on HAV is more complete than for HEV, a more recently discovered hepatotropic virus.

2.1 Epidemiology

2.1.1 Hepatitis A virus

This virus is found worldwide. The significant social and economic cost of hepatitis A is overshadowed by the burden of chronic liver disease associated with hepatitis B and C. Patterns of infection vary. High rates of exposure in infancy relate inversely to poor standards of sanitation, limited supplies of clean drinking water and overcrowded living conditions. In Asia and the East, Africa, India, certain Mediterranean countries and South America, seroprevalence rates for immunoglobulin G (IgG) antibodies, indicating previous infection, exceed 90% above 3 years of age and persist for life. Symptomless infection, especially in children under 7 years of age, encourages the spread of the virus. Improvements in hygiene delay exposure until older ages which, paradoxically, may result in more severe hepatitis.

HAV remains endemic in the West and is the main cause of acute viral hepatitis in the USA, although obvious epidemics are rare. Those at greatest risk, travelers to endemic areas, military personnel and individuals in institutions, may be protected by immunization. Adults infected, especially those aged more than 50 years, have increased morbidity and mortality. An outbreak of hepatitis A in Alaska in1992–3 involved 554 patients with jaundice; seven progressed to acute liver failure and four died.

In low prevalence regions, such as Northern Europe, North America and Australasia, the prevalence of IgG anti-HAV seropositivity rises with increasing age. This shift in herd immunity reflects recent improvements in sanitation (*Figure 2.1*). Consequently, increasing numbers of adults and children are susceptible to infection, especially when traveling to areas of high endemicity. Also, numbers of cases of hepatitis A show seasonal variation. The peak in autumn and early winter usually indicates importation from a higher prevalence area. Person-to-person transmission accounts for the majority of cases as HAV spreads through the community. Symptomless infection in children may obscure routes of spread. Vertical transmission from mother to child is rare but arises presumably from fecal contamination at delivery. Sporadic cases of hepatitis A in endemic areas maintain the reservoir in the population between epidemics.

2.1.2 Epidemic hepatitis A

Outbreaks arise from contamination of a common water or food source. Major epidemics have implicated uncooked shellfish, such as clams and oysters, and fruit and uncooked vegetables from soil fertilized with sewage. HAV can survive steaming and freezing processes. Spread by blood products and blood transfusion has been reported

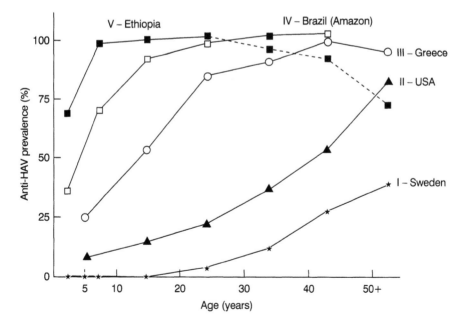

Figure 2.1. Patterns of seroprevalence of IgG anti-HAV. V, very high; IV, high; III, intermediate; II, low; I, very low. (Reproduced with permission from Hadler; Viral Hepatitis and Liver Disease; Hollinger, Lemon, Margolis (eds); © Lippincott Williams & Wilkins 1991.)

in multiply transfused individuals such as hemophiliacs and patients receiving hemodialysis but remains rare because viremia is transient. The solvent–detergent step used during preparation of Factor VIII for hemophiliacs, inactivates lipid-containing viruses but not HAV.

In the West, major epidemics of hepatitis A are uncommon. In the USA, these occur approximately every 10 years. The highest attack rates, but inapparent infection, occur in children aged 5–14 years; symptoms may occur in adults and older children. Clustering of cases is typical with poor sanitary hygiene and overcrowding (*Table 2.1*). Hepatitis A is endemic among North American Indians with equivalent rates between sexes. Immunization with hepatitis A vaccines of susceptible people, particularly children in schools and other institutions, (see section 2.10.2) can control small epidemics (*Table 2.2*).

2.1.3 Epidemiology of HEV

Originally, HEV, like HAV, may have been endemic worldwide. Improvements in sanitation and limited stability of the virion, especially in the gut, restrict the spread of HEV more than HAV, including to secondary contacts and by parenteral transmission. Seasonal variation with peak reporting in late summer and rainy seasons may reflect contamination of scarce water supplies.

Table 2.1. Risk factors for hepatitis A in USA (1989) (from Shapiro *et al.* 1992)

Risk factor	%
Personal contact with case	26
Employment/day-care center	14
Intravenous drug use	11
Recent international travel	4
Association with outbreak	3
Unknown	42

Table 2.2. Contrasting epidemiology of HAV and HEV in developing countries

	HAV	HEV
Major transmission route	Fecal–oral	Fecal–oral
Epidemics	Food-borne, waterborne	Waterborne
Exposure in childhood	Almost universal	Unknown frequency
IgG antibodies	Persist for life	Persist for years, may not be protective
Protective immunity	Yes, life-long	Not known
Reinfection	No	Possible
Mortality in pregnancy	Not increased	High during certain epidemics

Key notes Hepatitis A

Epidemiology

- Worldwide
- Epidemics and sporadic infections
- Fecal–oral transmission: contaminated food and water
- Blood transfusion—rare
- Stable virion; may survive freezing/steaming
- Spread with overcrowding, poor sanitation
- 40% unknown source in West

Infection

- Peak infectivity during incubation
- Infectivity wanes after jaundice
- High case-to-case (secondary) spread
- Severity of illness rises with age
- Relapsing and protracted illness—uncommon
- Acute liver failure—rare
- No chronic liver disease
- No excess severity in pregnancy

Reliable serological diagnosis

- IgM anti-HAV diagnostic of acute infection
- Natural immunity (IgG) probably life-long
- HAV RNA denotes current infection

Protection by

- Improved hygiene/sanitation
- Normal immunoglobulin
- Inactivated vaccines (parenteral route)
- Attenuated vaccines (oral route) under clinical trials

Our knowledge is evolving as serological tests become more reliable and available. Detection of IgG anti-HEV antibodies is unreliable for discriminating past from recent infection and their duration is unknown (see section 2.8.1). HEV causes more than 50% of cases of sporadic non-A, non-B hepatitis in India and Egypt. HEV RNA has been detected by reverse transcription–polymerase chain reaction (RT–PCR) in sewage and waste water and in serum and stool up to 52 days after jaundice. As with hepatitis A, vertical transmission can occur presumably from fecal contamination at delivery.

Hepatitis E is endemic in a zone stretching from Morocco to Hong Kong and in Central and South America. Population surveys of symptomless blood donors and volunteers suggest that between 5 and 25% have IgG antibodies (*Table 2.3*). The epidemiology in sub-Saharan Africa is under study. Sporadic cases in endemic regions, especially with protracted viremia and excretion of virus, may maintain the reservoir between epidemics. HEV has been detected in domestic swine in the USA, Nepal and the former Soviet Union, raising the question whether some human infections are zoonotic. Further, laboratory animals including pigs and rats and non-human primates have been infected experimentally with these isolates. Rats also may serve as a reservoir of infection for man, including in low prevalence regions (see also section 2.3.4). Furthermore, the surprisingly high (1.8–2.1%) seroprevalence of anti-HEV IgG among blood donors in the USA (*Table 2.3*) now is considered a true finding, because their antibodies show high specificity for the US isolate.

2.1.4 Epidemic hepatitis E

HEV is the main agent of enterically transmitted non-A, non-B hepatitis although additional viruses may be implicated. Epidemics are linked mostly to fecal contamination of water supplies. A distinct variety of epidemic, predominantly water-borne, non-A, non-B hepatitis was described in several cities in India between 1955 and 1982, the Middle East, South East Asia including Nepal and Burma (1973–1985) and Mexico (1987), Somalia, Ethiopia, Pakistan, Afghanistan, Bangladesh, Borneo, Pakistan and China.

Outbreaks resembled hepatitis A except for a longer incubation phase (around 45 days; range, 2–9 weeks) and low secondary spread. Also, attack rates were predominant among young adults previously exposed to hepatitis A in childhood, suggesting limited immunity. Several epidemics documented an excess severity and mortality in pregnancy. Subsequently, HEV was identified by immune electron microscopy in stool derived from these epidemics. Retrospectively, specific diagnostic assays confirmed HEV for epidemics in Kashmir and the former Soviet Union.

Table 2.3. Seroprevalence of IgG anti-HEV by EIA (after Dawson, *et al.*, 1992)

Sample	Origin	Number	Positive (%)
Volunteer BD	Egypt	102	25 (24.5)
Healthy people	Hong Kong	355	57 (16.1)
Volunteer BD	Saudi Arabia	861	79 (9.2)
Healthy people	Turkey	1367	76 (5.5)
Healthy people	Sudan	374	18 (4.8)
Healthy people	Singapore	87	3 (3.4)
Volunteer BD	Thailand	783	17 (2.2)
USA BD	Wisconsin	386	8 (2.1)
Volunteer BD	Europe[a]	7693	116 (1.5)
Volunteer BD	Texas (USA)	168	2 (1.2)

[a] Europe: UK, France, Germany, the Netherlands and Spain. BD, Blood donors.

2.1.5 Excess mortality with hepatitis E in pregnancy

Hepatitis E virus is the only major hepatotropic agent (see also herpesviruses, Chapter 7) associated with excess mortality in pregnancy in certain regions of high endemicity. Whether this severe outcome in developing countries reflects virus virulence, variants or host factors remains inconclusive.

Key notes Hepatitis E

Epidemiology

- Restricted distribution (developing countries)
- Epidemics and sporadic cases
- Fecal–oral transmission: contaminated water
- Blood transfusion—not reported
- Stability of virion—less than HAV

Infection

- Infectivity wanes after jaundice
- Low case-to-case (secondary) spread
- Severity of illness increases with age
- Relapsing and protracted illness—uncommon
- Acute liver failure—rare except in late pregnancy
- Excess severity in pregnancy
- No chronic liver disease

Serological diagnosis

- Significance of IgM anti-HEV, unclear, transient
- Antibody tests may not discriminate past from recent infection
- HEV RNA denotes current infection
- IgG persists following infection but may not be protective

Protection by

- Improved hygiene/sanitation
- No passive protection with immunoglobulin
- Vaccines under development

2.2 Biology of hepatitis A virus

The replication strategy of the *Picornaviridae* is based on poliovirus and other members that grow efficiently in cell culture. Some general features shared by HAV include an internal ribosomal entry site in the 5' untranslated region. However, the precise locations of some cleavage sites in the polyprotein remain unproven for HAV. Lack of an efficient cell culture system for HAV has hindered studies. Serial passage in cell culture leads to attenuation. The persistent, non-cytopathic growth of low titers of HAV in cell culture delayed vaccine development and may not be applicable to the high replication seen in man.

2.2.1 Classification

HAV was identified in the early 1970s using immune electron microscopy. Anti-HAV antibodies aggregated 27 nm spherical particles from stool. These particles banded in cesium chloride (density: 1.34 g/ml) and sedimented in sucrose gradients (coefficients: 150–160 S). The molecular mass of the contained RNA was around 2.5×10^6 kDa. In 1981, HAV was assigned to the family *Picornaviridae* that comprise the genera Enteroviruses, Rhinoviruses, Apthoviruses (foot and mouth disease viruses) and Cardioviruses (murine viruses). HAV previously was designated Enterovirus 72 for its close resemblance to the Enteroviruses; spread by the fecal–oral route and resistance to low pH and heat. Subsequently, sequencing of the HAV genome in the 1980s revealed a relatively low G+C content and little homology with the Enteroviruses. In 1992, HAV was assigned its own genus, Hepatovirus, of the *Picornaviridae*.

2.2.2 Structure of the hepatitis A virion

The genome is a single-stranded, positive-sense RNA of around 7.5 kb encapsidated in an unenveloped, icosahedral particle. The capsid comprises 60 copies of each of the four structural proteins VP1–4. A single copy of the genome-linked protein, VPg, is attached covalently to the 5' end of the RNA.

The morphogenesis of the virion is inferred from other picornaviruses with major differences from the enteroviruses and rhinoviruses. In these genera, the structural domain is cleaved *in cis* during translation from the remainder of the polyprotein (*Figure 2.2*) by the 2A protein. In the polioviruses, for example, the precursor, 1ABCD, is then cleaved by the 3C protease to proteins 1AB (VP0), 1C (VP3) and 1D (VP1), that remain associated as a protomer. Five protomers assemble to form a pentamer. Finally, 12 pentamers assemble around a molecule of genomic RNA with cleavage of the 1AB molecule (VP0 → VP4 + VP2). This final cleavage is believed to be auto-catalytic and several copies of VP0 remain uncleaved in the mature virion.

In HAV, the 2A protein lacks proteolytic activity. The primary cleavage event is between the 2A and 2B domains. The 2A protein is smaller than for other picornaviruses and remains covalently bound to VP1 as the intermediate pX, that accumulates in the infected cell. The precise role of pX is unclear. Small deletions of the central region of 2A do not affect viability. pX may be involved in the assembly of pentamers.

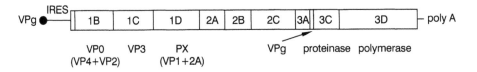

Figure 2.2. Organization of the HAV genome. The 7.5 kb RNA has a protein (VPg) attached covalently to the 5′ end and is polyadenylated at the 3′ end. The untranslated region (UTR, approximately 730 nt) at the 5′ end has extensive secondary structure and forms an internal ribosome entry site (IRES). There is a short UTR at the 3′ end. The remainder of the genome comprises a single ORF and is translated into a large polyprotein. Structural polypeptides (except VPg) are located at the amino-terminal end. The polyprotein is processed by the 3C proteinase. Modified from Harrison et al. (1999).

Unlike enteroviruses, maturation of the HAV provirion with auto-catalytic cleavage of the 1AB molecule (VP0 → VP4 + VP2), is slow, taking 1–2 h at 37°C *in vitro*. On infection, uncoating of newly synthesized hepatitis A virions requires completion of maturation. Furthermore, the VP4 of HAV is not myristilated. These differences may be essential for biliary secretion of HAV. Hepatitis A provirions, on release from the infected hepatocyte, may bind to the apical surfaces of other hepatocytes. However, the delayed maturation and uncoating favors transcytosis with release from the basal surface of the hepatocytes into the bile canaliculae. Detailed analysis of the virion structure awaits crystallization of hepatitis A virions.

2.2.2.1 Binding to the cell and uncoating

A candidate cell surface receptor for HAV has been identified using a monoclonal antibody (mAb) that blocks viral attachment, presumably by saturating the receptors. The putative protein, resembling the immunoglobulin superfamily that includes receptors for poliovirus and T cells, is mucin-like, with a globular apical domain. The cellular role is not known.

Binding of Enteroviruses to their receptors results in loss of VP4 and internalization of the virus particle. However, as the VP4 protein of HAV is not myristilated the process of internalization and uncoating is likely to be different.

2.2.3 Organization of the HAV genome

The genome comprises a single large open reading frame (ORF) that encodes a polyprotein of 2227 amino acid residues and is flanked by non-coding regions. The 5′ untranslated region has extensive secondary, and tertiary, structure essential for cap-independent translation and RNA replication. Base-pairing in the region 45–734 nt forms an IRES.

Incoming RNA acts as message for the translation of the polyprotein (*Figure. 2.2*). The structural domain is cleaved from the remainder of the molecule, probably by the protease activity in the 3C product (3C^pro); other cleavages (except VP0) appear to be carried out by 3C^pro.

Cleavage of the structural domain from the remainder of the molecule, between VP1 and 2A, for most picornaviruses is carried out *in cis* by the 2A product. However, for HAV, the initial cleavage between 2A and 2B seems to be carried out by 3C^{pro}. 2A appears to have a structural role, whereas the functions of the 2B product are unknown. The 2C region, in common with that of other picornaviruses, contains a marker for guanidine hydrochloride resistance. Guanidine hydrochloride blocks viral RNA synthesis at millimolar concentrations. The 2C product is believed to bind to nucleic acid and be involved in RNA replication.

The functions of the products of region 3 are better understood. Polypeptide 3B corresponds to the 23 amino acid, genome-linked protein VPg. Polypeptide 3AB probably is the precursor of VPg. 3C^{pro} has the features of a serine-like cysteine protease and can cleave itself from the polyprotein. Other cleavages of the polyprotein appear to be carried out by 3C^{pro} *in trans*. Finally, the 3D^{pol} protein contains the highly conserved Gly-Asp-Asp (GDD) motif characteristic of viral RNA-dependent RNA polymerases and replicates the viral genome.

2.2.4 Genetic variation

As all isolates of HAV are of a single serotype, diagnostic tests and vaccines have universal application. Seven genotypes are recognized on the basis of less than 85% nucleotide identity in selected regions of the genome (such as the VP1/2A junction, *Figure 2.2*). Four genotypes are associated with human disease and three were isolated from Old World monkeys.

2.2.5 Growth in cell culture

HAV grows poorly in cell culture, does not shut off host protein synthesis and can establish persistent infection without apparent cytopathic effects. The relatively low yields are sufficient, but expensive, for producing an inactivated vaccine.

2.2.6 Animal models

HAV, like HEV, has been passaged in non-human primates; the first cDNA clones were derived from virus after passage in marmosets. Three of the seven HAV genotypes

---

Key Notes The hepatitis A antigen, HAAg

- A single or overlapping set of epitopes comprising VP1 and VP3 sequences
- Defines the single serotype of HAV
- Neutralizing antibodies target this antigen
- Vaccines derived from any human strain of HAV should confer protection worldwide
- Is conformational (expression of individual proteins with hepatitis A antigenicity from recombinant DNA is problematic)

represent viruses initially isolated from Old World monkeys. Hepatitis A in non-human primates is mild. Caution is needed in controlling the attenuation of strains during development of live vaccines (see section 2.10).

2.3 Biology of hepatitis E virus

2.3.1 Classification

The HEV genome resembles those of the *Caliciviridae* but formal assignment to a family is awaited. Non-structural polypeptides are encoded towards the 5′ end. The structural polypeptide(s) towards the 3′ end presumably are translated from subgenomic mRNAs. Unlike authentic caliciviruses, HEV has a third ORF that overlaps the other two ORFs (*Figure 2.3*).

The HEV non-structural ORF contains the methyl transferase domain, suggesting that the 5′ end of the genome may be capped (caliciviruses have a protein linked to the 5′ end of their genomes). Furthermore, the sequence of the HEV non-structural ORF resembles more closely the 'alphavirus superfamily' such as rubella virus (Rubivirus) and certain plant viruses (Furoviruses), rather than the *Caliciviridae*.

2.3.2 Structure of the hepatitis E virion

HEV may be visualized as 28–35 nm particles by immune electron microscopy. Cup-like depressions, typical of caliciviruses, are sometimes visible on the surface. Immune electron microscopy reveals symmetrical, non-enveloped particles around 32 nm in diameter with cup-shaped depressions and spikes at the apex where two cups meet.

2.3.3 Cloning of the HEV genome

HEV RNA was cloned and sequenced from bile of infected monkeys. The genome is a polyadenylated, positive-sense RNA of around 7500 nt. Sequence analysis reveals

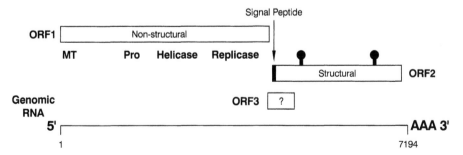

Figure 2.3. Organization of the HEV genome. The 7.5 kb RNA is polyadenylated at the 3′ end and has short untranslated regions (UTRs) at either end. There is no evidence for an internal ribosome entry site (IRES) and the 5′ end may be capped. ORF1 is non-structural and contains motifs suggestive of methyl transferase (MT), cysteine protease (Pro), helicase and RNA polymerase (Replicase) activities. No information is available regarding processing of this polyprotein. ORF2 encodes the major structural protein: this has a signal sequence at the amino-terminus and potential glycosylation sites (indicated by lolipops). (Reproduced with permission from Harrison, *Liver*, © 1999 Munksgaard International Publishers Ltd., Copenhagen, Denmark.)

three ORFs (*Figure 2.3*). ORF1, of approximately 5 kb, begins 28 nt from the 5′ end of the genome and contains the methyl transferase-like domain and encodes motifs associated with NTP binding, helicase and RNA-dependent RNA polymerase activities. The translation product yields proteins involved in replication. ORF2, of around 2 kb, begins 37 nt downstream of ORF1 and terminates 68 nt from the polyA tail. ORF2 encodes the major structural polypeptide(s). The third ORF is short (369 nt) and overlaps the others. ORFs 1 and 3 overlap by one nucleotide, the adenine forming the third base in the last codon of ORF 1 and beginning the initiation codon (AUG) for ORF 3. The function of the translation product of ORF 3 is unknown but may be a structural component of the virion. Subgenomic RNA species, detected by Northern hybridization of HEV cDNA probes to RNA isolated from liver of experimentally infected monkeys, have not been confirmed. Whether ORFs 2 and 3 are translated from subgenomic messenger RNAs, remains to be established.

2.3.4 Genetic variation

Complete nucleotide sequences first were reported for isolates from Burma ('Old World') and Mexico ('New World'). Refinements in PCR techniques and use of degenerate primers have identified HEV isolates in India, Pakistan, China, Italy and Greece and the USA. The overall sequence identities between the Burmese, Mexican and US isolates for nucleic acids are 74–76%, and for amino acids are 82–84% (ORF1), 90–93% (ORF2) and 79–87% (ORF3). Recent appreciation of the heterogeneity of nucleotide sequences has led to their classification within at least four genotypes; Burmese, Mexican, US and Chinese based on genetic distances of 0.25–0.45 substitutions per position (*Figure 2.4*). The Greek and Italian isolates probably represent additional genotypes and more regroupings, are anticipated. The degree of genetic diversity is less than for (HCV) (see section 4.2.4). In particular, much of the variation in nucleic acid sequences targets the third base of the triplet codon without effecting any change in the deduced amino acid sequence. How these changes impact on the design and sensitivity of serological tests for antibody and potential immunogens, remains unknown.

Burmese-like isolates are found throughout Asia and Africa. These sequences cluster closely and form genotype 1, whereas significant phylogenetic distance separates them from the sole isolate from Mexico (genotype 2) and from genotype 4 sequences from the People's Republic of China. The three USA sequences (two isolated from humans, one from a domestic pig) form genotype 3; their genetic closeness favors a zoonosis. Phylogenetic distances are sufficient to form separate genotypes for each of the three European isolates. The Italian isolate has been classified as genotype 5. The two sequences isolated in Greece are of uncertain origin and sufficiently diverse from each other to be separated as genotypes 6 and 7.

2.3.5 Growth in cell culture

HEV may not cause an obvious cytopathic effect in culture. Quantitation of extracellular virus using RT–PCR may provide a better assessment of virus replication.

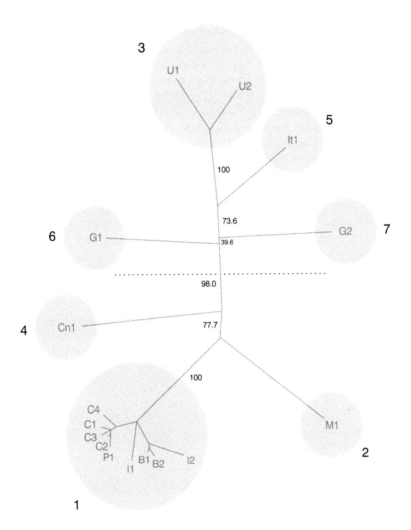

Figure 2.4. Genotypes of HEV (Reproduced with permission from Schlauder, *et al.*, 1999, *J. Med. Virol.* © 1999 John Wiley & Sons, Inc. Reprinted by permission of Wiley–Liss, Inc. a subsidiary of John Wiley & Sons, Inc.). Unrooted phylogenetic tree depicting the relationship of nucleotide sequences over a 242 nt region of ORF1. 1, Burmese-like viruses; 2, Mexican isolate; 3, US isolates (including murine HEV); 4, Chinese variants; 5, Italian isolate; 6, 7, Greek isolates.

2.3.6 Animal models

Experimental transmission initially was to cynomolgus macaques and other non-human primates. Virus particles were visualized by immune electron microscopy but attempts to purify large quantities from stool were unsuccessful due to significant degradation in the gut. Subsequently, bile from monkeys in the acute phase of infection was found to be a rich source of virus.

2.4 Clinical features of hepatitis A and E

The average incubation period is slightly shorter for hepatitis A (average 4 weeks; range, 1.5–7 weeks, *Figure 2.5*) than for hepatitis E (average 6 weeks; range 2–9 weeks). In epidemic hepatitis E, clinical disease occurs predominantly in young people aged 15–40 years although sporadic cases have been reported in children. High mortality rates (up to 20%) have been reported in the third trimester of pregnancy during certain epidemics of hepatitis E (see section 2.2).

Presenting symptoms often are non-specific with a flu-like illness of abrupt onset, headache, fatigue, arthralgias, nausea, anorexia, vomiting and diarrhea. Cholestatic jaundice with pale stools, dark urine and pruritus usually follows within days of the prodrome. Jaundice and overt symptoms appear to be more common with increasing age. Consequently, most cases have anicteric hepatitis, that may be overlooked without serological screening.

Hepatitis E was documented 30 days following ingestion of virus by a volunteer (*Figure 2.6*). Jaundice occurred between days 38 to 120. HEV RNA was detected in blood from day 22 to day 46 (peak alanine aminotransferase level) and in stool from day 34 until day 46 when collection of feces was discontinued. Virus particles were detected by electron microscopy in stool samples taken on days 34 and 37. Anti-HEV seroconversion occurred on day 41 and IgG anti-HEV persisted for at least 2 years.

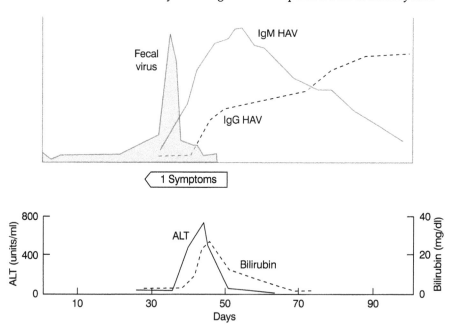

Figure 2.5. The course of acute hepatitis A. (Reproduced with permission from Sherlock, 1998 (eds, Zuckerman and Thomas) In: *Viral Hepatitis Scientific* 2nd edn, © 1998 Churchill Living-stone).

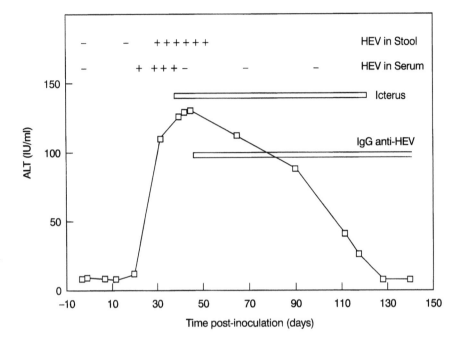

Figure 2.6. Acute hepatitis E in a volunteer. Alanine amino-transferase (ALT) levels are plotted as IU/ml. Collection of stool samples terminated on day 46. IgG was detectable at 2 years. (Reproduced with permission from Chauhan, *et al.*, 1993 *Lancet* **341**:149–150, © 1993 The Lancet Ltd.)

Key Notes Hepatitis A and hepatitis E

Hepatitis A and E are acute, self-limiting illnesses noted for their complete resolution. Symptoms and abnormal liver function tests typically resolve within 3 months. A chronic carrier state is not recognized in man. Persistent viremia accompanied by shedding of virus in stool has been documented up to 52 days from onset of jaundice in some instances of hepatitis E and in relapses of hepatitis A between 30 and 90 days after the primary episode

2.5 Complications

Occasionally, recovery may be delayed for months with protracted cholestasis and a biphasic course. Chronic sequelae have not been reported. Very rarely, relapse may progress to acute liver failure (see Chapter 5). The predeliction for excess mortality from epidemic hepatitis E during pregnancy in some regions, such as India, is unexplained.

The clinical severity of hepatitis A rises with increasing age and when superimposed on other liver diseases such as alcoholic hepatitis, dengue and HEV coinfection. A post-hepatitis syndrome may occur with malaise, fatigue, anorexia, failure to gain weight, intolerance to alcohol and anxiety. The liver may remain tender and palpable for several weeks; histopathology shows non-specific features of a resolving hepatitis.

2.6 Disease associations

Isolated cases report renal failure, pancreatitis and rhabdomyolysis complicating non-fulminant hepatitis A.

2.6.1 Immunological features

Rarely, a leukoclastic cutaneous vasculitis with purpura and cryoglobulinemia has been reported in hepatitis A (compare chronic hepatitis C, section 4.3.6.1). As with HCV, the cryoglobulins contain virus in immune complexes with IgM antibodies.

Immune complexes in association with a vasculitis and glomerulonephritis have been reported in hepatitis E but their prevalence and clinical significance are unknown. Disclosure of autoimmune liver disease with autoantibodies has been documented following hepatitis A but the role of HAV in triggering autoimmune events, remains unclear.

2.7 Histological features

Liver biopsy is not indicated in typical hepatitis A or E. A histological diagnosis may be useful for suspected underlying chronic liver diseases that require specific therapies, such as alcoholic hepatitis, Wilson's disease and autoimmune diseases. Hepatitis A can present with protracted cholestasis and itching and mimic primary biliary cirrhosis or sclerosing cholangitis.

In hepatitis A, typical histopathological features include marked periportal and perivenular inflammation with a mixed cell infiltrate of plasma cells, lymphocytes and polymorphonuclear leukocytes. Cholestasis may be prominent but bile ducts are pre-served. Fibrin-ring granulomata are reported rarely. Disruption of the limiting plate may simulate the interface hepatitis (formerly called piecemeal necrosis) of auto-immune hepatitis but IgM anti-HAV positivity confirms hepatitis A and autoantibod-ies are undetectable or of low titer (< 1 : 160).

Histological features for hepatitis E resemble hepatitis A based on limited information from autopsy specimens mostly in developing countries. Formation of pseudoductules was prominent in some reports.

2.8 Laboratory diagnosis

Hepatitis A and E should be excluded among many illnesses with non-specific presentations such as flu-like symptoms, malaise and jaundice especially in travelers returning from regions of high endemicity (*Table 2.3*). Serological diagnosis is required because the majority of cases are anicteric. Also, other, and additional, causes of jaundice and hepatitis include alcoholic hepatitis, drug-related hepatitis, biliary obstruction and chronic liver diseases such as primary biliary cirrhosis and sclerosing cholangitis.

2.8.1 Immunoassays

Diagnosis of recent hepatitis A infection relies on the detection of high-titer IgM-anti-HAV antibodies in serum. IgM anti-HAV antibodies may be detectable for months, particularly by research-based immune assays. IgG-anti-HAV is useful to demonstrate immunity following exposure to an index case and prior to travel to high prevalence areas. IgM anti-HAV levels fluctuate with relapse in 10–20% adults. Detection of anti-HAV antibodies following immunization requires specific testing assays (see hepatitis A vaccine, section 2.10.2).

2.8.2 Diagnostic tests for HEV

Enzyme immunoassays (EIA) detect IgG antibodies to immunoreactive epitopes identified in ORF2 and ORF3, based on recombinant proteins expressed from cloned HEV cDNA and synthetic peptides representing linear epitopes predicted from deduced amino acid sequences. One such EIA employs antigens from ORF2 and ORF3 gene products expressed in *Eschericia coli* as glutathione-S-transferase fusion proteins. Recombinant proteins derived from Mexican and Burmese isolates of HEV are included to allow for some antigenic variation of divergent strains but will have to be upgraded to include antigens representing the more recently discovered strains from Europe, the USA and China.

EIAs for IgM and IgA have been developed but have not been exploited commercially. IgM anti-HEV antibodies denote recent acute infection but may be undetectable during the first 1–3 weeks following jaundice. Current IgM and IgG assays peak around 6–10 days after the onset of jaundice. Duration of detection may be short lived (around 3–8 weeks); only 50% of cases have detectable IgM anti-HEV at 90 days after the onset of jaundice. IgG anti-HEV is detectable in over 90% of cases within 1 month of onset of jaundice. Consequently, IgG antibody positivity alone does not discriminate between recent acute and past exposure. Seroconversion from IgG anti-HEV negative to antibody positive is useful for diagnosing recent infection when results for IgM anti-HEV are unavailable or negative and if sampling occurs after seroconversion. The manufacturers recommend IgG anti-HEV to diagnose recent infection using a sample/cut-off ratio exceeding 3.0 during epidemics and outbreaks and below 3.0 for sporadic cases of hepatitis E. Tests for HEV antigen, including in stool, are not available commercially but the virus (HEV RNA) can be detected using RT–PCR.

The duration of persistence of IgG antibody, and whether this indicates immunity, remains unresolved. That symptomatic hepatitis E can recur in young adults despite probable prior exposure, suggests short-lived immunity for neutralizing antibody compared with life-long immunity following hepatitis A.

2.8.3 RT–PCR

Detection of HAV RNA and HEV RNA indicates current infection. Their detection in serum, indicating viremia, parallels positivity in stool samples, serum, liver and bile but typically only within the first 3 weeks following the onset of jaundice. Detection of HAV in fecal samples has been improved using an antigen capture technique. Virus is bound at 4°C, overnight, to tubes coated with a mAb prior to RT–PCR in the same tube.

2.9 Management

There are no specific recommendations. Bed-rest assists recovery; mobility can be resumed when the patient feels well. Most patients prefer a light diet based on low fat, high carbohydrate and protein. Supplements with vitamins are unnecessary except if malnourished, such as the alcoholic patient. Abstinence from alcohol is sensible despite lack of incriminating scientific data.

Corticosteroids do not improve the rate of clinical and histological resolution of hepatitis A. Their use occasionally for profound cholestasis and pruritus must be balanced against the risk of microbial infection and relapse precipitated by withdrawal.

Secondary spread of hepatitis A and E is uncommon especially once symptoms have developed. Handling of feces and bile can be minimized by washing hands thoroughly and disposing of contaminated clothes and fomites by autoclaving and incineration. Pregnant personnel should avoid exposure to hepatitis A and E. Although manufacturers do not recommend immunization against hepatitis A during pregnancy, susceptible pregnant women at high risk of hepatitis A should be counseled individually about the risks of catching hepatitis A relative to the remote risk of any adverse effects from an inactivated vaccine and immunoglobulin (see section 7.11.7). There are no forms of immunoprophylaxis against hepatitis E.

> **Key Notes** Management
> Significant progress towards the elimination of hepatitis A and E is achieved by making available clean water and safe sewerage facilities and practicing good hygiene, especially washing hands, when handling food

2.10 Prevention of hepatitis A

Strategies differ between low and high prevalence regions. In the West (*Table 2.1*) targeting motivated high-risk groups, such as international travelers, has limited the spread of infection. Other high-risk groups are difficult to reach. Almost one-half of the cases in the USA arise from an unknown source.

Universal immunization could eradicate HAV. Unresolved issues include duration of protection from a vaccine, that may not be life-long. Also, immunizing children requires justification given the generally mild (anicteric) illness and life-long protection following natural infection.

As herd immunity declines in developed countries, increased exposure through travel to endemic areas brings the risk of spread to a susceptible population especially via children who develop inapparent infection and transmit to others. Vaccine or immune globulin (IG) can control such outbreaks.

Practical recommendations to minimize fecal–oral transmission of viruses:

- Wash hands thoroughly after using the toilet

In endemic areas
- Cook food thoroughly, wash (in boiled water) and peel fruit
- Avoid raw and lightly cooked shellfish that may filter water contaminated with sewage
- Drink only bottled water, including for ice cubes
- Brush teeth in bottled or boiled water
- Avoid salads and foods washed in local water
- Avoid swimming in waters potentially contaminated with sewage

2.10.1 Immune globulin (passive immunization)

IG is pooled normal human immune globulin containing IgG anti-HAV antibodies (*Table 2.4*). IG is 80% effective in preventing hepatitis A infection in susceptible contacts. The declining seroprevalence of anti-HAV in the UK has reduced the availability of stocks of pooled immune globulin with high titers of anti-HAV antibodies. Introduction of a highly immunogenic vaccine has reduced significantly the demand for IG.

IG (0.02–0.06 ml/kg) is useful for rapid protection during an outbreak, or after exposure to hepatitis A when the vaccine is not specifically recommended, as during pregnancy.

Table 2.4. Passive immunization with IG

	Duration of potential exposure to HAV			
	<3 months		>3 months	
Body weight (kg)	IU	Vol. (ml)	IU	Vol. (ml)
<25	50	0.5	100	1.0
25–50	100	1.0	250	2.5
>50	200	2.0	500	5.0

Normal immune globulin contains anti-HAV antibodies (and anti-HBs antibodies) and should be considered in the pre-term neonate exposed to HAV in the intensive care unit. Doses typically are between 0.06–0.12 ml/kg or 250–500 mg depending on age (above and below 10 years) and offer some protection for around 2–3 months but may not be completely protective during outbreaks.

2.10.1.1 Safety

IG, prepared by ethanol fractionation (Cohn technique) of pooled human serum, and administered by intramuscular injection, has been used safely for more than 40 years. HBsAg, HCV and HIV are excluded.

2.10.2 Hepatitis A vaccines (active immunization)

The first vaccine was licensed for use in the UK in 1992 and in the USA in 1995. Three other vaccines are available in the West [SKB (Havrix), MSD (VAQTA®) and Pasteur Mérieux]. All are highly immunogenic: high levels of anti-HAV occur in over 95% of subjects 2–4 weeks after the first dose. A single dose of vaccine is preferable to IG prior to travel to endemic areas; the second dose may be given after return, 6–12 months later.

Protection is likely to be maintained for many years based on immunogenicity profiles but the duration of protective efficacy could be less than that achieved following natural infection, especially without the 'natural booster' effects once coverage of immunization becomes high.

Combined immunization with vaccine and IG, using different needles and syringes and sites, of contacts of index cases of hepatitis A may help to contain outbreaks. The slightly lower titers of antibody attained are not important in clinical practice.

2.10.2.1 Major steps in manufacture

These are conventional whole-virus 'killed' (formaldehyde-inactivated) preparations developed along similar lines to the Salk poliovirus vaccine. There is no recombinant formulation (unlike the linear peptides for hepatitis B vaccines) because structural conformation of the capsid proteins of whole, or near whole, virus is required to provide the neutralizing epitopes (immune determinants).

Low titers of HAV are produced from cell culture and large volumes are required to harvest virus and provide sufficient antigen. In 1979, Provost and Hilleman cultivated HAV in kidney cells (FRhK6) of the fetal rhesus monkey after serial passage in marmosets. Hepatitis A virus (SKB: an Australian isolate—HM175; Merck: the CR326 isolate; Pasteur Mérieux: the GBM isolate) has been adapted to grow in human diploid cells (such as MRC5). Vaccines stored appropriately at 2–8°C have a shelf-life of around 2 years.

2.10.2.2 Safety

The licensed vaccines are safe and do not transmit infection. They are conventional 'killed' vaccines with extended time (32 h) used for formaldehyde inactivation (*Figure*

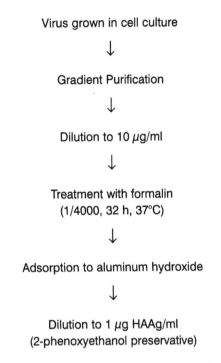

Virus grown in cell culture

↓

Gradient Purification

↓

Dilution to 10 μg/ml

↓

Treatment with formalin
(1/4000, 32 h, 37°C)

↓

Adsorption to aluminum hydroxide

↓

Dilution to 1 μg HAAg/ml
(2-phenoxyethanol preservative)

Figure 2.7. Production of an inactivated hepatitis A vaccine (Havrix).

2.7). Rigorous quality controls for every batch ensure safety, purity and sterility. The cell culture system is used for producing other vaccines and complies with World Health Organization safety requirements.

2.10.2.3 Combinations

The HAV vaccines are licensed for children aged more than 2 years. Accordingly, the combination vaccines (against hepatitis B and A viruses) are not licensed for use as immunoprophylaxis against vertical/perinatal transmission (see section 3.9).

2.10.2.4 Contraindications to immunization

None is specific excepting known allergies to the components (aluminum hydroxide, adsorbent and 2-phenoxyethanol, preservative) and febrile illness. Safety studies have not been carried out in pregnant women and manufacturers do not recommend giving the vaccine during pregnancy. However, there seems to be no specific reason for concern should pregnancy be discovered during immunization (see section 7.11.7).

2.10.2.5 Live, attenuated vaccines

Efforts continue to develop a stable attenuated live vaccine for oral administration. The potential for reversion to virulence that could lead to outbreaks of hepatitis A among vaccinees and their contacts remains a concern. Highly passaged strains are attenuated and show poor immunogenicity. Conversely, with less passage the propensity for reversion to wild-type is maintained; infection can occur with elevation in levels of aspartate aminotransferase.

Candidates under development include the HM175, CR326 and H2 strains. Preliminary studies have been carried out in non-human primates and humans, including children. Obvious advantages include ease of administration and capacity to mimic natural infection with development of local (mucosal) immunity and potential life-long protection. Whilst the value of IgA for protection for HAV is unproven, the parallel with the successful Sabin poliovirus vaccine (that induces IgG to high titer) makes this approach attractive.

2.10.3 Who needs hepatitis A vaccine?

Global eradication of hepatitis A is within reach given the high immunogenicity and clinical efficacy of the licensed vaccines. Highest rates of coverage are most likely to be achieved if hepatitis A vaccine is included within the expanded program of immunization (EPI) as well as for susceptible adults (catch-up program) (see caution on combination hepatitis B and A vaccines for children < 2 years). Unfortunately, high costs prohibit universal immunization, the strategy most likely to eliminate hepatitis A rapidly. Without a program of global eradication, the need for routine immunization of healthy children below the age of 10 years remains debatable for selected high endemicity regions despite good safety data on thousands of children immunized during clinical trials in the East, including Thailand. The vast majority of infections in young children are symptomless. As immunity from natural infection is life-long, this may outlast that anticipated following immunization with inactivated virus.

Universal immunization is needed in regions of low endemicity such as in the West to achieve global eradication. Immunization programs that target only high-risk groups are unlikely to succeed (compare the failure of targeted programs for hepatitis B) because risk factors cannot be identified in around 30–40% of cases. Coverage rates remain disappointingly low (< 30%) despite campaigns during outbreaks among motivated groups, such as men who have sex with men. Also, symptomless cases, especially among children, serve to maintain the reservoir of infection and spread hepatitis A within the community.

High-risk groups for hepatitis A:
- Care attendants with children in crèches, day-care centers, schools, residential homes
- Food-handlers
- Sanitation workers and plumbers exposed to sewage
- Personnel in closed and crowded situations:
- Prisons, residential homes, mental institutions, military personnel
- Intravenous drug users
- Health-care workers
- Homosexuals
- Travelers to areas of medium and high endemicity (All areas excluding Northern and Western Europe, North America, Australia and New Zealand)
- Contacts of cases

Only around one-half of the reported cases come from an identifiable source, such as a day-care center. Travelers are easy to target as high-risk but account for only around 5–10% of cases of hepatitis A in low endemicity regions such as the UK and USA. Selective immunization to this group alone is unlikely to reduce significantly the overall endemicity of HAV.

The licensed inactivated hepatitis A vaccines:

- For adults and children more than 2 years of age
- Safe
- Several manufacturers, pending widespread licensing—doses vary
- Very immunogenic
- Non-responders are rare (contrast hepatitis B vaccine)
- Anti-HAV antibodies: detectable within 2–4 weeks of first dose
- Detectable antibody signifies protection against infection
- Detectable antibody lasts years after second dose (at 6–12 months)

Although few data are available, active immunization with vaccine may play an important role in limiting outbreaks provided coverage is adequate. In the outbreak in Alaska in 1992–3, the incidence of new cases of hepatitis A fell to zero within 3 weeks of implementing the vaccine although coverage was only around 50% of the estimated population at risk (*Table 2.5*).

2.10.3.1 Cautions and contraindications

Hepatitis A vaccines are licensed for adults and children more than 2 years of age. Schedules and doses vary between the vaccines and are not comparable. Dosages are expressed as enzyme-linked immunosorbent assay (ELISA) units or Antigen units. HAV vaccines alone and as combination vaccines (against hepatitis B and A) are not licensed for use in new-born babies and cannot be used in children less than 2 years of age as part of EPI. Hepatitis vaccines are not recommended in pregnant women and are contraindicated for known hypersensitivity to vaccine components.

Table 2.5. Dose schedules for hepatitis A vaccine—Advisory Committee on Immunization Practice (USA), 1996

	Age (years)	Dose 0.5–1.0 ml	Schedule (months)
HAVRIX	2–18	720 EIU	0, 6–12
HAVRIX	> 18	1440 EIU	0, 6–12
VAQTA	2–17	25 U	0, 6–18
VAQTA	> 17	50 U	0, 6–18

There are no specific recommendations from CDC for testing immunogenicity (antibody levels). The vaccine should not be mixed with others in the same syringe. There is no contraindication to giving the hepatitis A vaccine at the same time, but at different sites, with other vaccines and IG.

2.10.4 Indications for immunization

High-risk groups for hepatitis A (CDC Guidelines):

- International travelers
- Military personnel
- People living in or relocating to areas of high endemicity
- Carers in child-care centers
- Laboratory and hospital staff handling hepatitis A virus
- Attendants looking after non-human primates
- Chronic liver disease
- Communities in which outbreaks have been reported:
  Alaskan Inuit, Native Americans, hemophiliacs,
  institutions for the mentally retarded, day-care centers
  sexually promiscuous groups, intravenous drug users

2.10.4.1 Side-effects

The vaccines are well-tolerated with few minor local reactions and fever. A few cases of transiently raised aspartate aminotransferase levels have been reported.

2.10.4.2 Post-immunization testing

The vaccines are very immunogenic. High levels of anti-HAV antibody can be detected 2–4 weeks after the first dose. Geometric mean titers fall with age above 40 years but non-response (contrast hepatitis B vaccine) is rare. There are no specific recommendations for testing for total anti-HAV antibodies following the basic two- or three-dose course.

2.10.4.3 Duration of protection

This should last years after the basic two-dose course. Anti-HAV antibody levels exceed those achieved after passive immunization with IG. Detectable antibody seems to equate with protection from infection. In clinical practice, the exact duration of protective efficacy has not been established. Antibody levels are likely to fall eventually. Long-term follow-up is necessary to show that protective efficacy outlives this fall (as for hepatitis B vaccine; see sections 3.9.3.3 and 3.9.3.5). Whether additional doses improve ('boost') immune memory or demonstrate its presence, requires study.

2.10.4.4 Pre-screening vaccinees

Up to 40% of adults in the UK and USA have IgG anti-HAV antibodies indicating past exposure and immunity to HAV. Pre-screening for IgG anti-HAV antibodies is cost-effective for high-risk individuals such as regular travelers and persons who have lived in developing countries. Pre-screening is essential in pregnancy to avoid unnecessary immunization especially during outbreaks and prior to working in high-risk situations and travel abroad.

2.11 Protection against hepatitis E

No immunoprophylaxis is available for HEV. Controversial results following passive immunization of cynomolgus macaques highlight the limitations of animal models. In one study, although three out of four animals showed no histological or serological evidence of hepatitis after challenge, all four became viremic and shed HEV in the stool. Vaccines are under development (see section 2.13.1).

2.12 Future issues for HAV

We await the crystallization of the virion and resolution of its structure and more information on the cellular receptor to enable interpretation of the early events in infection. The precise role of the VP1/2A fusion pX also remains to be determined, as does the mechanism of biliary excretion.

The development of live attenuated vaccines continues despite concern over the inverse relation between attenuation and immunogenicity. Questions remain over the appropriateness of testing (in clinical trials) new vaccines in areas of high endemicity. High cost remains the major barrier to global immunization programs that could lead to eradication of HAV.

2.13 Future issues for HEV

Many remain to be resolved, including confirmation that the genomic RNA is capped, whether subgenomic transcripts are synthesized and the mechanism of their transcription. Questions remain also concerning assembly of the capsid, including the signal sequence and membrane localization of the ORF2 product and the role of the ORF3 product.

Serological tests for HEV antigens require further development to allow discrimination between acute and past infections and to accommodate the heterogeneity of isolates. The mechanisms of liver damage and involvement of the immune response have not been determined, particularly the propensity for severe disease in late pregnancy.

2.13.1 Prospects for HEV vaccines

As with HAV, development of a killed vaccine may be superior to the recombinant approach in terms of immunogenicity. However, attempts to propagate HEV efficiently in cell culture have been unsuccessful.

A universally effective vaccine seems a tangible prospect although the issue of cross-protection between the Asian, Mexican and other isolates remains a challenge. Progress remains slow because the immunology of HEV is poorly understood and data on cross-protection yield conflicting results. Protection following recovery from natural infection seems short-lived despite high titers of antibodies against the structural protein ORF2. In contrast, data from an epidemic in Pakistan supported the conclusion that IgG anti-HEV antibodies are protective: the clinical cases in young adults resulted from primary infections. Single cases indicate that IgG anti-HEV can persist for at least 12 years.

The most direct approach to developing a vaccine utilizes recombinant HEV antigens as immunogens. Cynomolgus macaques immunized with a protein expressed from ORF2 (capsid region) were completely protected following challenge with the homologous (Burmese) isolate. However, challenge with a Mexican isolate resulted in signs of virus replication with HEV RNA in the liver and HEV antigen shed in the stool, although there was no histopathological evidence of hepatitis. Fusion proteins may form suboptimal antigens because they contain *E. coli* or other protein domains, whereas some native proteins, for example those produced by recombinant baculoviruses, may present conformational epitopes. The role of conformational epitopes in eliciting a protective immune response remains to be determined.

Further reading

Alter, M.J., Kruszon-Moran, D., Nainan, O.V. *et al.* (1999) The prevalence of hepatitis C virus infection in the United States, 1988 through 1994. *New Engl. J. Med.* **341**: 556–562.
NHANES serological study of approximately 40 000 people aged 2 months to 89 years. Anti-HCV prevalence was 1.8% (95% CI 1.5–2.3%) viz 3.1–4.8 million people in the USA are infected with HCV. Of all positives 65% were in the age range 30–49 years. Black men aged 40–49 years had the highest rate of seropositivity for HCV (9.8%).

Brown, E.A., Day, S.P., Jansen, R.W., Lemon, S.M. (1991) The 5' nontranslated region of hepatitis A virus RNA: Secondary structure and elements required for translation in vitro. *J. Virol.* **65**: 5828–5838.
Structural model for the HAV IRES.

Centers for Disease Control and Prevention. (1993) Hepatitis E among US travellers, 1989–1992. MMWR **42**: 1–4.
Hepatitis E among US travelers.

Chauhan, A., Jameel, S., Dilawari, J. B., Chawla, Y. K., Kaur, U., Ganguly, N. K. (1993) Hepatitis E virus transmission to a volunteer. *Lancet* **341**: 149–150.
Clinical and serological features of hepatitis E in a volunteer who swallowed infected stool.

Dawson, G.J., Chan, K.H., Cabal, C.M., *et al.* (1992) Solid-phase enzyme-linked immunosorbent assay for hepatitis E virus IgG and IgM antibodies utilizing recombinant antigens and synthetic peptides. *J. Virol. Methods* **38**: 175–186.
Seroprevalences of antibodies to hepatitis E virus typically ranging 1.1–7.6% among blood donors and general populations in New Zealand, Thailand and Mexico.

Emerson, S.U., Huang, Y.K., McRill, C., *et al.* (1992) Mutations in both the 2B-gene and 2C-gene of hepatitis A virus are involved in adaptation to growth in cell culture. *J. Virol.* **66**: 650–654.
Role of 2B/2C mutations in adaptation of HAV to cell culture.

Hadler, S. C. (1991) Global impact of hepatitis A virus infection changing patterns. In: *Viral Hepatitis and Liver Disease* (eds F. B. Hollinger, S. M. Lemon, H. S. Margolis) Williams and Wilkins, Baltimore, pp. 14–20.
World-wide prevalence of hepatitis A.

Harrison, T.J., Dusheiko, G.M., Zuckerman, A.J. (1999) Hepatitis viruses. In: *Principles and Practice of Clinical Virology.* 4th edn. (eds A. J. Zuckerman, J. E. Banatvala, J. R. Pattison) John Wiley and Sons Ltd, Chichester, pp. 187–233.

Innis, B.L., Snitban, R., Kunasol, P., *et al.* (1994) Protection against hepatitis A by an inactivated vaccine. *JAMA* **271**: 1328–1364.
Efficacy of immunization against HAV, including in children in high endemicity regions.

Jansen, R.W., Siegl, G., Lemon, S.M. (1990) Molecular epidemiology of human hepatitis A virus defined by an antigen-capture polymerase chain reaction method. *Proc. Natl Acad. Sci. USA.* **87**: 2867–2871.
Improved sensitivity of RT–PCR for HAV using an antigen capture technique.

Lemon, S.M. (1992) Hepatitis A virus: current concepts of the molecular virology, immunobiology and approaches to vaccine development. *Rev. Med. Virol.* **2**: 73.
A good, general review on HAV.

Lemon, S.M., Shapiro, C.N. (1994) The value of immunization against hepatitis A. *Infect. Agents Dis.* **1**: 38–49.
Recommendations by the Advisory Committee on Immunization Practices of the US Public Health Service. Review of hepatitis A epidemiology with emphasis on potential vaccination strategies.

Mast, E.E., Alter, M.J., Holland, P.V., Purcell, R.H. (1998) Evaluation of assays for antibody to hepatitis E virus by a serum panel. Hepatitis E Virus Antibody Serum Panel Evaluation Group. *Hepatology* **27**: 857–861.
Comparison of assays for anti-HEV, including recombinant protein and synthetic peptide formats.

Nanda, S.K., Ansari, I.H., Acharya, S.K., *et al.* (1995) Protracted viremia during acute sporadic hepatitis E virus infection. *Gastroenterology* **108**: 225–230.
Hepatitis E detection in stools and serum, 7–14 weeks of illness.

Purdy, M.A., McCaustland, K.A., Krawczynski, K., Spelbring, J., Reyes, G.R. (1993) Preliminary evidence that a trpE-HEV fusion protein protects cynomolgus macaques against challenge with wild-type hepatitis E virus (HEV). *J. Med. Virol.* **41**: 90–94.
Recombinant approaches to hepatitis E vaccine development.

Robertson, B.H., Jansen, R.W., Khanna, B., *et al.* (1992) Genetic relatedness of hepatitis A virus strains recovered from different geographical regions. *J. Gen. Virol.* **73**: 1365–1377.
Hepatitis A virus may be classified into seven genotypes.

Schlauder G.G., Dawson, G.J., Erker, J.C., et al. (1998) The sequence and phylogenetic analysis of a novel hepatitis E virus isolated from a patient with acute hepatitis reported in the United States. *J. Gen. Virol.* **79**: 447–456.
The US genotype of HEV.

Schlauder, G.G., Desai, S.M., Zanetti, A.R., et al. (1999) Novel hepatitis E virus (HEV) isolates from Europe: Evidence for additional genotypes of HEV. *J. Med. Virol.* **57**: 243–251.
Evidence for at least seven genotypes.

Schultz, D.E., Honda, M., Whetter, L.E., McKnight, K.L., Lemon, S.M. (1996) Mutations within the 5' nontranslated RNA of cell culture-adapted hepatitis A virus which enhance cap-independent translation in cultured African green monkey cells. *J. Virol.* **70**: 1041–1049.
Association of mutations in the HAV IRES with adaptation to growth in cell culture.

Shaffer, D.R., Brown, E.A., Lemon, S.M. (1994) Large deletion mutations involving the first pyrimidine-rich tract of the 5' non-translated RNA of human hepatitis A virus define two adjacent domains associated with distinct replication phenotype. *J. Virol.* **68**: 5568–5578.
Attenuation of hepatitis A viruses with deletions in the 5' UTR.

Shapiro, C.N., Coleman, P.J., McQuillan, G.M. et al. (1992) Epidemiology of Hepatitis A. Seroepidemiology and risk groups. *Vaccine* **10** (Suppl. 1): S59–S62.
A review of hepatitis A surveillance in the USA. The whole volume reports the international symposium on active immunization against hepatitis A (Vienna 1992).

Sherlock, S. (1998) Clinical features of hepatitis (eds, Zuckerman, A.J., Thomas, H.C.) In: *Viral Hepatitis Scientific.* 2nd edn. Churchill Livingstone, London, pp. 1–13.

Tsarev, S.A., Tsareva, T.S., Emerson, S.U., et al. (1994) Successful passive and active immunization of cynomolgus monkeys against hepatitis E. *Proc. Natl Acad. Sci. USA* **91**: 10198–10202.
Immunization may prevent disease, but not replication of HEV.

Wang, Y., Ling, R., Erker, J.C., et al. (1999) A divergent genotype of hepatitis E virus in Chinese patients with acute hepatitis. *J. Gen. Virol.* **80**: 169–177.
The Chinese genotype of HEV.

Yarbough, P.O., Tam, A.W., Fry, K.E., et al. (1991) Hepatitis E virus: identification of type-common epitopes. *J. Virol.* **65**: 5790–5797.
Identification of the HEV ORF 2 and 3 epitopes used in EIAs.

Chapter 3

Hepatitis B (HBV) and Hepatitis D (HDV, delta) Viruses

3.0 Introduction

HBV and HDV share with hepatitis C virus a major route of transmission (parenteral), risk factors for infection and the propensity for persistent infection leading to chronic hepatitis and cirrhosis. Individuals in high-risk groups (*Table 3.1*) may be infected with all three viruses. Epidemiological differences, including high rates of sexual and vertical transmission for HBV, but low for HDV and HCV, may reflect differing efficiencies of transmission and levels of viremia. HCV is reviewed in Chapter 4.

Aspects of hepatitis B and D relating to acute liver failure, organ transplantation, pregnancy and pediatric issues are dealt with in Chapters 5, 6 and 7.

3.1 Epidemiology

HBV and HDV share the same epidemiology and dual infection accounts for significant morbidity and mortality. HBV is endemic worldwide. An estimated 350 million

Table 3.1.   Risks for transmission of HBV

High prevalence areas
- Infants of carrier mothers
- Household contacts of carriers
- Inadequately sterilized instruments for percutaneous use

Low prevalence areas
- Sexual contact with carriers
- Intravenous drug use
- Multiply transfused and recipients of blood products

---

people carry hepatitis B surface antigen (HBsAg) and around 5% (at least 5 million) of these are superinfected with HDV (*Table 3.2*).

HBV is a major cause of liver disease worldwide, including chronic hepatitis, cirrhosis and hepatocellular carcinoma (HCC or primary liver cancer, PLC). Worldwide, HCC is one of the most common cancers, at least in males, and HBV has been implicated in over 80% of cases. The 750 000 deaths per annum from HCC are expected to rise with population growth and falling infant mortality. The long-term risk for a HBsAg seropositive carrier male dying from cirrhosis and other complications may exceed 40%.

Almost 90% of the world's population live in geographic regions of high (>8% HBsAg) and intermediate (2–8% HBsAg) endemicity for HBV. However, the prevalence varies greatly (*Figure 1.1*). Infection is most common in sub-Saharan Africa, China and South-East Asia. In these regions, 10–25% of the population are seropositive for HBsAg and the majority of the remainder have serological markers of past infection (anti-HBs, anti-HBc). In these high prevalence areas vertical transmission from the mother, and horizontal transmission from household members, to unimmunized children maintain the virus in the population.

Areas defined by the World Health Organization as of intermediate prevalence rates (2–8%) for HBsAg seropositivity include the Mediterranean littoral, Eastern Europe and the Indian subcontinent.

In northern Europe, the USA and Canada, the chronic carrier rate is below 1% in the general adult population. However, this may exceed 2% in inner cities that concentrate high-risk groups, such as immigrants from higher prevalence countries, intravenous drug users and men who have sex with men. Horizontal transmission predominantly is between young persons by sexual contact and intravenous drug use. In the USA, 30 000 new cases of hepatitis B occur annually, with 70% in the 15–39 year age group and 15% in teenagers.

3.1.1 Subtypes of HBV

All isolates of HBV share a common group determinant, designated *a*. Four major sub-types, *adw*, *ayw*, *adr* and *ayr*, are recognized on the basis of mutually exclusive pairs of allelic subdeterminants, *d* or *y* and *w* or *r*. The molecular basis of this variation is

Table 3.2. Superinfection and coinfection with HDV

HDV superinfection

>   Superinfection of an HBsAg seropositive individual results in persistent infection, typically following a flare of acute-on-chronic hepatitis

HDV coinfection occurs *de novo* with HBV

>   Simultaneous dual infection is common especially in intravenous drug users and their sexual contacts and may result in severe acute hepatitis, even evolving into acute liver failure, before clearance of both viruses

discussed below (see section 3.2.5.3). These subtypes show varying geographical distribution and have been used in epidemiological studies. For example, the *ayr* subtype occurs across northern and western Africa, the Mediterranean littoral and northern and eastern Asia, whereas *adr* predominates in South-east Asia.

Serological subtyping has been surpassed by DNA sequencing in modern epidemiological studies and in implicating the source of infection in individual transmission events.

Key Notes Hepatitis B

Epidemiology
- Worldwide
- High endemicity in China, South-east Asia, sub-Saharan Africa
- Parenteral and sexual routes

Infection
- Symptomless illness in babies and young adults
- Acute liver failure—uncommon
- Chronic infection (HBsAg seropositive > 6 months): common, > 90% babies and 1–10 % adults
- Up to 40% of long-term carriers may die from liver-related causes

Reliable serological diagnosis
- HBsAg—the frontline assay for infection
- IgM anti-HBc—acute infection or exacerbation
- HBeAg indicates viremia
- Anti-HBs indicates recovery or response to immunization

Protection by
- Hepatitis B immune globulin (HBIG)
- Plasma-derived and recombinant vaccines
- Natural immunity probably long-lived

3.1.2 Hepatitis D virus

High prevalence areas where up to 50% of HBsAg seropositive individuals have HDV include the Mediterranean littoral, especially Southern Italy, Israel and the Middle East, South America and sub-Saharan Africa (*Table 3.3*). In Northern Europe and the USA, dual infection predominantly is confined to high-risk groups, specifically intravenous drug users, immigrants from high prevalence areas and the multiply transfused. For transmission, HDV requires helper function from the surface protein of HBV specifically for entry into, and egress from, the hepatocyte.

Evidence for horizontal transmission of HDV comes from Italy. Clustering of cases is seen within HBsAg positive families without obvious risk factors. The relative roles of

Table 3.3. Prevalence of anti-HDV antibodies in HBsAg seropositive groups (1980–1995)

Selection criteria	Country or region	Positive (%)
Patients with liver disease	Somalia	50.0
	Senegal	44.0
	Hungary	13.6
	Italy	23.4
	Kenya	42.0
	India	4.0
	Yemen	2.0
	Iran	2.4
	Japan	0.6
Prostitutes	Taiwan	16–55
IVDU	Taiwan	73
	USA	21

intimate contact between sexual partners and spread among siblings are unclear. Sexual transmission of HDV between individuals, excepting intravenous drug users, is uncommon for reasons not clearly understood. Vertical transmission of HDV also seems to be uncommon and may require high levels of HBV replication. Hepatitis B immune globulin (HBIG) and/or hepatitis B vaccine protects against HDV as well as HBV infection.

Key Notes Hepatitis D (delta hepatitis)

Epidemiology
- Worldwide
- High endemicity in Mediterranean, South America
- Prevalence declining in some countries
- Risk factors, natural history and transmission linked with HBV
- Parenteral and sexual routes of transmission

Infection
- Symptomless illness in babies and young adults
- Acute liver failure—uncommon (co-infection with HBV)
- Chronic liver disease common (superinfection with HBV), cirrhosis

Serological diagnosis
- Specific serological tests for antigen and antibody

Protection by
- Preventing HBV (HBIG and vaccines)
- No specific vaccine to protect HBsAg carriers from superinfection
- Natural immunity probably long-lived

3.2 The biology of HBV

Research is hampered by lack of a suitable cell culture system. HBV infects only man and higher primates. The discovery of related viruses in rodent and avian species, as well as the development of transgenic animals, provided alternative models with which to study the virology and pathogenesis. These viruses share with HBV host specificity and hepatotropism and are classed as hepadnaviruses (family *Hepadnaviridae*, for *hepa*tropic *DNA* viruses).

Woodchuck hepatitis virus (WHV) is closely related to HBV and causes chronic hepatitis and HCC (see section 3.5.3). HDV also can infect woodchucks infected with WHV. Californian ground squirrels may be infected with a virus (GSHV) closely related to WHV. Duck hepatitis B virus (DHBV) infection is endemic in many domestic flocks of Pekin ducks and related viruses are found in wild ducks and herons. The DHBV model proved valuable in elucidating mechanisms of hepadnavirus DNA replication and evaluating potential antiviral compounds (*Table 3.4*).

Divergent genotypes of HBV may be found in chimpanzees and gibbons. In 1998, a distinct new virus, woolly monkey HBV (WMHBV), from a New World non-human primate, was added to the *Hepadnaviridae* following discovery of persistent infection in a colony of captive animals.

3.2.1 Organization of the HBV genome and replication

Some features of the HBV genome are unusual. The complete strand is only around 3200 nucleotides in length. This circular DNA molecule is partially single-stranded and associated with a DNA polymerase that can complete the short strand (*Figure 3.1*). Circularity is maintained by base-pairing of the overlapping 5′ ends of the strands

Table 3.4.   Comparative analysis of the hepadnaviruses

	(Human) HBV	Woodchuck WHV	Ground Squirrel GSHV	(Pekin) Duck DHBV	Woolly monkey WMHBV
Host range	Higher primates	Other *Sciurridae*	Other *Sciurridae*	Other ducks	Other primates
Genome size	3.2	3.3	3.3	3.0	3.2
X gene	Yes	Yes	Yes	No	Yes
Infection with HDV	Yes	Yes	NR	NR	NR
Persistent infection	Yes	Yes	Yes	Yes	Yes
Cirrhosis	Yes	Yes	Less common	NR	NR
HCC	Yes	Yes	Less common	No[a]	NR

[a]DHBV-infected Pekin ducks develop HCC only with co-factors such as dietary carcinogens (aflatoxin). NR, Not reported.

[cohesive end region of 224 base pairs (bp)]. DR1 and DR2 are directly repeated motifs of 11 bp near to the 5' ends involved in template switching during genome replication. The 5' ends of the strands are modified; the minus strand has a protein moiety covalently attached whereas the plus strand has a capped oligoribonucleotide. The minus strand has a terminal redundancy of around eight nucleotides. The position of the 3' end of the plus strand is variable, that strand being usually 60–80% full length.

Four open reading frames (ORFs) are conserved and encoded by the short (plus) strand. The largest ORF overlaps the other three and encodes the viral polymerase. The second ORF has three in-phase initiation codons (*Figure 3.1*) and encodes the

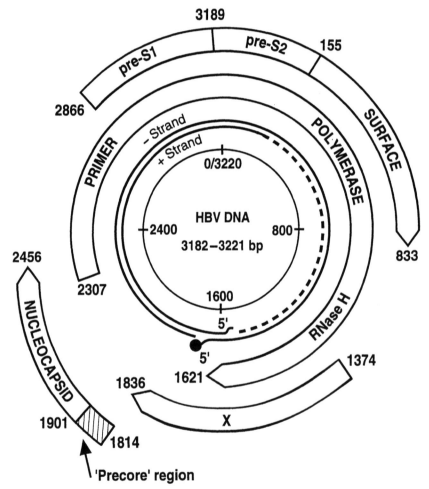

Figure 3.1. Structure and organization of the HBV genome showing open reading frames and multiple initiation sites in the core and surface regions. Note the compact organization: all nucleotides encode protein and around half are located where ORFs overlap, all *cis*-acting elements are embedded within coding sequence. 3182–3221 bp refers to the extremes of observed genome size. (Reproduced from Arrand and Harper, *Viruses and Human Cancer*, 1998; BIOS Scientific Publishers Ltd.)

various forms of the surface antigen, HBsAg. The region between the first and second initiation codons is known as pre-S1, and between the second and third, as pre-S2. The third ORF has two in-phase ATG codons that define the precore region and encode the nucleocapsid protein, HBcAg, and precursor of the secreted antigen, HBeAg. The product of the final ORF, originally designated X for its unknown function, trans-activates transcription.

3.2.2 Binding to the hepatocyte and virus entry

In the 1980s, a domain within the pre-S1 region of the large surface protein (see section 3.2.5.1) that attaches to the hepatocyte receptor was located. However, identification of that receptor remains elusive. Candidates include a soluble (M_r = 50) binding factor, IgA, the asialoglycoprotein receptor, interleukin (IL)-6 and carboxypeptidase D. That the pre-S2 domain may bind to hepatocytes through polymerized human serum albumin, largely has been discredited. However, some authors postulate binding of the major [226 amino acid (aa)] form of HBsAg to apolipoprotein H or annexin V on the hepatocyte. Receptor binding by HBV is complex and probably involves several ligands and co-receptors. The process of virus entry and delivery of the genome to the nucleus is poorly understood.

3.2.3 The HBV polymerase and genome replication

The hepadnaviruses appear to be related evolutionarily to the retroviruses, as the replication strategy of this DNA genome is unusual for proceeding through an RNA intermediate. The HBV polymerase comprises several domains (*Figure 3.1*). The amino-terminal region acts as a primer for synthesis of the minus strand as described below. There follows a spacer (tether) region and the RNA-, and DNA- dependent DNA polymerase. The domain at the carboxyl-terminus has an RNase H activity that degrades the RNA pregenome during synthesis of the minus strand. Homology of the amino acid sequence with retroviral polymerases (reverse transcriptases) is noted particularly around a highly conserved motif, Tyr-Met-Asp-Asp (YMDD), that forms part of the active site of these polymerases. Resistance of HBV to some nucleoside analogs is associated with mutations affecting this domain (see section 3.8.2.3).

After binding of HBV to the hepatocyte and uptake and uncoating, the genome is detectable in the nucleus in a covalently closed, circular (ccc; supercoiled) form. This supercoiled cccDNA is the template for transcription of the HBV pregenomic, and messenger, RNAs. All of these transcripts are 3′ co-terminal, being polyadenylated in response to a signal in the core ORF (*Figure 3.2*). Transcription from the core promoter yields the pregenomic RNA that also acts as message for HBcAg and the polymerase, and the precore mRNA, the precursor of HBeAg.

Pregenomic RNA is packaged with a molecule of the polymerase into precores (surrounded by HBcAg) in the cytoplasm. A stem–loop structure (ε) near to the 5′ end acts as the packaging signal. The requirement to maintain this structure limits the emergence of viable mutants with change in this region. This segment of RNA also encodes the precore sequence and the emergence of precore mutants (see section 3.2.4.1) is constrained for some genotypes.

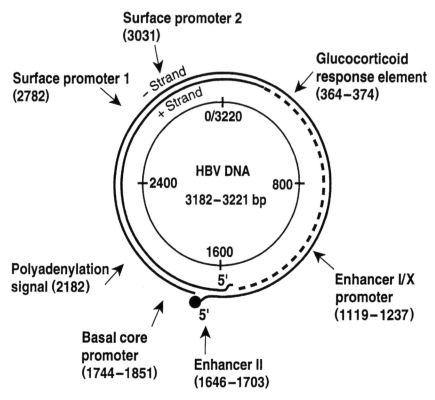

Figure 3.2. *Cis*-acting elements in the HBV genome. (Reproduced from Arrand and Harper, *Viruses and Human Cancer*, 1998; BIOS Scientific Publishers Ltd.)

The amino-terminal domain of the polymerase primes synthesis of a 4 nt nascent DNA strand at a bulge in the ε stem. The primer and nascent strand undergo a template switch to a complementary (τ) sequence near to the 3′ end of the pregenome. Synthesis of minus strand proceeds by reverse transcription of the pregenome, with concomitant degradation of the template (RNase H-like activity). The remaining capped oligoribonucleotide (which was the 5′ end of the pregenome) is at the position of the direct repeat, DR1. This is believed to translocate to the other copy of the direct repeat, DR2, on the minus strand and to prime synthesis of the plus strand. The short terminal redundancy of the minus strand permits circularization of the genome, with another template switch by the polymerase, as the plus strand is synthesized. Completion of the core may starve the polymerase of precursor nucleoside triphosphates, leaving the plus strand incomplete. The cores are then coated with HBsAg to form mature virus particles.

Early in infection of a hepatocyte, some progeny genomes cycle back to the nucleus to build up and maintain the pool of cccDNA template. Elimination of hepatocytes containing such cccDNA is essential to successful antiviral therapy. Nucleoside analogs that inhibit the HBV polymerase will not act on established cccDNA.

3.2.4 The core and e antigens

These are translated from the same ORF but from different RNA transcripts. The core antigen (HBcAg) forms the nucleocapsid, an essential component of the virion. HBcAg may be detected in the infected liver and its antibody (anti-HBc) in serum. HBeAg is a soluble protein secreted from infected hepatocytes into the circulation. HBeAg indicates viremia but its antibody (anti-HBe) does not necessarily indicate clearance of virus as HBeAg-negative viruses are viable.

Pregenomic RNA molecules act as mRNA for HBcAg and the polymerase in addition to their role in DNA synthesis. These mRNAs lack the translation initiation codon at the beginning of the precore region and are translated from the second, in-phase initiation codon in the core ORF to the 22 kDa nucleocapsid protein. The carboxyl-terminal portion of this protein is highly basic, containing a large number of arginine residues, and is believed to interact structurally with the nucleic acid in the HBV core.

Translation of the entire core ORF from the upstream initiation codon of precore mRNAs yields a precursor protein, p25. The precore region (29 aa) contains a signal sequence at the amino terminus, that directs p25 to the endoplasmic reticulum where the first 19 aa are removed by the cellular signal peptidase. The protein is processed further through the Golgi apparatus with proteolytic removal of the arginine-rich,

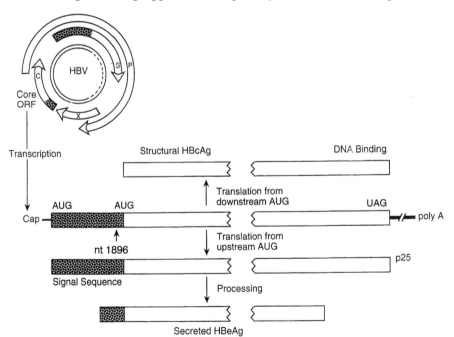

Figure 3.3. Generation of HBcAg and HBeAg from the core ORF. The location of the most common precore mutation (G1896A, which converts a tryptophan codon, UGG, to a termination codon, UGA, is shown). (Reproduced with permission from Harrison TJ (1998) Genetic variants of HBV, pp 93–110; In: Koshy and Caselman (eds), *Hepatitis B Virus*, © 1998 Imperial College Press.)

carboxyl-terminal domain before secretion as HBeAg (see *Figure 3.3*). HBeAg is anti-genically distinct from the nucleocapsid protein. However, proteolysis *in vitro* of native or recombinant HBcAg reveals HBe antigenicity.

3.2.4.1 HBeAg-negative infections

Patients seropositive for HBsAg may remain viremic despite absence of HBeAg. Such patients were first described from Mediterranean countries but their distribution is worldwide. A number of mutant forms of HBV DNA are associated with HBeAg-negative, persistent infections. In the most common prototype, a point mutation (G1896A) in the precore region of the genome forms a stop codon (codon 28, TAG; normally TGG, tryptophan) preventing translation of the precore region and production of HBeAg. Transcription and translation of HBcAg and production of infectious virus particles are unaffected. Other mutations may inactivate the precore initiation codon or interfere with proteolysis of the p25 precursor, with a similar outcome.

Some mutations in the core promoter region down-regulate synthesis of the precore mRNA, whilst up-regulating the pregenomic RNA. Such mutants also may infect patients without detectable HBeAg—any small amount of HBeAg secreted by the infected hepatocytes is masked by antibody.

In mixed infections with wild-type virus, the precore mutant may be selected by the process of seroconversion from HBeAg to anti-HBe positivity when hepatocytes with surface expression of HBeAg are eliminated by the immune response of the host and during treatment with interferon therapy.

Precore variants have been associated with the development of severe chronic liver disease, some cases of fulminant hepatitis B and fibrosing cholestatic hepatitis (FCH) that can develop following transplantation for chronic hepatitis B (see section 6.3.3). However, they have been found also in patients entering clinical remission and in symptomless contacts of acute liver failure. Early studies focused on sequencing relatively small regions of the HBV genome, particularly with respect to codon 28 of the precore region. Mutations in other regions of the genome, including affecting core promoter and/or the amino acid sequences of HBcAg and the X gene product, may have been overlooked.

3.2.5 The surface antigen

In addition to forming the surface envelope of HBV, HBsAg is shed in excess from infected hepatocytes as subviral particles. The antibody response to HBsAg is protective and subviral particles harvested from the blood of persistently infected individuals formed the basis of the first (plasma-derived) hepatitis B vaccine.

3.2.5.1 Expression of the surface proteins

Various forms of HBsAg are expressed by translation from the three in-phase initiation codons in the surface ORF. Control of the synthesis of each product is principally at the level of transcription. The most active of the two surface promoters (surface promoter

2, *Figure 3.2*) is in the pre-S1 region and lacks a TATA box, so that the cap sites vary. Some species of the 2.1 kb RNA synthesized from this promoter lack the pre-S2 initiation codon. Translation of these mRNAs from the third in-phase initiation codon in the ORF yields the 226 aa residue major surface protein (SHBs). This protein may be non-glycosylated (p24) or glycosylated at the asparagine residue at position 146 (gp27). This primary 226 aa sequence is common to all forms of HBsAg and includes the major antigenic determinants.

Additional surface proteins (pre-S proteins) are essential components of the virion. Translation from the pre-S2 initiation codon results in an additional 55 aa residue (pre-S2 domain) at the amino terminus of HBsAg. This contains a further site for N-linked glycosylation, resulting in variably glycosylated forms, gp33 and gp36 (depending upon whether the 146 site is glycosylated). These proteins are known as the pre-S2 or middle surface proteins (MHBs). The quantity of MHBs proteins synthesized is at least 10-fold less than SHBs proteins. Glycosylation (the addition of sugar moieties) may serve to mask immunogenic epitopes on the protein.

A 2.4 kb RNA is transcribed from the less active surface promoter 1 (*Figure 3.2*). The entire surface ORF is translated from this mRNA yielding pre-S1 proteins (or large surface proteins) with around 128 additional residues at the amino terminus. Pre-S1 proteins are found in non-glycosylated and glycosylated forms (p39 and gp42), depending upon glycosylation of the residue 146 asparagine. The pre-S2 site is not glycosylated in the pre-S1 proteins and may be masked by the pre-S1 domain. The pre-S1 region seems also to act as a signal for virion assembly. The virion surface contains pre-S1, pre-S2 and S proteins in the ratio 1: 1: 4. Tubular forms of subviral particles also are rich in pre-S1-containing proteins.

The pre-S domains constitute important additional targets for antibody and vaccines incorporating pre-S epitopes are undergoing clinical trials (see section 3.9.2.3). As noted above, a region of the pre-S1 domain binds to the hepatocyte receptor and antibodies to this region are neutralizing.

3.2.5.2 The virion and subviral particles

Subviral particles also are secreted by infected hepatocytes. These vastly outnumber the circulating virions and can swamp and tolerize the immune system. They are composed of HBsAg embedded in lipid derived from the hepatocyte, but lack HBV DNA and HBcAg. The 22 nm spherical particles are most abundant and lack pre-S1 proteins. The 22 nm tubules are aberrant forms resulting from incorporation of pre-S1 proteins into subviral particles. Patients who have cleared their viremia may remain HBsAg positive because 22 nm spherical particles are synthesized and secreted by hepatocytes with integrated HBV DNA. These subunit particles were harvested to manufacture the plasma-derived vaccines.

3.2.5.3 Antigenicity

The major surface protein is highly hydrophobic and embedded mostly in lipid or internal in the virions and subviral particles (*Figure 3.4*). Anti-HBs antibodies that

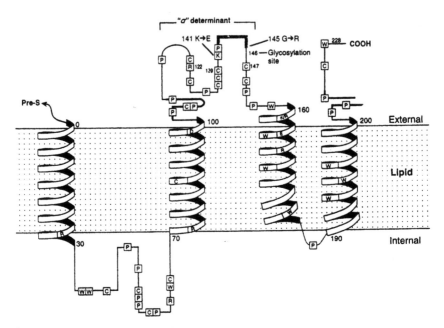

Figure 3.4. Proposed structure for the major HBV surface protein showing the location of some antibody escape substitutions (residues 141 and 145). Modified from Stirk *et al.*, 1992, *Intervirology*.

recognize the major surface protein react principally with a hydrophilic region, the *a* determinant, exposed on the surfaces of the virions and subviral particles. Anti-*a* antibodies are neutralizing and protective.

The *a* determinant is found in all wild-type strains of HBV and contains conserved cysteine residues at positions 121, 124, 137, 138, 139, 147 and 149. Disulfide bridging between these residues maintains a highly conformational structure, probably a double loop. The remainder of the *a* determinant also is highly conserved, particularly between residues 139 and 147 (the 'second loop'). Other epitopes, or sub-determinants, define the subtypes of HBV and map close to or overlap the *a* determinant. Subtype variation, *d/y* and *w/r*, correlates with the amino acids at residues 122 (*d* = lysine, *y* = arginine) and 160 (*w* = lysine, *r* = arginine) of the 226 aa major surface protein.

3.2.5.4 Anti-HBs escape variants

Natural or vaccine-induced immunity to HBV is attributable mostly to neutralizing antibodies to epitopes within the *a* determinant. Among different isolates of HBV, variants with mutations affecting residues known to be highly conserved have been reported in infants infected perinatally despite combined immunoprophylaxis. The HBIG may have served to select variant HBV present as minor species in virus populations in the mother. Similar variants have been selected in liver transplant patients given HBIG to prevent HBV infection of the graft (see section 6.6.1). HBsAg with the arginine 145 substituted for glycine is most common and shows altered reactivity with monoclonal antibodies that should bind to the *a* determinant. Mutations affecting other residues, such as 141, 142, 144 have been described. The asparagine at position 146 remains a conserved glycosylation site and any mutation here may be lethal to the virus. Mutations affecting the less conserved region between cysteines 124 and 137 have been reported less often and their significance is unclear.

During an immunization program begun in Singapore in 1983, 41 out of 345 infants of HBsAg- and HBeAg-positive mothers became infected despite combined immunoprophylaxis, probably from intrauterine transmission. Analysis of HBV DNA, encoding the *a* determinant and surrounding region revealed only wild-type sequences in 25. Arginine 145 substitutions were detected in 12 out of the remaining 16, in mixed combinations with another mutation or mutations in the same genome.

In some published reports, individuals infected with surface variants, especially those with multiple amino acid substitutions, failed to test positive for HBsAg using commercial assays. New screening assays, particularly those based on monoclonal antibodies, should be evaluated for their ability to detect variant HBsAg to minimize likely transmission, especially from blood donors.

3.2.6 The x protein (HBx)

The product of the fourth ORF is expressed from its own mRNA (X gene), a transcript of around 0.8 kb expressed from the X promoter/enhancer I complex (*Figure 3.2*). The protein functions as a transcriptional transactivator and up-regulates *in vitro* the activities of all HBV promoters and various cellular promoters. The x protein is not required for expression of hepatitis B virions in cell culture but is required to establish infection in the woodchuck model. The role of the X gene product in hepatocarcinogenesis is discussed below (see section 3.6.3).

Transactivation by the X gene product does not involve binding to DNA. Instead, the protein seems to act by disrupting various cellular signalling cascades. All four promoters in the HBV genome, including the X promoter itself, are upregulated by the X gene product *in vitro*, as are many cellular genes.

3.2.7 HBV at extrahepatic sites

In HBsAg seropositive individuals, HBV-specific nucleic acid sequences have been reported most commonly in peripheral blood leukocytes (PBLs) and occasionally in pancreas, kidney, skin, spermatozoa, bile duct epithelium and in Kaposi's sarcoma. The occasional reports of viral RNA and virus-specific proteins suggest HBV can replicate in certain extrahepatic sites, such as PBLs. These observations, if confirmed, could help to explain infection of the graft in transplant recipients and difficulties in eradicating HBV from extrahepatic reservoirs.

3.3 Biology of hepatitis D virus

The HDV particle comprises the RNA genome surrounded by the nucleocapsid protein (hepatitis D antigen, HDAg) and the hepatitis B envelope. The virus is considered defective because production is limited to cells that synthesize HBsAg. The genome resembles elements of infectious RNA (viroids) in plants but, unlike viroids, which are not encapsidated, HDV encodes a protein, HDAg.

3.3.1 The defective nature of HDV

HBV provides helper function for virion assembly and transmission of HDV but not genome replication and expression. HBsAg is required for assembly of infectious virus particles, egress from the infected cell and adsorption and entry to target hepatocytes. The surface antigen of WHV (WHsAg) can substitute for human HBV in experimental infection.

3.3.2 Replication of HDV

The HDV genome (*Figure 3.5*) is a circular RNA molecule of around 1700 nucleotides that is folded into a rod-like structure. Around 70% of the residues are base-paired, in common with infectious RNA molecules, such as plant viroids. Similarly, HDV RNA contains consensus motifs implicated in auto-catalytic self-cleavage and ligation (RNA circularization). Unlike the viroids, HDV encodes protein (HDAg) via an ORF in the

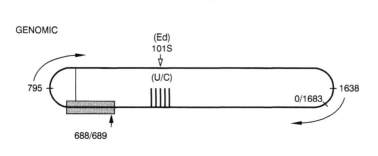

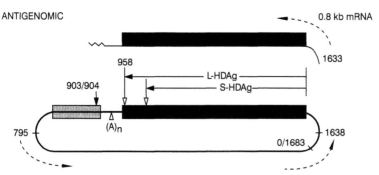

Figure 3.5 Organization of the HDV genome. The antigenomic RNA and mRNA are detected only in infected cells. The genomic RNA is represented in clockwise orientation and the antigenomic RNA anticlockwise; both are numbered in the genomic orientation. The hatched boxes represent the ribozyme domain and nucleotides 688/689 and 903/904 are the ribozyme cleavage sites. With permission, from the *Annual Review of Biochemistry*, Volume 64, © 1995, by Annual Reviews www.AnnualReviews.org

antigenomic sequence. This protein is translated from a polyadenylated mRNA of around 800 nt, not the antigenome.

The genome is replicated by the host RNA polymerase II rather than by a virus-encoded RNA-dependent RNA polymerase. Infected hepatocytes contain around 300 000 copies of the genome, 50 000 copies of the genome complement and 600 copies of the HDAg mRNA. Replication involves a rolling circle mechanism with self-cleavage (ribozyme activity) and subsequent circularization of unit length molecules. How the mRNA for HDAg is generated remains unclear.

3.3.3 The delta antigen, HDAg

This is detectable in two forms in the infected hepatocyte. The smaller (195 aa) form is required for HDV RNA replication and binds to the rod-like structures of the genome and genome complement. The larger (214 aa) structural form is required for virion assembly and seems to be synthesized following RNA editing. This process converts the termination codon at the end of the ORF for the short form to a tryptophan codon, resulting in a 19 aa carboxyl-terminal extension.

3.3.4 The structure of the hepatitis D virion

The virion is smaller than HBV—around 36 nm. The surface protein envelope resembles more the 22 nm HBsAg particles than the hepatitis B virion (see section 3.2.5.1) but pre-S1 proteins seem to be essential for attachment to the hepatocyte receptor. The nucleocapsid, of around 19 nm in diameter, comprises the RNA genome surrounded by HDAg. Large and small forms of the delta antigen are detectable in varying proportions in the virion.

3.4 Clinical features

Outcomes of acute HBV infection depend on the age of acquisition and, inversely, on the severity of the immunological response of the host (*Figure 3.6*). Mild and symptomless infections are common especially in childhood and adolescence. At the other extreme, acute liver failure (see section 5.2.5) is rare but the mortality exceeds 50%.

Clinical features are non-specific. Serological tests are required to distinguish acute from chronic hepatitis B and these from hepatitis C and other causes of hepatitis. Patients with viral hepatitis and chronic liver disease cannot be distinguished clinically from the many other causes of chronic liver disease, in particular alcoholic hepatitis, autoimmune liver disease and genetically determined chronic liver diseases, such as Wilson's disease and alpha-1-anti-trypsin deficiency.

3.4.1 Acute resolving hepatitis B and D

The patient may remain symptomless and anicteric. Any symptoms are non-specific, such as a 'flu-like' illness, malaise, anorexia, nausea and vomiting. Occasionally, a patient may present with consequences of immune complex disease and cryoglobu-

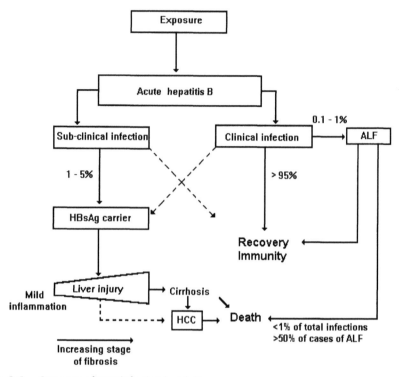

Figure 3.6.   Outcome of HBV infection in adults.

linemia (Chapters 1 and 7) such as a serum sickness-like syndrome (see section 3.4.5), Gianotti–Crosti's disease, glomerulonephritis, (see also section 4.3.6.1), and, rarely, aplastic anemia (see section 5.2.7.1).

Clinical recovery from uncomplicated infection may take months. A biphasic illness with resurgence of serum levels of transaminases and bilirubin can occur, particularly with coinfection (B and D). Co-infection can be severe and may lead to acute liver failure.

3.4.2 Fulminant hepatitis B

Acute liver failure due to HBV is discussed in Chapter 5.

3.4.3 Chronic hepatitis B and cirrhosis

Following acute infection, between 1 and 5% of presumed healthy adolescents and adults, and more than 90% of infected babies, fail to clear the virus. As chronic carriers, they remain HBsAg seropositive for more than 6 months. Prior to the advent of testing for HBV DNA, patients with acute hepatitis B typically had to wait a minimum of 6 months to demonstrate lack of seroconversion from HBsAg to anti-HBs before being considered eligible for antiviral therapy.

The term 'healthy carrier' has been used incorrectly. A symptomless carrier already may have chronic hepatitis and cirrhosis.

Pointers to underlying chronic liver disease include cutaneous stigmata such as spider naevi and palmar erythema. Cirrhosis may be silent for many years before the patient presents with complications such as portal hypertension that manifests itself as ascites and is prone to becoming infected (spontaneous bacterial peritonitis).

Key Notes: Chronic carrier state: definitions

HBsAg seropositive for a minimum of 6 months

This classical definition focused on disease outcome. Such patients were likely to remain HBsAg seropositive for years and risk complications of chronic hepatitis and especially liver cancer

HBV DNA is persistently detectable on serial testing by molecular techniques

This indicates virus replication. HBV DNA becomes undetectable within weeks of acute infection in the presence of an adequate immune response

The 'carrier state' can be predicted earlier than 6 months and the patient selected to receive antiviral therapies

---

3.4.4 Mechanisms of acute and chronic liver damage

The classical view emphasizes that clearance of HBV from the liver depends on lysis of infected cells by CD8+ T lymphocytes. Also, in the immunocompetent host, the HBV-related acute liver damage (cell lysis) is attributable mostly to a virus-specific, HLA-restricted (MHC class I) cytotoxic T-lymphocyte (CTL) response directed against processed peptides displayed on the surface of infected hepatocytes. This cellular response involves multiple epitopes that may serve to limit the possibility of 'escape variants'. An appropriate immune response leads to non-critical, reversible liver damage with clearance of virus from liver and extrahepatic sites and development of neutralizing (anti-HBs) antibodies. However, HBV can persist at very low levels for decades in individuals who recover from acute infection. This persistence maintains a constant stimulation of the CTL responses that keep the virus under control.

The classical view has been challenged by findings in two acutely infected chimpanzees where around a 10-fold reduction in viral DNA levels preceded the peak of T-cell activity and consequent rise in transaminases. This non-cytolytic clearance of viral DNA may be attributable to the action of cytokines. Any liver damage also could be related to release of necroinflammatory cytokines such as tumor necrosis factor, interferon gamma (IFN-γ) and IL-2 by cytotoxic T cells (CD8+) following recruitment of antigen non-specific inflammatory cells, especially of the monocyte/macrophage lineage. Helper T-cells (CD4+ lymphocytes) specific for HBcAg also may be involved in clearance of HBV by inducing the HBcAg-specific cytotoxic T-cell response either directly or via their release of cytokines.

Conversely, failure to clear the virus (HBsAg seropositive) and progression to chronic hepatitis is attributed to an ineffective CTL response. Anti-HBc immunoglobulin (Ig)G replaces IgM but does not ameliorate the infection. The severity of chronic damage and risk of cirrhosis reflect continuing viral replication (HBeAg seropositivity and/or HBV DNA) and repeated episodes of necroinflammatory activity aimed at clearing virus from infected hepatocytes.

3.4.5 Disease associations

As with HCV, acute and chronic hepatitis due to HBV can be associated with several extrahepatic manifestations (see also section 1.11), typically related to deposition of immune complexes and mixed cryoglobulinemia (see also section 4.3.6.1), although the association is more obvious with HCV than with HBV. These include a serum sickness-like syndrome, glomerulonephritis, skin rash (papular acrodermatitis: Gianotti–Crosti disease, see section 7.6.2.1). Such manifestations are more common in children than adults with hepatitis B. Other associations involving immune complexes include poly-arteritis nodosa (PAN) and, rarely, aplastic anemia.

3.4.5.1 Polyarteritis nodosa

Up to 50% of patients diagnosed with PAN are found to be chronically infected with HBV. This complicated condition can present with diverse clinical manifestations in numerous organs due to vasculitis affecting all sizes of arteries. The vasculitis is related to deposition in the arterial wall of immune complexes comprising viral antigens and antibodies. Reduction in the viremia with antiviral therapies can ameliorate the condition but the outcome remains poor despite the numerous therapeutic strategies that have been tried.

3.5 Hepatocellular carcinoma

This is one of the 10 most common cancers in man. More than 250 000 new cases are reported worldwide each year. In regions of sub-Saharan Africa, China and South East Asia, the age-adjusted incidence of HCC is over 30 new cases per 100 000 population each year; whereas it is less than five cases per 100 000 per year in Western Europe and North America. HCC is more common in males than females and the incidence of the tumor increases with age, reaching a peak in the 30–50 year age group.

Concerning the epidemiology of HCC:

- In the world's 'top 10' most common cancers
- Diagnosis to death typically < 1 year
- 80% association with HBV (in Africa)
- 80% associated with HCV (in Japan)
- 80% associated with cirrhosis
- 80% alpha-fetoprotein seropositive

Risk factors for HCC are:

- HBV, HCV, especially early infection
- Cirrhosis
- Male sex
- Smoking

Geographical areas of high incidence of HCC concur with high prevalence of HBsAg. A considerable interval occurs between time of initial infection (typically in childhood) and presentation with HCC. An elegant prospective study performed in Taiwan by Beasley and colleagues of over 22 000 male factory workers in Taiwan followed more than 3000 HBsAg carriers per 75 000 man-years. The relative risk of developing HCC was over 200-fold for the HBsAg carrier over matched, non-carrier, controls. More than half of the deaths in the carrier group were liver-related.

3.5.1 Oncogenic mechanisms

These are not fully understood. Multiple factors probably operate such as the propensity to cause cirrhosis, viral integration (covalent linkage of viral and chromosomal DNA in the hepatocyte, which may be mediated by topoisomerase I), male gender, infection at an early age and disruption of expression of tumor suppressor genes.

3.5.2 HCCs frequently contain integrated HBV DNA

Primary liver tumors from HBsAg-positive patients frequently stain positive for HBsAg. The first successful attempt at long-term culture yielded a cell line (PLC/PRF/5) that secretes HBsAg. Southern hybridization analysis of this line shows multiple sites of integration of HBV DNA (*Figure 3.7*, lane 1).

At least 80% of tumors from HBsAg-positive patients contain viral integrants detectable by Southern hybridization (*Figure 3.7*, lane 5). The more sensitive polymerase chain reaction (PCR) can detect viral DNA sequences undetectable by hybridization. Tumors typically contain viral DNA integrated randomly throughout the human genome. Usually, tumors show clonality for HBV DNA integration and arise from a single cell after the integration event. Multiple tumors in the same liver may show different clonalities and patterns of integration.

Analysis of liver biopsies from patients with chronic hepatitis B frequently reveals evidence of integration (*Figure 3.7*, lane 3). In fact, integration may occur during the acute phase of infection. Detection of specific hybridization bands in non-tumorous tissue implies either that integration has occurred at the same site(s) in many cells or that a single cell has expanded clonally following an integration event. There is no evidence for specific targets for integration of viral DNA in the human genome, and these data seem to indicate expansion of what may be potentially pre-neoplastic clones.

3.5.3 The search for evidence of cis-activation

Sequencing of the HBV genome did not reveal an obvious viral oncogene. Furthermore, attempts to transform cells *in vitro* with HBV DNA were unsuccessful and the long latency between HBV infection and development of HCC argued against the direct action of a viral gene product. An early hypothesis maintained that the neoplastic process involved insertional mutagenesis but this preceded our understanding of the transactivating functions of the X gene product and other proteins expressed from integrated HBV DNA. For example, viral integrants may contain *cis*-acting sequences, other than viral promoters, such as enhancers and the glucocorticoid-responsive element (*Figure 3.2*). Studies to test this hypothesis did not provide evidence of the frequent activation or inactivation of cellular genes. In isolated examples, typically from tumors arising on non-cirrhotic liver, HBV DNA was integrated into sequences such as a novel retinoic acid receptor gene and the cyclin A gene. It is extremely unlikely that a cellular gene has escaped detection, the expression of which is influenced directly by HBV DNA integration in a significant proportion of HCC. In contrast, *myc* oncogenes are frequently activated by integrated WHV DNA in the woodchuck with HCC, the most actively studied animal model of virus-associated HCC.

3.5.4 Transactivation by viral gene products

Failure to find evidence of *cis*-activation in the majority of HBV-associated tumors prompted a reconsideration of the role of HBV gene products. In particular, the x protein was shown to function as a transcriptional transactivator, consistent with a role in the oncogenic process. Furthermore, patients with HCC were frequently found to be seropositive for antibodies that reacted with recombinant X gene products or syn-

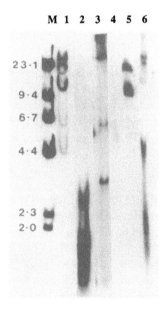

M 1 2 3 4 5 6

23·1
9·4
6·7
4·4
2·3
2·0

Figure 3.7. Composite illustrating the state of the HBV genome in the liver determined by Southern hybridization. Lane 1, integration in the PLC/PRF/5 cell line; lane 2, replicative intermediates in the biopsy of an HBeAg-positive patient; lane 3, integration in the biopsy of an anti-HBe positive patient; lanes 4 and 5, non-tumorous and tumor tissue from an anti-HBe positive patient (partial hepatectomy); lane 6, biopsy from an anti-HBe positive patient showing replicative forms and integrated HBV DNA. All DNA samples were digested with *Hind*III which does not cut HBV DNA. Lane M, molecular weight markers, sizes in kb to the left. Modified from Harrison *et al.*, 1986.

thetic peptides derived from the predicted amino acid sequence. *In vitro*, growth stimulation and tumorigenic transformation of NIH 3T3 cells by the HBx gene have been successful. Other studies suggest that HBx may be sufficient to transform immortalized cell lines to the ability to grow in soft agar and induce tumors in nude mice.

Reports of the effect of HBx as a transgene are contradictory. Transgenic mice in which the X gene was under the control of the α-1-antitrypsin regulatory region alone did not develop serious liver damage or HCC but levels of expression of the x protein were variable. Transgenics with the HBx gene under the control of its own regulatory elements developed multifocal areas of altered hepatocytes progressing to benign adenomas and then carcinomas. Expression of the X gene product also seems to interfere with cellular DNA repair processes, again consistent with a role in oncogenesis. Truncated pre-S/S proteins expressed from integrated HBV DNA in tumors also have been shown to function as transactivators *in vitro*.

3.5.5 Other factors

Development of human cancer likely is multifactorial and involves several (mutagenic) hits. In addition to HBV, dietary factors, smoking, other hepatotropic viruses [especially hepatitis C virus (HCV)] and other factors may play a causal role in oncogenesis. In some areas of the world with a high incidence of HCC, contamination of foodstuffs with mycotoxins such as aflatoxin is common. Aflatoxin can cause HCC when fed experimentally to rats and may contribute to the development of HCC in man, for example, by causing *p53* mutations.

Mutations in the *p53* tumor suppressor gene, in particular, a G to T transversion in codon 249, are the most commonly detected genetic alterations in human malignancies and have been observed in HCCs from high risk areas such as China and Africa. Exposure to aflatoxin may be associated with the mutation. Codon 249 mutations seem to be uncommon in areas with a low incidence of HCC, even in HBV-associated tumors.

Any association between HBV and p53 remains unclear. Although the X gene product interacts directly with p53, the insolubility of the x protein and its tendency to aggregate *in vitro* suggest that non-specific binding to other proteins may occur. HBV DNA can integrate near to the *p53* gene in chromosome 17p but the consequences remain unclear.

There are no data to support any specific association of HDV infection with development of HCC. HDAg may be detected in the livers of chronically co-infected patients with HCC, but seems to be confined to non-tumorous tissue.

3.6 Serological diagnosis

The principal screening assay for acute and chronic hepatitis B and for screening blood donors detects HBsAg in serum. Testing should include all the main agents as multiple infections (such as B, B with D and C) are common among high risk groups especially intravenous drug users and the multiply transfused.

3.6.1 Acute hepatitis B

Diagnosis relies on detecting IgM anti-HBc in the presence of HBsAg (see also *Table 1.8*). The former may be falsely negative with concurrent HDV infection. Most current assays for HBsAg are based on an enzyme immunoassay format. Anti-HBs antibodies bound to a solid substrate capture HBsAg from the sample under test. The surface antigen is then detected with a second, labeled antibody.

Key Notes: Significance of IgM anti-HBc

- Relatively reliable in recent hepatitis B
- May be only marker in severe hepatitis B
- Correlation with inflammatory activity
- Potentially, false-negative with HDV co-infection

Recovery from acute hepatitis B with viral clearance usually is associated with seroconversion to anti-HBs. Anti-HBs assays also are used to monitor the response to immunization. Antibody levels are expressed as milli-International Units per milliliter (mIU/ml), based upon a WHO anti-HBs reference standard. Assays that detect antibodies to pre-S determinants (see section 3.2.5.1) are available for research purposes only.

Hepatocytes infected with wild-type HBV secrete HBeAg, a marker of viremia (virus replication) that may be monitored using quantitative assays. However, seroconversion to anti-HBe does not necessarily indicate cessation of virus replication because variants such as precore mutants may replicate without synthesis of HBeAg (see above). Direct detection of HBV DNA (see section 1.4) is required to determine whether the anti-HBe-positive patient is viremic.

Acute infection typically leads to the detection of HBsAg in blood 2–8 weeks before development of abnormal serum liver biochemistry and onset of symptoms. Serum HBsAg usually remains detectable until the convalescent phase. HBeAg becomes detectable soon after HBsAg.

In uncomplicated acute hepatitis B, recovery is associated with rapid clearance of serum HBV DNA and detection of anti-core antibodies (IgM and, soon after, IgG), shortly followed by anti-HBe, with decline and disappearance of HBe and HBs antigenemia within 3 months. The subsequent 'window' phase, defined by the serological presence of anti-HBc in the absence of HBeAg and anti-HBs, persists for 2–16 weeks. This phase terminates with detection of anti-HBs, signalling recovery, viral clearance and immunity from future HBV infection. The window phase probably is an artefact. HBsAg can be detected using monoclonal antibodies and HBV DNA by the PCR technique. As noted above, extremely low levels of HBV replication (controlled by CTL responses) may persist in anti-HBs positive individuals who have recovered from infection. Hepatitis B may reactivate in such individuals in the face of immunosuppression.

3.6.2 Predicting chronicity

Clinical indicators do not predict likely chronic infection. Paradoxically, those with a mild (anicteric) clinical illness are more likely to become chronic carriers than those with jaundice and symptoms reflecting a severe illness (*Table 3.5*). Also, babies and young children are less likely to clear the infection than immunocompetent adults (see section 7.6.1).

This paradox reflects the inverse impact of the immune response of the host on outcome. An aggressive immune attack is more likely to result in liver damage (and symptoms), but with clearance of the virus, than a mild immune response. Chronic infection is more common in the immunosuppressed than immunocompetent individual.

Historically, the word carrier often was interpreted as being benign and applied particularly to potential blood donors who were symptomless at routine screening. There was reluctance to biopsy seemingly well individuals. Subsequently, longitudinal follow-up and histological assessments showed that chronic carriers were at risk of progressive liver disease, including development of cirrhosis and HCC.

Chronic hepatitis B is likely following acute hepatitis if HBV DNA in serum (dot-blot hybridization) remains detectable after the peak elevation of liver enzymes [alanine aminotransferase (ALT), aspartate aminotransferase (AST)]. Serial analyses of HBV DNA over 1–3 weeks should give an indication of the likelihood of clearing hepatitis B infection.

HBV DNA levels in acute versus chronic hepatitis B:

- Low, clearing: self limited acute hepatitis
- Persistent in chronic infection
- Early consideration for antiviral therapies

Table 3.5. Consequences of acute hepatitis B

Symptoms	Risk of persistent infection
None	High (> 10%)
Jaundice	Low (1–10%)
Acute liver failure	Low (< 10%)

Spontaneous seroconversion from HBeAg to anti-HBe occurs in 1–10% of chronically infected individuals annually. However, seroconversion from HBsAg to anti-HBs with clearance of virus is uncommon, especially following vertical transmission.

IgG-anti-HBc is present in the serum of HBsAg positive chronic carriers and may play an important role in immunomodulation of fetal serological responses to HBV infection following vertical transmission from the carrier mother. Carriers of HBsAg are usually positive for antibodies to the nucleocapsid protein, anti-HBc, determined by assaying for total antibody. These antibodies are not neutralizing and do not appear to influence virus replication (*Table 3.6*).

Occasionally, anti-HBc may be the sole marker of hepatitis B. Such findings usually are interpreted as indicative of past infection without an anti-HBs response or following a decline in anti-HBs levels below detectable limits. Anti-HBc was used as a surrogate marker for parenteral risk for donated blood for non-A non-B hepatitis prior to the availability of anti-HCV assays.

Most patients with underlying inflammatory activity in the liver have serological evidence of virus replication: HBV DNA is detectable by molecular techniques. In clinical practice, the majority are seropositive for anti-HBe rather than HBeAg and, in some of these patients, progressive liver disease has been associated with the presence of HBV mutants. Levels of HBV DNA may correlate inversely with inflammatory activity reflecting the attempt by the host to eradicate the virus.

Table 3.6.   Hepatitis B virus infection: typical serological profiles

	HBsAg	HBeAg	HBV DNA	IgM anti-HBc	IgG anti-HBc	anti-HBe	anti-HBs
Acute	+	+	+	+	–	–	–
Acute rapid clearance	–/+	–	–	+	–	–	–
ALF	+/–	–	–	+	+	–/+	–/+
Chronic[a] no virus replication	+	–	–	–	+	+	–/+
Chronic virus replication	+	+	+	–	+	–	–
Chronic virus replication (HBeAg-ve)[b]	+	–	+	–	+	+	–

[a] Chronic with no detectable virus replication: HBV may persist for more than 6 months without serological evidence of virus replication (undetectable levels of HBV DNA by hybridization). [b] Typically with the precore stop (G1896A) variant.

3.6.3 Occult hepatitis B

HBV infections which are not detected by routine assays for HBsAg in serum have been termed 'occult'. Typically, low levels of viremia are detectable by PCR for HBV DNA in serum. Low level HBV replication may be attributable to virus variants or suppression of replication during co-infection with another virus, such as HCV. In a recent study, HBV DNA was detected by PCR in 66 out of 200 (33%) patients with chronic hepatitis C liver disease, but in only seven out of 50 patients with liver disease who were negative for HBsAg and markers of HCV. HBV infection correlated significantly with cirrhosis in the HCV-infected patients.

3.6.4 Hepatitis D virus

As hepatitis D occurs with hepatitis B, the patient with dual chronic hepatitis typically is seropositive for HBsAg and anti-HBe (rather than HBeAg) as well as IgG anti-HD antibody. Specific assays are available for detection of anti-HD antibodies of the IgM and IgG classes. Serum-based assays to detect HDAg and HDV RNA are not widely available. Testing must include HDAg, HDV RNA and anti-HDV, markers of HBV infection, and be repeated in suspected cases.

3.6.4.1 Hepatitis B and D co-infection

Coexistence of IgM anti-HBc with markers of HDV infection indicates co-infection. HBsAg, HBeAg and HBV DNA become detectable in serum along with HDAg and HDV RNA. IgM anti-HDV becomes detectable, followed by IgG anti-HD. IgM anti-HDV is not as reliable for diagnosing acute HDV infection as is IgM anti-HBc for acute HBV infection. Also, IgM (and IgG) antibodies may be detectable only transiently, or later than anticipated using current assays. This probably accounts for some of the under-reporting of acute infection. Titers of anti-HDV antibodies may be low in co-infection and undetectable until several weeks after the onset of illness (*Table 3.7*).

3.6.4.2 HDV superinfection on chronic hepatitis B

Most patients acquire HDV infection superimposed on chronic hepatitis B. Typically they are seropositive for anti-HD antibodies (IgG and IgM) and HBsAg but negative

Table 3.7. Markers of HDV infection (modified from Polish *et al.*, 1993)

Marker	Tissue	Early acute	Acute	Convalescent (co-infection)	Chronic (symptomatic)	Chronic (asymptomatic)
HDAg	Serum	++	+	–	+/–	–
	Liver	++	++	–	++	–
Anti-HD						
IgM	Serum	+/–	++	+/–	++	–
IgG	Serum	+/–	++	+/–	+++	++
HDV RNA	Serum	+	++	–	+/–	+/–

for IgM anti-HBc. Anti-HD antibodies should be sought in serial samples (acute and convalescent), and despite undetectable HBsAg, which may have cleared before clinical presentation. The diagnosis of chronic hepatitis D is more reliable when HDAg is detectable in the nuclei of hepatocytes using immunohistochemical techniques. HDAg is more prominent and widespread in chronic than in acute HDV infection. Hepatitis D viremia is followed by anti-HD IgM, then IgG. Anti-HD IgM persists along with HDAg and HDV RNA in chronic hepatitis D. Progression of chronic HDV infection may lead to rising titers of anti-HD. HDV can suppress replication of HBV, measured by falling titers of serum HBsAg, HBV DNA and IgM anti-HBc, leading to misdiagnosis.

IgM anti-HDV:

- Transient, less reliable than IgM anti-HBc
- Low titers in co-infection
- Serial testing
- Prefer histology—HDAg in liver
- IgG and IgM anti-HDV may coexist

3.7 Histological features

Liver biopsy is not essential to the management of acute hepatitis B and D.

3.7.1 Acute hepatitis B

Histological features add little to serological markers of acute infection and do not predict the probability of developing chronic hepatitis and cirrhosis. In acute hepatitis B, damage is maximal in periportal areas and plasma cells are few. The inflammatory infiltrate comprises mostly lymphocytes and macrophages and a few scattered polymorphonuclear cells and eosinophils. CTLs and macrophages are prominent, especially adjacent to hepatocytes, reflecting the immunological nature of damage. Cytochemical staining for HBsAg and HBcAg is weak and confined to a few hepatocytes. Large numbers of positive cells suggest chronic hepatitis B. In patients transplanted for chronic hepatitis B, many hepatocytes may become positive for HBsAg and HBcAg following infection of the graft, especially in FCH (see section 6.3.3).

3.7.2 HBV and HDV

Co-infection may result in a more severe illness than with HBV alone. Histological features without special stains do not discriminate between acute hepatitis B and acute B with D co-infection. HDAg may be detected in the nuclei of some hepatocytes; large numbers of positive nuclei suggest chronic hepatitis D. Microvesicular fat may be noticeable.

Features suggestive of chronicity include the presence of bridging fibrosis (*Figure 1.4*) and HBsAg detectable in several hepatocytes as ground-glass cells or as antigen (orcein or Victoria blue stains) in the cytoplasm of hepatocytes using immunohistochemical stains. HBcAg is detectable in nuclei of hepatocytes using immunohistochemical stains. HDAg may be detectable in the nuclei of hepatocytes in HDV superinfection of chronic hepatitis B.

3.7.3 Multiple infections

Concomitant infection with HCV and human immunodeficiency virus (HIV) (as well as HDV) are common among intravenous drug users and other high risk groups that share transmission routes (see section 4.3.5). Clearance of HBV (and HCV) is difficult to achieve with antiviral therapies.

3.7.3.1 Impact of HBV on other infections

HBV infection impacts adversely on the natural history of HCV (see section 4.3.5.1) but not HIV infection. Liver disease is more aggressive with concomitant HBV and HCV infection.

3.7.3.2 Impact of HIV on HBV infection

Recrudescence of HBV and flares of hepatitis B can occur following restoration of the CTL repertoire and apparent remission of the HIV infection, especially with successful highly active antiretroviral therapy. Occasionally the anti-HBc and anti-HBe seropositive patient with HIV may revert to a state of virus replication with detectable HBeAg and HBV DNA levels. Also, flares in hepatitis B have been documented with the long-term use of lamivudine (3TC) in HIV infection due to emergence of resistant variants (see sections 3.8.2.3 and 8.1.2).

3.8 Management

General principles are discussed in section 1.12.

3.8.1 Acute hepatitis B

Management is non-specific and focuses on encouraging rest until liver enzymes and bilirubin (jaundice) return toward normal ranges. There are no specific dietary recommendations that seem to alter the clinical course. Corticosteroids are contraindicated in acute viral hepatitis: they do not improve recovery and may promote sepsis. The efficacy of antiviral therapies in acute hepatitis B has not been studied in large-scale clinical trials. Traditionally antiviral agents, such as α-IFNs, have been withheld for fear they may overstimulate the immune response and precipitate massive hepatic necrosis resulting in acute liver failure (ALF). Early instigation of antiviral therapy before the 6-month time period that defines the chronic carrier seems a sensible option for uncomplicated acute hepatitis B with slow clearance of HBV DNA. No large-scale data are available to test the efficacy of this strategy in preventing chronic infection and other sequelae such as integration of HBV DNA into hepatocyte DNA.

3.8.1.1 Severity of illness

Clinical indicators do not predict early those most likely to pursue an uncomplicated course. Patients with acute hepatitis B should be monitored with serial levels of prothrombin time (PT) or International Normalized Ratio and traditional liver biochemical profile (see section 1.6.2), as well as HBsAg and HBV DNA to predict early those most likely to develop ALF or persistent infection. Falling levels of serum ALT and AST do not necessarily herald recovery. Instead, falling liver enzyme levels accompanied by rising PT/INR usually indicates deterioration from reduced functioning liver mass.

Patients should be followed-up to confirm clearance of virus and seroconversion to anti-HBs. Those likely to achieve clearance of the virus and seroconversion to anti-HBs typically show rapid reduction in levels of HBV DNA within the time-frame of maximum elevation in liver enzymes (AST, ALT). Antiviral therapy should be considered when high levels of serum HBV DNA are maintained in serial samples over 1–2 months as this profile suggests chronic infection.

3.8.2   Chronic hepatitis B

General management of chronic viral hepatitis is regardless of etiology (see section 1.12.2). As with hepatitis C (see section 4.7.8.2), the decision to treat depends on the severity of the liver inflammation and long-term concerns over development of cirrhosis and HCC. Data are insufficient to determine whether treating long-term carriers slows down the development of fibrosis and, ultimately reduces the risk of developing HCC.

3.8.2.1 Approaches to antiviral therapies

The mainstay of treatment is with IFN-α-2b, and, more recently with the nucleoside analogs (see section 1.13 for discussion of IFNs and antiviral strategies).

IFN-induced seroconversion is favored with recent HBV infection, interface hepatitis, elevated serum transaminase (AST and ALT) levels and female sex. Patients with normal transaminases typically are poor at responding to antiviral therapies. Such patients typically are from the East and Far East, acquired their infections perinatally or early in infancy and are tolerant to viral antigens. Conversely, clearance rates of HBV DNA with IFN-α therapy are comparable to those seen in Caucasians with elevated serum transaminase levels indicative of a CTL response to infection (*Table 3.8*).

Typical regimes for IFN-α-2b in recombinant form, are 2.5–10 MU/m^2 (million units per square meter), three times a week by subcutaneous or intramuscular injection (depending on the product) for 6–12 months. Note these doses are higher than those used typically for treating hepatitis C (see section 4.7.9). Oral formulations of IFNs are not available but patients may be taught to self-administer the treatment. Many clinical trials reported successful seroconversion from HBeAg to anti-HBe seroconversion in 15–40% of patients. Complications of interferon therapy are covered in section 1.13.2.

Table 3.8.   HBeAg to anti-HBe seroconversion following interferon therapy

Positive predictors[a]	Negative predictors
• High serum ALT	• Asian (vertical transmission)
• Recent (adult) infection	• Neonatal infection
• History of acute hepatitis	• Immunosuppressed (HIV)
• Inflammation on liver biopsy	• No inflammation
• Female sex	• Male sex
• Low HBV DNA	• High HBV DNA

[a] The positive predictors reflect some integrity of the cytotoxic immune response.

The IFNs have many actions *in vivo* in addition to inhibiting virus replication and can stimulate the host CTL response (see section 1.13). IFN-α may upregulate class I HLA display of viral antigens on hepatocytes. Consequently, clearance of virus frequently is preceded by a flare of hepatitis (*Figure 3.8*) Accordingly, success most often is achieved with a low viral load and recent (adult) infection (*Table 3.8*).

Limitations of current approach to antiviral regimes:

• Seroconversion from HBeAg to anti-HBe (end-point) < 30–40% patients
• Optimum doses of interferons uncertain
• Optimum duration of therapy uncertain
• No evidence for efficacy in reversing integration of HBV DNA

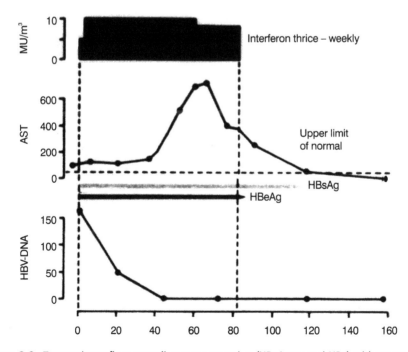

Figure 3.8. Transaminase flare preceding seroconversion (HBeAg to anti-HBe) with successful interferon α therapy. (Reproduced with permission from Alexander *et al.*, 1987 *Lancet* **ii**: 66–69, © 1987 The Lancet Ltd.)

3.8.2.2 Nucleoside analogs

Nucleoside analogs such as lamivudine (3TC), famciclovir (the prodrug of penciclovir), abucovir and the nucleotide analog adefovir dipivoxil reduce HBV replication but, unlike IFNs, have no known direct immunomodulatory impact to clear the virus. HBV DNA levels typically return to baseline levels on discontinuing therapy because most chronically infected individuals mount an insufficient immune cytotoxic attack against infected hepatocytes. Their main advantages are oral administration once daily and lack of serious side-effects.

In one study of Chinese patients aged 16 years or more, lamivudine (100 mg/day, orally for 1 year) was associated with improved histology, 16% (compared with 4% on placebo) seroconversion from HBeAg to anti-HBe and loss of detectable HBV DNA. Further seroconversions are anticipated in the year subsequent to antiviral therapy (*Table 3.9*).

The combination therapy with a nucleoside analog to reduce virus replication and IFN-α-2b to boost the immune attack seems a logical advance in therapeutic strategy. Although results of large-scale studies are not available, preliminary data using this combination, including in patients who failed IFN-α monotherapy, do not show an overwhelming advantage. More research is required to optimize the timing of these antiviral agents.

3.8.2.3 Emergence of drug-resistant variants

Mutations in the polymerase gene, such as those involving substitution of isoleucine (I) or valine (V) for methionine (M) in the YMDD motif, arise in 10–30% of patients treated long-term (beyond 6 months) with the nucleoside analog 3TC. This motif is conserved among retroviral polymerases (see section 3.3.2) and equivalent changes have been observed in the reverse transcriptase of 3TC-resistant HIV. Clinical and histological deterioration are not inevitable but severe exacerbations of hepatitis B with death have occurred, particularly following liver transplantation. Combination therapies with multiple nucleoside analogs may inhibit selection of resistant variants. However, consecutive use of nucleoside analogs (such as 3TC followed by famciclovir) that target the same region of the DNA polymerase may result in rapid emergence of resistance to both agents.

Table 3.9.   Nucleoside analogs (e.g. 3TC: lamivudine)

Advantages
Oral and once daily administration
Side-effects usually not serious:
• Headache
• Pancreatitis
Disadvantages
No immune immunomodulatory effects
Emergence of resistant variants after > 6 months of therapy

3.8.2.4 Other antivirals

Prior to the advent of the nucleoside analogs, drugs such as ganciclovir, acyclovir and adenine arabinoside monophosphate were used to reduce virus replication, including following transplantation (see section 6.5). Unfortunately, decreases in HBV DNA levels were temporary and use was limited by need for intravenous infusions and significant side-effects such as renal toxicity.

3.8.2.5 Other immunomodulatory approaches: corticosteroid withdrawal (priming) therapy

Corticosteroids are known to have adverse effects in viral hepatitis, especially by precipitating sepsis and decompensated liver disease. However, their withdrawal can precipitate a flare (acute-on-chronic hepatitis) with elevations in the liver enzymes, presumably due to priming (albeit temporarily) of the immune response. This observation rationalized their use with withdrawal timed prior to instigating IFN therapy. Most authorities consider that the risks of using corticosteroids outweigh any potential benefit from seroconversion to anti-HBe.

3.8.2.6 Other immunomodulatory approaches: T-cell stimulants

Reports of seroconversion from HBeAg to anti-HBe with α thymosin and levamisole have not been confirmed in controlled studies.

Cytokines, including IL-2 and IL-12, can inhibit virus replication but side-effects are significant. Interest continues with the use of IL-12 as this promotes differentiation of T-helper cell (Th1) responses that are activated during spontaneous clearance of HBV.

Clearance of HBsAg has been documented following transfer of bone marrow from immune (anti-HBs) donors to patients with chronic HBV infection and despite the inevitable requirement for immunosuppressive therapy.

3.8.3 Cirrhosis and hepatocellular carcinoma (HCC)

Management principles are similar for cirrhosis regardless of etiology (see section 1.12).

3.8.4 Hemophilia

The seroprevalence of HBsAg remains high in hemophiliacs and other multiply transfused patients (thallassemics). As for hepatitis C, the constraints on liver biopsy make monitoring these patients difficult. However, responses to antiviral therapies are similar to the general population in patients without evidence of multiple infections and HIV.

3.8.5 HBV in renal disease

The seroprevalence for HBsAg remains high in patients with renal failure, especially those receiving hemodialysis (see also section 6.11.2). Segregation of patients into dialysis units that specialize in handling hepatitis B has helped to reduce the spread of hepatitis B. Patients with renal failure requiring dialysis are immunosuppressed with respect to HBV. Consequently, infected patients tend to have high levels of HBV DNA and show poor responses to antiviral therapies. Many infected patients have significant liver disease (chronic hepatitis B) and this is prone to exacerbation especially following renal transplantation. In fact, the HBV-related liver disease can show accelerated progression to fibrosis and cirrhosis following renal transplantation of the HBsAg seropositive chronic carrier.

Unfortunately fewer than 50% of patients with renal failure respond to the hepatitis B vaccine, including in the young (<40 years) age group (see section 3.9.3). Use of immune stimulants such as thymosin, and additional doses ('booster shots') has led to a temporary and modest elevation of anti-HBs levels in the minority of patients. Whether any improved response equates to improved immune memory and protective efficacy against HBV infection, remains unknown.

Hepatitis C is especially common in renal failure and adds to the complexity of managing these patients.

3.9 Prevention

The hepatitis B vaccine was the first vaccine produced using recombinant techniques to target a human infectious disease and the first anti-cancer vaccine. The licensed plasma and recombinant licensed vaccines are safe and effective (see also section 7.11).

3.9.1 Strategies for immunization

Global eradication of HBV is feasible with current licensed vaccines. The lack of an animal reservoir makes possible the eventual eradication of HBV. The problem of lifelong persistent infection and need to break the chain of mother-to-infant transmission (see section 7.5.2) necessitate an immunization program spanning several generations.

The WHO recommended universal immunization of babies and infants be in place by 1994 in areas of high endemicity defined by a prevalence of HBsAg exceeding 8%, and in all countries by 1997, with integration of hepatitis B vaccine into the Expanded Programme of Immunization (EPI). As part of the EPI, other vaccines, such as against polio, diptheria, tetanus and pertussis, can be given at the same time using separate sites, needles and syringes. Multivalent vaccines (vaccine combinations) are under development.

By 1999, more than 100 countries had adopted this recommendation, particularly in Asia and the Pacific regions. In China, with its large population and high prevalence of HBsAg, up to 90% of children in the cities, but fewer in rural areas, are being immunized. In India, with a carrier rate of around 8%, few individuals were immunized up to 1996 but local production of a plasma-derived vaccine is planned. Certain regions, such as sub-Saharan Africa, that rely on overseas aid have poor coverage within EPI and the hepatitis vaccines.

Many countries with a low prevalence of infection, including the UK, opted for a strategy of immunizing high-risk individuals only. Such strategies have not decreased the incidence of acute hepatitis B.

Universal immunization of infants in Taiwan since 1984 has reduced the prevalence of chronic infection from around 10 to below 2%. Further, the incidence of HCC in children aged 6–14 years was reduced by almost 50% between 1990 and 1994 compared with between 1981 and 1986. Successful immunoprophylaxis against HBV infection also should protect against HDV regardless of route of transmission.

> **Key Notes:** Universal immunization is the key to global eradication of HBV
>
> Despite the availability of safe and effective vaccines for more than a decade, there has been little epidemiological impact
>
> Targeting of 'high risk' groups such as healthcare workers in the West has been ineffective in reducing the global burden of HBV
>
> Recommendation by the WHO to immunize universally all infants in all countries is the most effective way to facilitate global eradication of HBV

3.9.2 Hepatitis B vaccines

Individuals from around the world who recover from acute hepatitis B produce antibodies (anti-HBs) to HBsAg and have life-long immunity to re-infection. This observation suggested that anti-HBs represented a neutralizing antibody and that HBsAg would elicit a protective immune response. The lack of a cell culture system for HBV necessitated the use of the huge reservoir of HBsAg in the carrier population as a novel source of antigen.

3.9.2.1 Plasma-derived hepatitis B vaccines

First-generation vaccines, comprising HBsAg harvested from donated plasma of hepatitis B carriers, remain in use in some countries. HBV-infected hepatocytes secrete HBsAg in the form of non-infectious, subviral particles (22 nm spheres and tubules) in vast excess over the 42 nm virions. Procedures for purification of HBsAg included several steps to inactivate HBV and other pathogens (*Figure 3.9*).

The resulting preparations were shown to be safe and immunogenic and protected chimpanzees against challenge with HBV, including cross-protection against different virus subtypes. Subsequently, immunization was shown to protect volunteers against

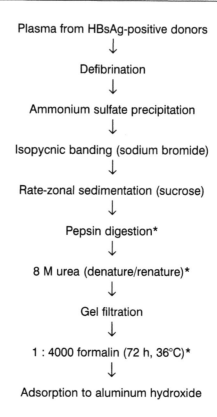

Plasma from HBsAg-positive donors
↓
Defibrination
↓
Ammonium sulfate precipitation
↓
Isopycnic banding (sodium bromide)
↓
Rate-zonal sedimentation (sucrose)
↓
Pepsin digestion*
↓
8 M urea (denature/renature)*
↓
Gel filtration
↓
1 : 4000 formalin (72 h, 36°C)*
↓
Adsorption to aluminum hydroxide

Figure 3.9. Typical production process for a plasma-derived hepatitis B vaccine. Steps marked with an asterisk will inactivate any viruses present in donated plasma.

HBV infection, most notably in a trial amongst high-risk individuals in New York. The vaccine also prevents perinatal transmission. The advent of the HIV pandemic caused unfounded concern over the safety of first generation vaccines and coincided with the development of recombinant DNA vaccines. Plasma-derived vaccines remain an option where carrier rates are high, provided safety and quality control measures are adhered to strictly.

3.9.2.2 Recombinant hepatitis B vaccines

Second-generation vaccines contain recombinant HBsAg synthezised in eukaryotic (yeast or mammalian) cells. Yeast systems give higher yields and are used more commonly than mammalian cell systems. Most recombinant plasmids encode only the 226 aa major surface protein but can include pre-S2 and pre-S1 sequences (*Figure 3.1*). HBsAg synthesized in mammalian cells is secreted into the culture fluid as 22 nm particles resembling those circulating in carriers. The protein is not secreted from yeast but particles may be detected on disruption of the cells. An example of a purification process for a yeast-derived recombinant hepatitis B vaccine is shown in *Figure 3.10*.

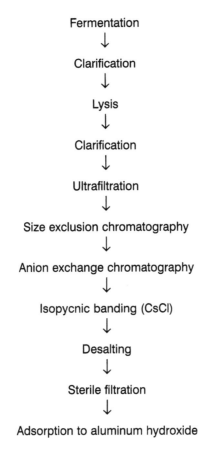

Fermentation
↓
Clarification
↓
Lysis
↓
Clarification
↓
Ultrafiltration
↓
Size exclusion chromatography
↓
Anion exchange chromatography
↓
Isopycnic banding (CsCl)
↓
Desalting
↓
Sterile filtration
↓
Adsorption to aluminum hydroxide

Figure 3.10.   Production process for a licensed, yeast-derived hepatitis B vaccine.

3.9.2.3 Pre-S vaccines

Inclusion of pre-S epitopes in vaccine preparations could improve non-responsiveness and diminish immune pressure on the *a* determinant that has led to vaccine escape mutations. Most current plasma-derived vaccines do not contain pre-S epitopes because the proteolytic treatment (*Figure 3.9*) removes these domains from middle and large surface proteins. The most common licensed recombinant vaccines contain only the major surface protein. Detection of anti-pre-S antibodies remains a research tool. What levels are protective, remains unclear.

3.9.2.4 Combined hepatitis A and B vaccines

A combined vaccine (Twinrix, SKB) is licensed in Europe and the USA. Each dose contains 720 ELISA Units HAAg and 20 µg HBsAg. These amounts are reduced 50% for the pediatric dose. The antigens are adsorbed separately to aluminum hydroxide (as for Havrix and Engerix B), before mixing and diluting with saline. Phenoxyethanol is added as preservative. Clinical trials, using a 0, 1 and 6-month immunization schedule, showed equivalent immunogenicity to either vaccine for seroconversion rates and geometric mean titers. There was no evidence of an increase in side-effects or local reactions compared to the individual vaccines.

3.9.2.5 DNA vaccines and T-cell vaccines

Injection of plasmid DNA, for example, into muscle, leads to expression of proteins encoded therein and, consequently, to an immune response. This method of immunization also stimulates MHC class I-mediated CTL responses. Levels of circulating anti-HBs induced in volunteers immunized with DNA encoding HBsAg do not approach those typical of conventional hepatitis B vaccines. However, DNA-based vaccines may prove useful for therapeutic immunization of hepatitis B carriers because of their potential for inducing T-cell responses.

Repeated immunization with peptides encoding relevant T-cell epitopes may lead to a cellular immune response and such T-cell vaccines also may prove valuable for therapeutic immunization.

3.9.3 Clinical practice

The immunization schedule for licensed vaccines of 0, 1 and 6-month doses was found to be effective. The third dose is important in increasing the seroconversion rate in immunocompetent adults from 75–85% to over 90% and in increasing the geometric mean titer of antibody.

Key Notes: Dosing schedules for hepatitis B vaccines

The schedule of 0, 1 and 6 months is widely used

The accelerated schedule of 0, 1, 2 and 12 months usually gives more rapid seroconversion, longer duration of detectable antibody (anti-HBs) levels and is preferred in high-risk situations (e.g. following a definite exposure and for babies born of HBsAg seropositive mothers)

3.9.3.1 Contraindications to the hepatitis B vaccines

There is none excepting known allergy to aluminum hydroxide and thiomersal preservative (see section 7.12.4.4 for pediatric considerations) and the usual contraindications such as history of fevers, current fever and infection and seizures. All manufacturers state that the hepatitis B vaccine is not recommended in pregnancy. However, there have been no excess adverse effects noted on the fetus for the many hundreds of women who began immunization without knowledge of their early pregnancy.

3.9.3.2 Side-effects

These are notably uncommon given the many million doses administered safely to new-born babies, children and adults (*Table 3.10*).

Reports of neurological abnormalities following immunization include individual cases of Guillain–Barré syndrome, aseptic meningitis, transverse myelitis and seizures and later development of multiple sclerosis. As with the reports of an increased incidence of diabetes mellitus, these findings are considered to occur by no more than chance (random association).

3.9.3.3 Post-immunization testing

Follow-up testing of anti-HBs antibody is essential to assess good (>100 mIU/ml = 100 U/l), poor (< 100 mIU/ml) and non-response (< 10 mIU/ml = < 10 U/l). Evaluation should be carried out around 1 month after the three-dose basic schedule. Poor and non-responders cannot assume protection from infection.

As HBsAg is the sole antigen in the vaccine, only anti-HBs is detectable after successful immunization. In contrast, anti-HBc as well as anti-HBs, including anti-pre-S antibodies, become detectable with recovery from natural infection.

3.9.3.4 Immunogenicity

The hepatitis B vaccines are very immunogenic and stimulate high levels of anti-HBs, especially in children. However, up to 20% of otherwise healthy adult vaccinees do not make an adequate (assumed protective) response to the vaccine antigen (see good responder below). The data sheets provided by some manufacturers report seroconversion rates of up to 95%. Usually, this refers to detectable levels of anti-HBs (above 10 mIU/ml) and does not equate necessarily with protection from infection (protective efficacy). From childhood, response rates fall with increasing age. Only around half of vaccinees aged more than 45 years can anticipate a good response. Females show higher anti-HBs levels than males. Response is better with a deep intramuscular (deltoid or femoral) site rather than into fat (buttock).

Up to 10% of immunocompetent recipients, especially those above 50 years, fail to make an antibody response. Infants of infectious mothers may become infected,

Table 3.10.   Side effects of HB vaccine (adapted from Lemon and Thomas, 1997)

Local and systemic	Incidence
Soreness	10–15%
Redness and swelling	10–15%
Fever (low-grade)	<3%
Headache, malaise, nausea, rash	<1%
Arthralgias and myalgias	<1%

sometimes despite a seemingly adequate antibody response, and viral variants which seem to escape neutralization have been implicated in some cases.

3.9.3.5 Protective efficacy

The minimum level of anti-HBs that guarantees protection from infection is unknown. Hepatitis B infection is unlikely in good responders who achieve levels above 100 mIU/ml when tested 1 month following the basic three-dose course. In the good responder, protection from clinical infection seems to persist for at least 10 years, based on follow-up data. Protection from infection outlives the circulating level of anti-HBs, which falls in all vaccinees after the basic three-dose course.

Poor responders to hepatitis B vaccine achieve less than 100 mIU/ml of anti HBs 1 month after the third dose and are more common with:

- Rising age
- Immunosuppressed (e.g. HIV, liver transplant recipient)
- Renal failure

The definition of poor responder is arbitrary but seems sensible because the maximum level of anti-HBs does not rise rapidly and substantially following an additional dose of vaccine. This indicates a sluggish immune memory to HBsAg and presumably an impaired response to the virus. Studies in mice suggest that the immune responses to the major surface protein and pre-S domains are regulated separately and pre-S2 epitopes may augment the response to the major surface protein. This begs the question whether inclusion of pre-S epitopes in vaccine preparations will reduce the rate of non-responsiveness. The results of clinical trials suggest that a subset of those who fail to respond adequately to a vaccine containing only the major surface protein may respond to vaccines containing pre-S epitopes.

The poor responder should be given an extra dose of vaccine and re-tested 1 month later. This overcomes any doubt over inadequate administration, faulty technique and batch of vaccine. The safest policy is to assume no protection. Seroconversion to anti-HBc can occur but the poor responder may be protected from developing persistent infection (HBsAg seropositivity).

Many authorities recommend an additional, 'booster', dose after 5 years as prudent clinical practice. Scientific evidence for this recommendation is lacking. The USA Advisory Committee on Immunization Practice specifically has not recommended additional doses over the 12-year period for which follow-up data are available.

3.9.3.6 Non-responders

Most cases of non-response and poor response seem to be genetically determined as responses to other vaccines are not impaired.

Non-responders achieve less than 10 mIU/ml of anti HBs 1 month after the third dose and may be:

- Genetically determined
- HBV infected

Key Notes: Pre-screening for anti-HBs

This is not cost effective in the UK or USA. Pre-screening should be considered for immigrants and travelers from high-prevalence countries, the sexually promiscuous, and intravenous drug users, amongst others. Savings on pre-screening should be channelled into testing after completion of the basic course of hepatitis B vaccine

3.9.3.7 Problems of vaccine failure in infants

See section 7.12.4.5

3.9.3.8 Approaches to improving immunogenicity

Immunostimulants, such as α-thymosin and IL-2, have accompanied hepatitis B immunization in the immunosuppressed and renal dialysis. The few detectable anti-HBs levels typically have remained low and short-lived, raising doubts over long-lasting protective efficacy.

Further reading

Alexander G., Brahm, J., Fagan, E.A., *et al.*, (1987) Loss of HBsAg with interferon therapy in chronic hepatitis B virus infection. *Lancet* **ii**: 66–68.
Landmark paper showing clearance of HBsAg as well as seroconversion of HBeAg following antiviral therapy.

Alter, M.J., Hadler, S.C., Margolis, H.S., *et al.* (1990) The changing epidemiology of hepatitis B in the United States. Need for alternative vaccination strategies. *JAMA* **263**: 1218–1222.
Failure of targeting high risk groups for immunization to impact on incidence of disease.

Alter, M.J., Margolis, H.S., Krawczynski, K., *et al.* (1992) The natural history of community-acquired hepatitis C in the United States. *N. Engl. J. Med.* **327**: 1899–1905.
Follow-up of patients with non-A, non-B hepatitis in Sentinel Counties studies, risks of persistent viremia and chronic hepatitis.

Cacciola, I., Pollicino, T., Squadrito, G., Cerenzia, G., Orlando, M.E., Raimondo, G. (1999) Occult hepatitis B virus infection in patients with chronic hepatitis C liver disease. *N. Engl J. Med.* **341**: 22–26.
66% of patients with hepatitis C liver disease were HBV DNA-positive by PCR despite undetectable HBsAg.

Centers for Disease Control. (1991) Hepatitis B virus. A comprehensive strategy for eliminating transmission in the United States through universal childhood vaccination. Recommendations of the Immunization Practices Advisory Committee (ACIP). *MMWR* **40 (RR-13)**: 1–20.
Current recommendations for use of hepatitis B vaccine.

Centers for Disease Control. (1991) Public Health inter-agency guidelines for screening donors of blood, plasma, organs, tissue and semen for evidence of hepatitis B and hepatitis C. *MMWR* **40: (RR-4)**: 1–17.
US Public Health Service recommendations for counselling and medical management of persons with hepatitis B and anti-HCV antibodies.

Centers for Disease Control. (1999) Notice to readers update: recommendations to prevent hepatitis B virus transmission—United States. *MMWR* **48 (02)**: 33–34.
Recommendations of the Advisory Committee on Immunization Practices (ACIP) since October 1997 to expand hepatitis B immunization from 0–18 years (including 'catch-up' in ages 11–12 years) for all children and older adolescents and adults in defined risk groups.

Chang, M.H., Chen, C.J., Lai, M.S., et al. (1997) Universal hepatitis B vaccination in Taiwan and the incidence of hepatocellular carcinoma in children. Taiwan Childhood Hepatoma Study Group. *N. Engl. J. Med.* **336**: 1855–1859.
Landmark paper showing reduction in HCC by almost 50% in children aged 6–14 years within 15 years of implementing universal immunization.

Chazouillères, O., Mamish, D., Kim, M., et al. (1994) 'Occult' hepatitis B virus as source of infection in liver transplant recipients. *Lancet* **343**: 142–146.
HBV infection following liver transplantation may be attributable to previous occult infection of the recipient or donor.

Chen, J.-Y., Harrison, T. J., Lee, C.-S., Chen, D.-S., Zuckerman, A. J. (1986) Detection of hepatitis B virus DNA in hepatocellular carcinoma. *Brit. J. Exp. Path.* **67**: 279–288.
More than 80% of HCCs from HBsAg carriers contain integrated HBV DNA, detectable by Southern hybridization.

El-Serag, H.B., Mason, A.C. (1999) Rising incidence of hepatocellular carcinoma in the United States. *New Engl. J. Med.* **340**: 798–799.
The age-adjusted incidence of liver cancer from 1976–1995 increased from $1.4/10^5$ between 1976–1980 to $2.4/10^5$ from 1991–1995 with over a 40% increase in mortality. The rises were especially high in black men and persons aged 40–60 years.

Guidotti, L.G., Rochford, R., Chung, J., Shapiro, M., Purcell, R., Chisari, F.V. (1999) Viral clearance without destruction of infected cells during acute HBV infection. *Science* **284**: 825–829.
Non-cytolytic viral clearance in two chimpanzees.

Harrison, T.J., Anderson, M.G., Murray-Lyon I.M., et al. (1986) Hepatitis B virus DNA in the hepatocyte. A series of 160 biopsies. *J. Hepatol.* **2**: 1–10.

Heathcote, J., McHutchinson, J., Lee, S., et al. (1999) A pilot study of the CY-1899 T cell vaccine in subjects chronically infected with hepatitis B virus. *Hepatology* **30**: 531–536.
Disappointing response to single epitope vaccine.

Hoofnagle, J., Shafritz, D.A., Popper, H. (1987) Chronic type B hepatitis and the 'healthy' HBsAg carrier state. *Hepatology* **7**: 758–763.
Landmark review of natural history.

Koshy, R., Caselman, W.H. (1998) *Hepatitis B Virus.* Imperial College Press, London.

Lai, C.-L., Chien, R.-N., Leung, N.W.Y., et al. (1998) A one-year trial of lamivudine for chronic hepatitis B. *New Engl. J. Med.* **339**: 61–68.
Lamivudine for 1 year at 100 mg p.o./day for Chinese HBsAg seropositive carriers aged above 16 years resulted in improved histology (reduced progression of fibrosis), seroconversion of HBeAg to anti-HBe (16% compared with 4% placebo group) and loss of detectable HBV DNA. YMDD resistant variants were detected in 14%. Follow up data are awaited.

Lai, M.M. (1995) The molecular biology of hepatitis delta virus. *Ann. Rev. Biochem.* **64**: 259–286.

Lau, G.K., Liang, R., Wu, P.C. et al. (1998) Use of famciclovir to prevent HBV reactivation in HBsAg-positive recipients after allogeneic bone marrow transplantation. *J. Hepatol.* **28**: 359–368.
Famciclovir reduced the number of cases and severity of reactivation of hepatitis B following bone marrow transplantation compared with controls. Interestingly, two patients (anti-HBe with undetectable HBV DNA by bDNA) cleared HBsAg following famciclovir and grafting with HLA identical bone marrow from immune (anti-HBs and anti-HBc) donors.

Lemon, S.M.,Thomas, D.L. (1997) Vaccines to prevent viral hepatitis. *New Engl. J. Med.* **336**: 196–204.
Review of recombinant and plasma vaccines with strategies for the future.

Ling, R., Mutimer, D., Ahmed, N., et al. (1996) Selection of mutations in the hepatitis B virus polymerase during therapy of transplant recipients with lamivudine. *Hepatology* **24**: 711–713.
First description of YMDD substitutions in HBV.

Locarnini, S., Birch, C. (1999) Antiviral therapy for chronic hepatitis B infection: lessons learned from treating HIV-infected patients. *J. Hepatol.* **30**: 536–550.
Comprehensive review of HBV replication and nucleoside analogs.

Margolis, H.S., Alter, M.J., Hadler, S.C. (1991) Hepatitis B: evolving epidemiology and implications for control. *Sem. Liver Dis.* **11**: 84–92.
Landmark paper on epidemiology and control.

Marusawa, H., Osaki, Y., Kimura, T., et al. (1999) High prevalence of anti-hepatitis B virus serological markers in patients with hepatitis C virus related chronic liver disease in Japan. *Gut* **45**: 284–288.
Anti-HBc in 49.9% of 2014 Japanese patients with HCV related chronic liver disease including HCC.

Mutimer, D., Naoumov, N., Honkoop, P., et al. (1998) Combination alpha-interferon and lamivudine therapy for alpha-interferon resistant chronic hepatitis B: results of a pilot study. *J. Hepatol.* **28**: 923–929.
Disappointing results in 20 patients treated with both drugs for 16 weeks; sustained response for HBV DNA and for HBeAg seroconversion in only one patient.

Norder, H., Hammas, B., Lee, S.D., et al. (1993) Genetic relatedness of hepatitis B viral strains of diverse geographical origin and natural variations in the primary structure of the surface antigen. *J. Gen. Virol.* **74**: 1341–1348.
Correlation of HBsAg antigenic sub-determinants with amino acid sequence.

Oon, C.J., Lim, G.K., Ye, Z., *et al.* (1995) Molecular epidemiology of hepatitis B virus vaccine variants in Singapore. *Vaccine*, **13**: 699–702.
Summary of the Singapore vaccine study.

Polish, L.B., Gallagher, M., Fields, H.A., Hadler, S.C. (1993) Delta hepatitis. Molecular biology and clinical and epidemiological features. *Clin. Microbiol. Rev.* **6**: 211–229.
Overall review of hepatitis D virus and infection.

Rehermann, B., Ferrari, C., Pasquinelli, C., Chisari, F.V. (1996) The hepatitis B virus persists for decades after patients' recovery from acute viral hepatitis despite active maintenance of a cytotoxic T-lymphocyte response. *Nature Med.* **2 (10)**: 1104–1108.
Individuals who seroconvert to anti-HBs following an acute infection may not clear the virus completely but harbor a low level infection which is controlled by the CTL response.

Stirk, H.J., Thornton, J.M., Howard, C.R. (1992) A topological model for hepatitis B surface antigen. *Intervirology* **33**: 148–158.
Prediction of the structure of the major surface protein.

Summers, J., Mason, W. S. (1982) Replication of the genome of a hepatitis B-like virus by reverse transcription of an RNA intermediate. *Cell* **29**: 403–415.
Discovery that hepadnaviruses replicate with reverse transcription of an RNA pregenome.

Chapter 4

Hepatitis C Virus (HCV)

4.0 Introduction

Worldwide, hepatitis C virus (HCV) is a major cause of sporadic (community-acquired) hepatitis. Parenteral exposure via injection of recreational drugs remains a major risk factor for acquiring HCV infection (*Figure 4.1*). HCV also was a major cause of post-transfusion non-A, non-B hepatitis prior to the advent of screening blood donors for anti-HCV antibodies in the 1990s. Chronic hepatitis with persistent virus replication is the most common outcome of acute hepatitis C. Cirrhosis and primary liver cancer are associated with long-term infection. Vertical and sexual transmissions are much less common for HCV than for hepatitis B virus (HBV).

4.1 Epidemiology of hepatitis C virus

HCV is endemic worldwide with around 170 million chronic carriers, of whom an estimated 5 million are in Western Europe and 4 million in the USA. Prevalence rates vary and data are incomplete. The global prevalence based on antibody tests is around 3% and ranges between 0.5 and 5% in most countries. The first-generation enzyme immunoassays (EIAs) gave high reactivity rates—false positives—and overestimated the prevalence of infection. Conversely, antibody prevalences among volunteer blood donors, selected for low-risk factors, underestimate the general prevalence of infection. In the UK, less than 1 in 1000 volunteer potential blood donors test positive for anti-HCV antibodies, compared with around 1–2% in the USA (*Table 4.1*).

HCV is the major cause of HBsAg seronegative chronic liver disease worldwide. In the USA alone, HCV accounts for more than 10 000 deaths annually from chronic liver disease. End-stage chronic hepatitis C, often with concurrent alcoholic liver disease, is the most common referral for liver transplantation in the West. Paradoxically, many individuals diagnosed by screening with chronic hepatitis C seem to have mild liver disease over long-term follow-up.

131

Table 4.1. Seroprevalence of hepatitis C in US populations before 1992 (various authors, Kelen *et al.*, 1992; after Schiff *et al.*, 1999)

Hepatitis C in high-risk groups	Anti-HCV positivity (%)
Patients presenting to Casualty (ER) Departments in USA	18
• IVDU[a]	>80[a]
• Cocaine users	>10
Healthcare personnel	0.6–6.6[b]
• Dentists	1.75
• Oral surgeons	9.3
Hemodialysis	6–38
Organ transplantation	
• Cadaveric donors	4.2
• Renal transplant candidates	10–30
• Liver transplant candidates	20–40
Bone marrow candidates transfused before 1990s	
• Leukemia	5
• Thalassemias	50

[a] Screened in Casualty (ER) Departments: falling seroprevalence since 1995. [b] Seroconversion after sustained needlestick exposure.

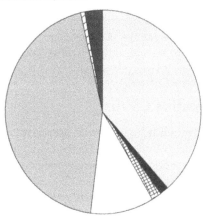

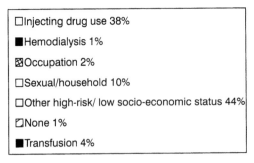

□Injecting drug use 38%

■Hemodialysis 1%

▨Occupation 2%

□Sexual/household 10%

□Other high-risk/ low socio-economic status 44%

▨None 1%

■Transfusion 4%

Figure 4.1. Risk factors for acute hepatitis C in the USA, 1990–1995 (CDC Sentinel Counties Study). Note that recent studies (*Table 4.2*) have identified several of the factors grouped here as 'other high-risk/low socio-economic status'. (Data available from www.cdc.gov/ncidod.diseases/hepatitis/slideset/hep49.gif)

HCV is a common cause of inapparent acute hepatitis. Studies on acute hepatitis C are limited because the majority of patients who present with symptoms have chronic hepatitis. (see also section 1.2).

4.1.1 Risk factors for HCV infection

In the USA around 50% of anti-HCV seropositive individuals admit to intravenous drug use (IVDU) and 13% declare other risk factors such as a seropositive household/sexual partner. Only 3% recall a blood transfusion and 1% have potential occupational exposure. Although, no high-risk behavior is declared in the remaining 32%, on repeat questioning some admit previous intravenous drug use and snorting cocaine. In the West, HCV infection is more common in males than females, blacks than whites and middle aged (30–50 years) adults than other age groups. The high rates of infection (5–10%) in Egypt, various Mediterranean countries, Japan and elsewhere may reflect inadvertent transmission, including that associated with parenteral administration of medicinal agents.

Key Notes Hepatitis C

Epidemiology
* Worldwide
* High endemicity in East and Far East
* Parenteral and sporadic (unknown) routes
* Sexual transmission uncommon
* Vertical transmission uncommon

Infection
* Symptomless illness in majority
* Acute liver failure—rare
* Chronic infection
      common (>80%) after acute infection
      chronic hepatitis, cirrhosis and hepatocellular carcinoma may ensue

Reliable serological diagnosis
* Antibody—by enzyme immunoassays
* Antigen—assays not available
* RT–PCR for viremia (HCV RNA)
      for delayed seroconversion

Protection
* Protective immunity ill-defined
* Immune globulin not available
* No immediate prospects of a vaccine

Table 4.2. Age adjusted prevalence of anti-HCV antibodies according to risk factors (NHANES III study: after Alter MJ *et al.*, 1999)

	Number	Total population	Non-hispanic whites	Non-hispanic blacks	Mexican Americans
Ever healthcare	769	1.4%	1.4%	2.1%	0.6%
Never healthcare	15 625	2.0%	1.6%	3.7%	2.7%
Cocaine > 10 times	519	17.8%	11.6%	25.3%	26.0%
Marijuana > 100 times	1025	9.5%	6.6%	15.3%	23.4%
> 50 sexual partners	454	9.4%	11.8%	9.1%	6.1%
Herpes simplex type 2	3453	3.4%	2.8%	5.0%	3.6%

A recent survey for risk factors for anti-HCV seropositivity, between 1988 and 1994, was conducted by the Centers for Disease Control and Prevention (CDC) using the NHANES III population database of over 20 000 individuals. Interestingly, these data showed no correlation with healthcare occupation, ever being in the armed forces or previous surgery involving transfusion and dental visits. Instead, there was a strong correlation with use of cocaine, use of marijuana and promiscuous life-style as well as race (*Table 4.2*). Other positive correlations included male sex, low level of education, being below the poverty level and being divorced or separated. This study estimates that around 2.7 million individuals in the United States are chronically infected with HCV.

4.1.2 Vertical and sexual transmission

Perinatal transmission is sufficiently uncommon for the CDC and the American Academy of Pediatrics to consider unnecessary specific recommendations surrounding the mode of delivery and breast-feeding (see also section 7.5.3).

4.1.3 Sporadic, community-acquired infections

Routes of transmission remain unknown in up to 30% of cases of chronic hepatitis C making difficult the control of the spread of infection.

4.2 Biology of hepatitis C virus

Prior to the advent of molecular techniques, results of physiochemical and transmission studies favored an enveloped virus but electron microscopy, immunological techniques and all other efforts failed to characterize the elusive non-A, non-B agent. HCV, like HBV, is difficult to propagate in cell culture. The diagnostic tests developed are based on recognition by circulating antibodies of recombinant proteins and synthetic peptides representing the gene products of various regions of the virus genome.

4.2.1 Cloning of the HCV genome

The HCV genome was cloned in 1989 and published with an antibody test (first generation). The starting pool of serum from the CDC was implicated in post-transfusion non-A, non-B hepatitis and had high titer of agent based on transmission studies. High-speed centrifugation pelleted putative virus particles. Nucleic acid was isolated and complementary DNA synthesized. An expression cDNA library was screened with pooled sera from chronically infected patients anticipated to contain anti-viral antibodies.

A clone was isolated that was complementary to a small segment of the viral genome that bound to antibodies in serum from individuals with chronic non-A, non-B hepatitis. This clone, in turn, was used as a probe to detect others ('walking the genome'), leading to the assembly of overlapping sequences covering the viral genome. These sequences hybridized to a positive-sense RNA molecule of around 10 000 nucleotides (nt) present in the livers of infected chimpanzees but not in uninfected controls.

4.2.2 Organization of the HCV genome

The HCV genome comprises a positive (mRNA)-sense, single-stranded RNA molecule of around 9.4 kb that resembles closely the pestiviruses and flaviviruses that are enveloped RNA viruses of the family *Flaviviridae*. The 5' and 3' untranslated regions (5' UTR and 3' UTR) flank a single, long open reading frame (*Figure 4.2*). The 5' UTR comprises 342 nt preceding the codon that initiates translation and is highly conserved among diverse HCV sequences. The genome is not capped but the 5' UTR is predicted to have extensive secondary structure and contains an internal ribosome entry site for initiation of translation. A hairpin at the extreme 5' terminus may be involved in replication of the genome.

As for other positive-sense RNA viruses, the 3' end is important for RNA replication and packaging into virus particles. The 3' UTR contains a poly U tract originally thought to constitute the 3' terminus of the genome. However, a length (98 nt) of novel sequences detected downstream is highly conserved between subtypes and includes a terminal hairpin that may be involved in replication of the genome.

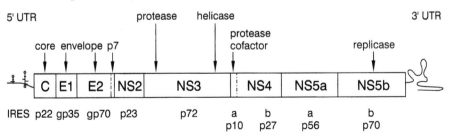

Figure 4.2. Organization of the HCV genome. Boxes show the major products of the polyprotein with the untranslated regions at either end. Known functions of these products are shown above the boxes and the predicted sizes of the polypeptides are given below.

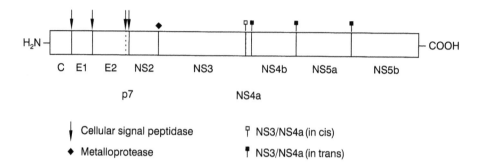

Figure 4.3. Processing of the HCV polyprotein by host and viral proteases. Further details of the products may be found in *Figure 4.2*.

Following translation, the polyprotein is cleaved into 10 or more polypeptides via a series of host and viral peptidases (*Figure 4.3*). The structural proteins are encoded in the amino-terminal portion, the core or nucleocapsid protein preceding two envelope glycoproteins, E1 and E2, and are processed in the endoplasmic reticulum by the host signal peptidase. The glycoproteins vary considerably among HCV sequences, particularly a hypervariable region at the amino terminus of E2. There is no non-structural glycoprotein equivalent to NS1 in the flaviviruses. However, cleavage by the cellular signal peptidase at the carboxyl-terminus of E2 may yield a small, membrane-bound protein called p7. The E1, E2 (and E2–p7 fusion) proteins show N-linked glycosylation and are localized to the endoplasmic reticulum. E1 and E2 may complex as a heterodimer. Hypervariability in this region reflects continued selection of variants that escape neutralizing antibody.

The non-structural region (NS) is processed to NS2, NS3, NS4a, NS4b, NS5a and NS5b. Further processing of NS2 to NS2a and NS2b, analogous to the flaviviruses, has not been reported. The NS2 protein has a zinc-dependent metalloproteinase activity and is responsible for cleavage (in *cis*) of the NS2/NS3 site. As with the flaviviruses, the NS3 protein functions as a serine protease in the amino-terminal region and RNA helicase with ATPase activity towards the carboxyl terminus. The RNA helicase acts to relax RNA duplexes during replication. Most cleavages in the non-structural region, NS3/NS4a (in *cis*) and NS4a/NS4b, NS4b/NS5a and NS5a/NS5b (in *trans*), are attributed to the serine protease activity NS3. NS4a binds to NS3 and acts as a co-factor for the *trans* cleavages. Binding of the NS3/NS4a protease to its substrates is being studied in detail and the crystal structure of the NS3 protease domain, complexed with a synthetic NS4a cofactor peptide, has been determined. Inhibitors of NS3 enzymatic activity are being developed as potential antiviral agents. However, NS3/NS4a lacks the distinct, substrate-binding pocket seen in other serine proteases, confounding the design of inhibitory molecules. The Gly-Asp-Asp (GDD) motif, common to the RNA-dependent RNA polymerases, is located in NS5b, that has replicase activity. The function of NS5a is unclear but may be involved in replicase activity.

4.2.3 In search of the receptor—binding of HCV to the cell surface

The E2 glycoprotein has been shown to bind to CD81, a molecule expressed on the surface of diverse cell types. E2/CD81 binding may be required for virus entry to host cells and CD81 may constitute the primary receptor or, more likely, a co-receptor. Binding of E2 to CD81 can be inhibited by serum from immunized chimpanzees, but this may result from steric effects rather than direct neutralization of a receptor-binding domain.

CD81 is a member of the tetraspannin superfamily of molecules, with four transmembrane domains and two external loops. The second extracellular loop (EC2) is highly conserved across species, with only one amino acid difference between humans and chimpanzees (the only hosts for HCV). There are numerous sequence differences between human and rodent EC2 and HCV E2 does not bind to murine CD81.

CD81 with CD19 and CD21 form a signalling complex on the surface of B cells that is important for their activation and proliferation. Binding of HCV to CD81 on B cells may account for some aspects of its pathology, such as the association of HCV with cryoglobulinemia and certain B-cell lymphoproliferative disorders (see sections 4.3.6.1 and 4.3.6.3).

It has been reported recently that HCV, and other members of the *Flaviviridae* including GBV-C/HGV, enter cells via the low density lipoprotein receptor.

4.2.4 Classification of HCV

HCV has been classified provisionally as the prototype of a third genus, the hepaciviruses, in the family *Flaviviridae*. Other genera are the flaviviruses (prototype, Yellow Fever virus), pestiviruses and, potentially, the GB viruses (*see Table 8.6*).

4.2.4.1 HCV genotypes and clinical significance

The genetic organization of HCV is conserved among all isolates but these may differ in sequence at as many as 45% of nucleotide positions. Several candidate schemes to classify this genetic diversity emerged in early publications. The term 'genotype' describes groups of genetically related HCVs. Genotypes were numbered in chronological order of discovery rather than by degree of divergence or any pathological and clinical significance. So far, no non-pathogenic genotypes have been reported.

Simmonds *et al.* (1994) proposed a more complex but rational scheme involving subtypes within genotypes. HCVs sharing more than 88% nucleotide identity fall within the same subtype. Viruses of different subtypes within one genotype share 75–86% nucleotide identity, whilst those with only 55–72% identity with known sequences may constitute new genotypes. Eleven genotypes and more than 70 subtypes are recognized. The first six genotypes account for more than 90% of isolates found worldwide and are designated by an Arabic numeral, with subtypes of a lower case letter: 1a, 1b, 2a etc.

Genotypes of HCV are characterized after polymerase chain reaction (PCR) amplification by comparisons of nucleotide sequences and, indirectly by restriction fragment length polymorphisms (restriction enzyme digestions of PCR amplicons from the 5′ UTR), line probe assay (hybridization to immobilized, specific oligonucleotides) and by serotyping assay (detection of genotype-specific antibodies by EIA). Sequencing of more than one subgenomic region [usually, NS5 and nucleocapsid (core) regions] remains the gold-standard for identification of new genotypes and subtypes. Whereas, sequencing of diagnostic reverse transcription (RT)–PCR products that often target the more highly conserved 5′ UTR (see section 1.4.5) may indicate the genotype but not, necessarily, the subtype.

4.2.4.2 Geographic distribution of HCV genotypes

As with other parenterally transmitted viruses, spread of HCV was favored before the 1960s by increase in intravenous use of recreational drugs, multiple transfusions for major surgery, recycling and repeated use of sharp instruments without adequate sterilization, and international travel.

The major genotypes vary in geographic distribution but genotypes 1 and 2 are distributed worldwide. Genotype 1b is the most common (> 50% of isolates) in the USA, southern Europe and Japan. More heterogeneity is found in Europe than the USA, presumably reflecting migration of peoples. Genotype 1a is prevalent in Western Europe and more common than genotype 2a. Genotypes 2a and 2b tend to be found in older individuals who deny parenteral risk factors. Genotype 3a is found in India and the Far East, and, in the West. The preponderance of genotypes 1a and 3a among IVDU is consistent with their more recent divergence from ancestral isolates and spread through Europe. Genotype 4 is prevalent in the Middle East, especially Egypt and Yemen, and sub-Saharan Africa. Genotypes 3 and 4 are extremely divergent with a large number of subtypes. Genotypes 5–11 seem to be more restricted geographically. Genotype 5 was described first in South Africa and genotype 6 in Hong Kong, Thailand, Taiwan and Vietnam. The rare genotypes 7–11 are found across South-East Asia.

So far, data remain limited on the clinical correlation between viral genotypes, disease severity and outcome. Several publications correlate genotype 1b with adverse clinical outcomes such as development of severe chronic liver disease before and after liver transplantation, hepatocellular carcinoma (HCC) and poor sustained virological response to α-interferons (IFNs). In one study from Japan, the presence of multiple mutations within NS5a in variants of genotype 1b correlated with clearance of virus following α-IFN therapy. Responses to IFN therapy seem especially poor for genotypes (1a and 1b) in patients with liver enzymes in the normal range but comparative and large-scale data are lacking for the rarer genotypes.

Any correlation of genotype with pathogenicity deserves caution given the basis for allocating genotype, the bias as predominant genotype, tendency for high levels of viremia and diversity of quasispecies eliciting different immune responses, duration of infection with rising age and other confounding factors. Also, early versions of the branched chain DNA assay (see section 1.4.4) used oligonucleotides complementary to genotype 1 sequences leading to suboptimal quantification of other genotypes. Cirrhosis seems as prevalent with genotypes 1 and 4 as compared to genotypes 2 and 3.

Moreover, the universal classification has problems. There is no restraint over assignment of new numbers. The delay between submission and publication of papers results in simultaneous 'claims'. More critically, the distinction between genotype and subtype, based on pair-wise comparisons of HCV sequences available in 1993, has broken down with identification of more sequences.

4.2.4.3 Quasispecies of HCV and their clinical significance

The quasispecies effect is discussed in Chapter 1 (see section 1.3.1). Minor quasispecies have been implicated in the reinfection of chimpanzees challenged, after recovery, with the same inoculum of HCV. Such studies provide evidence that the evolution of viral quasispecies involves selection by host factors, such as the immune system. Similarly, a great diversity of quasispecies reduces the likelihood of clearing HCV during antiviral therapy and favors relapse with selection of resistant species following cessation of treatment.

4.2.5 Mechanisms of liver damage and immunopathogenesis

Evidence for a direct cytopathic effect is lacking, except in severe disease following transplantation (see fibrosing cholestatic hepatitis, section 6.3.3). However, examination of liver biopsies from patients with chronic hepatitis C occasionally reveals dying hepatocytes without adjacent inflammation. Conversely, in most patients, markers of liver damage such as transaminase levels do not correlate with viral load, arguing against a direct cytopathic effect.

Evidence for immune-mediated damage is increasing and resembles that seen in acute and chronic HBV infection. Histopathological features of lymphoid aggregates and the reduction in serum alanine aminotransferase (ALT) levels observed in the short-term with corticosteroid therapy suggest an immunopathological effect of cytotoxic T cells. CD8+ T cells may be detected in infected livers. These may lyse HCV-infected hepatocytes directly and also release antiviral cytokines such as tumor necrosis factor (TNF)-α and IFN-γ.

Studies on acute infection focus on non-human primates as data for man are lacking. Following transmission to chimpanzees, HCV-specific cytotoxic T lymphocytes (CTLs) become detectable and numbers of CD8+ cells rise with the acute elevations in liver enzymes. There is an inverse relationship between the levels of viremia and CTL response, as measured *in vitro*, suggesting regulation of the infection by the immune system. In the acute phase of infection, the CTL response targets several structural and non-structural proteins of HCV. Several CTL epitopes have been defined within the HCV core protein, such as amino acids 88–96.

As with chronic hepatitis B, proliferation assays using synthetic peptides to viral core epitopes show that T-cell responses remain active long after resolution of acute infection. Similarly, in the chronic phase of infection, numbers and activity of the CTL response in peripheral blood mononuclear cells are low. CTLs have been detected in the lymphoid aggregates of liver biopsies in man and chimpanzees but no specific HLA alleles have been identified that predict clearance of HCV infection.

Extrapolations are limited from comparable studies in chronic hepatitis B. Different strategies must have evolved for persistence of HCV in the host. CD4-mediated T-cell responses are more intact in chronic hepatitis C than B and cytokine profiles differ. Also, the role of CD4+ helper T lymphocytes is controversial; they form aggregates in the liver but seem unable to clear HCV from infected hepatocytes by release of cytokines and cell lysis. Cytokine profiles of clones of intrahepatic T cells in chronic hepatitis C show a predominance of the non-cytolytic T helper phenotype: Th1 (CD4+, CD8−) that produces IFN-γ, interleukin-2 and TNF-β. This profile should stimulate cell-mediated immune responses that protect against intracellular pathogens and favor clearance of virus.

4.2.5.1. Neutralizing antibodies

Certain non-human primates with high titer antibodies following immunization with recombinant E1 and E2 proteins can be protected from infection after challenge with an autologous inoculum. Neutralizing antibodies may recognize and block domains on the envelope glycoproteins that bind to receptors or co-receptors (potentially; CD81; see section 4.2.3) on the hepatocyte surface.

The development of neutralizing antibodies serves to drive the evolution of the HCV quasispecies in persistently infected individuals and may account for the extreme variability of the HVR-1 of E2. Serum taken from one well-studied patient (patient H) contained antibodies that neutralized the infectivity of his acute phase plasma for chimpanzees at 2, but not 11, years after acute hepatitis C.

However, the capacity to neutralize and protect does not extend to heterologous challenge. Anti-E2 antibodies are detectable in most viremic individuals with chronic hepatitis C. Here, there seems to be a dynamic equilibrium between maintaining the antibody response to the surface glycoproteins of HCV and the emergence and selection of mutated viruses that escape neutralization by antibody. Peaks of viremia in chronically infected patients may be followed by emergence of variants with mutations and which escape neutralization by antibody. As with HBV, mutations also have been detected in immunodominant CTL epitopes such as residues 88–96 of the nucleoprotein of HCV. These mutations may facilitate escape of the CTL response and favor persistent infection, especially in the presence of high viral loads.

The variability of critical regions of the surface glycoproteins may account for the limited success in developing vaccines and failure of immune globulin (passive immunization) to protect against HCV infection.

4.2.6 Cell culture

Few laboratories claim to be able to grow the virus consistently in cell culture.

4.3 Clinical features

These do not distinguish acute from chronic hepatitis C or from any other causes of hepatitis. Most presentations are with chronic hepatitis C because the majority of acute infections are inapparent, subclinical, episodes.

The clinical features of hepatitis C infection:

- Acute hepatitis C goes unnoticed in most patients
- Most patients present with chronic hepatitis
- 50% have risk factors for parenteral transmission
- 75% without obvious risk factors are in low socio-economic groups
- Concomitant alcoholic liver disease is common
- Non-specific presentations include skin rashes, fatigue, itching and arthralgia
- Cryoglobulinemia is symptomless in 80% of patients
- Vasculitis, pupura and renal dysfunction may occur in 2–5%

4.3.1 Acute hepatitis C

This is diagnosed infrequently except for post-transfusion hepatitis with unscreened blood. The incubation period is around 6–8 weeks (range, 2–26 weeks) in the few patients who develop symptoms. Malaise and jaundice occur in around 20–40%. Elevations in liver enzymes (ALT > 5–15 times normal range) occur in the majority of patients. The ALT levels characteristically fluctuate with intermittent normalization making prediction of clearance difficult without testing serially HCV RNA levels. For treatment of acute hepatitis C, see section 4.7.7. Acute liver failure attributable to HCV is extremely rare and is covered in Chapter 5.

4.3.2 Chronic hepatitis and cirrhosis

Persistent viremia follows acute infection in more than 85% of patients. Most presentations suggesting acute hepatitis, with fever, malaise and elevations of ALT, are discovered later to be acute exacerbations of chronic hepatitis C (acute-on-chronic).

The distribution in severity of disease differs between that seen in community-based practice and referral centers. In screening programs such as for blood donors, around 40% of symptomless anti-HCV seropositive individuals have serum ALT levels within the normal range. Most have mild hepatitis on liver biopsy; cirrhosis is uncommon in this setting. Conversely, significant numbers of patients presenting to hepatology centers are referred for abnormal elevations in liver enzymes and liver damage that may progress to cirrhosis. Others will present with, or be referred for, complications of cirrhosis such as portal hypertension with bleeding varices and ascites and HCC (see section 4.4). Typically, in these, HCV infection is diagnosed retrospectively after presentation.

4.3.2.1 Long-term prognosis

Follow-up beyond 10 years of patients with post transfusion chronic hepatitis C has yielded conflicting data. Reports from non-specialized centers of patients typically who acquired HCV during cardiac surgery prior to the advent of screening, have failed to show an increase in mortality over uninfected matched controls. The high mortality rates (around 50%) in both groups reflected deaths from common causes in an older cohort that had survived major surgery.

Conversely, after 17 years follow-up of a cohort of 704 Irish women who in 1977–1998 received anti-D globulin contaminated with HCV, only 2% had developed cirrhosis; the vast majority have remained symptomless with histologically documented, mild hepatitis. In contrast, reports from liver transplant centers tend to emphasize the severity of chronic liver disease, cirrhosis and HCC especially in patients infected after 50 years of age.

4.3.2.2 Risk factors for progression

Progression of chronic liver disease is slow and may take 20–30 years from infection to cirrhosis and HCC. Several co-factors, especially duration of infection and rising age, play an important role in the development of progressive liver disease and development of cirrhosis and may explain the disparity in clinical outcomes. The prevalence of cirrhosis does not differ between virus genotypes.

Risk factors for developing progressive liver injury:

- Age > 40 years at acquisition of infection
- Alcoholism
- Co-infection with human immunodeficiency virus, HBV
- Male sex

4.3.3 Hepatitis C and alcohol

The prevalence of HCV infection and related severe disease is significantly higher in alcoholics than in the general population.

4.3.4 Hepatitis C after liver, other organ and bone-marrow transplantation

These aspects are dealt with in Chapter 6.

4.3.5 Multiple infections

Concomitant infections with HBV, hepatitis D virus (HDV) and human immunodeficiency virus (HIV) are common among IVDU and other groups at high-risk of parenteral transmission. GBV-C/hepatitis G virus RNA has been detected in up to 20% of anti-HCV seropositive individuals but the pathogenic significance remains unclear (see section 8.9).

Concomitant hepatitis C and B infections:

- Patients share common risk factors
- HCV can suppress HBV replication
- HCV RNA may be undetectable in serum but positive in liver
- Increased risk of severe disease and early HCC

Hepatitis B may play an important role in liver disease previously attributed solely to HCV. In a study of 200 Italian patients with HCV and liver disease; 33% had HBV DNA detectable in liver compared with 14% controls without HCV. HBV DNA was detected twice as frequently among those who had failed IFN-α-2b monotherapy and more often associated with cirrhosis than milder liver disease. No specific gene rearrangements were noted on sequencing but the low level viremias and undetectable antigens suggest possible suppression of gene expression by HCV. Further, in a large study of 2014 patients from Japan, almost half (49.9%) with HCV infection also were seropositive for anti-HBc, including 59% of those with HCC. Virological and histopathological mechanisms require further study.

4.3.5.1 Impact of HCV on other infections

HCV infection *per se*, does not seem to impact adversely on the natural history of HBV or HIV infection. No clear correlation has been demonstrated between the levels of immunodeficiency, assessed by reduction in CD4+ T-lymphocyte counts and HCV viremia.

4.3.5.2 Impact of HBV and HIV infections on hepatitis C

The natural history of the liver disease is accelerated with concomitant hepatitis B and/or HIV infection (see Chapter 8). Limited data suggest that occult HBV infection (HBV DNA in liver biopsy tissue despite undetectable HBsAg in serum) in patients with hepatitis C may be found more frequently with cirrhosis than less severe liver disease. Rearrangements in the S gene may account for some, but not all, cases of undetectable HBsAg. The corresponding generally low levels of viremia and viral antigens (undetectable HBcAg and HBsAg by immunohistochemical techniques) in liver suggest that occult HBV infection results from suppression of virus replication and gene expression. Also, a sustained virological response to IFN and ribavirin (see section 4.7.19) is less likely with, than without, concomitant infections.

4.3.6 Disease associations

Many are associated with chronic hepatitis C especially with immunological features such as mixed cryoglobulinemia (see section 1.11).

4.3.6.1 Mixed cryoglobulinemia

HCV infection has been documented in 30–90% of patients diagnosed with mixed cryoglobulinemia (*Table 4.3*). Conversely, cryoglobulinemia (types 2 and 3) can be detected in 30–50% of symptomless patients with chronic HCV infection. Levels of viremia, distribution of genotypes, severity of liver disease and response to IFNs, do not discriminate those with, and without, cryoglobulinemia. Only longer duration of diagnosis and, consequently, more advanced liver disease are associated with cryoglobulinemia. The finding of mixed cryoglobulinemia is not diagnostic for chronic

Table 4.3. Cryoglobulinemias and association with HCV infection

Cryoglobulinemia	HCV	IgG	IgM	Rheumatoid factor
Type 1	No	Neg	Monoclonal	−
Type 2 (mixed)	Yes	Polyclonal	Monoclonal	+
Type 3 (mixed)	Yes	Polyclonal	Polyclonal	+

hepatitis C and occurs in chronic liver diseases regardless of etiology. Cryoglobulinemias accompany many diseases especially systemic lupus erythematosus.

HCV RNA predominates in the cryoprecipitates rather than the serum, suggesting a direct role in their formation. HCV RNA associates with the fraction of very low density and low density lipoproteins but underlying mechanisms remain obscure.

Most (80%) patients with hepatitis C and cryoglobulinemia remain symptomless. Typical symptoms are excess fatigue, pruritus and also weakness, arthralgias and palpable purpura (systemic vasculitis) that form the classic triad in mixed cryoglobulinemia. Some present to departments other than hepatology with these extrahepatic features. Raynaud's phenomenon, dry eyes, vasculitis with purpura, membranoproliferative glomerulonephritis and neuropathy occur in less than 5% of patients with cryoglobulinemia, and usually advanced liver disease.

Clinical consequences reflect hypergammaglobulinemia with deposition of immune complexes, cryoglobulinemia and complement in small vessel walls, as well as development of various autoantibodies including SSA, SSB and Rheumatoid factor (Rf). The vasculitis is leukocytoclastic: neutrophils are attracted to the vessel walls with extravasation of erythrocytes. Lymphocytes infiltrate around the vessels during the healing process. HCV RNA (positive strand) has been detected around vessel walls. Immunoglobulin (Ig)M and IgG are deposited within the walls. The renal disease typically is a membranoproliferative glomerulonephritis with cryoglobulins integral to the pathological process; IgG and Rf are deposited in the subendothelial space. Dry eyes and dry mouth can occur but, unlike sicca syndrome, CD8+ T cells predominate in infiltrates in lacrimal glands. The vasculitis and renal dysfunction (creatinine clearance) may improve in up to 50% of patients who respond to IFN-α and ribavirin with reduction in HCV RNA levels. Long-term antiviral therapy may be required as relapse is common on cessation. Immunosuppressive therapy may improve the vasculitis but any short-term benefits must be balanced against the long-term adverse effects on virus replication.

The high (10–20%) seropositivity for HCV in patients with autoimmune thrombocytopenic purpura may reflect the repeated exposure to blood products.

4.3.6.2 Autoimmune disorders, autoantibodies and coexistent diseases

Hypothyroidism, in association with anti-thyroid autoantibodies, is the most common (up to 5%) autoimmune association with hepatitis C. IFN therapy may aggravate thyroid disorders but usually these resolve on cessation of therapy.

Table 4.4. Autoantibodies in autoimmune hepatitis and chronic hepatitis C

Autoimmune hepatitis	HCV infection	ANA	SMA	LKM-1
Type 1	Rare	+	+	−
Type 2	Common	+	+	+

Various liver-associated autoantibodies are found in patients with chronic hepatitis C (*Table 4.4*). Usually, titers of anti-nuclear antibodies (ANA) and smooth muscle antibodies (SMA) are lower (<1 : 160) than those considered diagnostic for Type I autoimmune hepatitis except in regions where HCV is highly endemic. Diagnosis can be problematic with additional serological features suggestive of autoimmune diseases such as cryoglobulins and elevated levels of globulins. Some patients with chronic viral hepatitis have additional histological and serological features suggesting an overlap syndrome with autoimmune hepatitis including interface hepatitis (piecemeal necrosis around portal tracts) and rosette formation of hepatocytes.

Type II autoimmune hepatitis, defined by the detection of liver, kidney, microsomal (LKM-1) autoantibodies, can be associated with anti-HCV positivity. Most patients are male, whereas most with autoimmune liver diseases are female.

Treatment options should be considered carefully as IFNs can aggravate autoimmune hepatitis. Conversely, immunosuppressive therapies, especially corticosteroids, can enhance virus replication. Also, the immunosuppressed host is at increased risk of bacterial and fungal infections. Sepsis is a major cause of morbidity and mortality in patients with liver disease, regardless of etiology.

4.3.6.3 B-cell activation and lymphomas

Oligoclonal activation of B cells may be attributable to the envelope glycoprotein, E2, binding to the surface receptor, CD81 (see section 4.2.3) and seems to be a feature of chronic hepatitis C. HCV has been associated with various non-Hodgkin's lymphomas, including mucosa associated lymphoid tissue tumors.

4.3.6.4 Porphyria cutanea tarda

This is the most common form of porphyria worldwide and around 50–80% of patients have detectable HCV RNA. These share several features such as alcohol excess, iron overload in the liver and risk of HCC. Their interrelation remains unclear but alcohol excess and HCV may favor expression of this inherited disorder.

4.3.6.5 Lichen planus

Around 10–40% of patients with lichen planus are seropositive for anti-HCV antibodies. Similarly, liver biopsies in small numbers of patients with lichen planus have shown chronic hepatitis C. The immunopathogenesis is not well-defined but exacerbation of the skin condition can occur with administration of α-IFNs.

4.3.6.6 Iron overload

This is found more often in chronic hepatitis C than other types of chronic viral hepatitis. Chronic hepatitis C is common in alcoholics. Differences in the intensity of staining of the hepatocytes and distribution of iron should allow discrimination from alcoholic liver disease. An inverse correlation exists between the amount of iron deposited in the liver and reduction in serum ALT levels following antiviral therapy. Some authorities recommend venesection for iron overload prior to antiviral therapy. Iron tends to accumulate in the liver with ribavirin and may reflect hemolysis, a side-effect of this drug. (See also section 4.7.20.3).

4.4 Hepatocellular carcinoma

HCV is the major cause of HCC in HBsAg-negative patients in the world and specifically in regions with high endemicity for this virus, such as Japan and Southern Europe. Regions of intermediate prevalence for anti-HCV antibodies and HCC (20–50% of tumors) include Greece, Austria, Switzerland and the Middle East. Low correlation rates (< 20% HCC associated with anti-HCV) are noted in the West, including the USA and in Northern Europe, parts of sub-Saharan Africa and China. In these regions, chronic HCV infection accounts for the doubling of cases of HCC observed since the 1980s especially among migrants from regions of high endemicity for HCV. Furthermore, these changes have coincided with improvements in techniques for clinical diagnosis of small tumors and more accurate documentation in cancer registries.

4.4.1 Risk factors for HCC

The incidence of HCC is 1–4% per year in cirrhosis related to HCV but is rarer without cirrhosis than for HBV (see section 3.6). Prevalences of anti-HCV as high as 60–80% have been reported in HCC patients from areas of the world such as Italy, Spain and Japan where HBV infection is relatively uncommon. In Japan, where 88% of primary liver tumors are associated with HCV infection, the risk of developing cirrhosis and, eventually, HCC rises by 5–10% annually, and may reach 75% after 20 years. Conversely, for chronic hepatitis B these risks rise from 7% annually to a plateau of around 20–30% after 15 years. In the USA and Europe, these risks are less than in Japan but may reach around 20–30% over 20 years. The impact of HCV and HBV on development of HCC may be underestimated. Occasionally, HCV RNA may be detected in the liver in the absence of serum markers.

At least 20 years of chronic HCV infection precede the development of HCC based on documented time of infection from cases of symptomatic acute post-transfusion hepatitis. Malignancy may develop more rapidly with late (after 50 years of age) acquisition of HCV infection and concomitant alcoholic liver disease; from 60% in non-alcoholics to over 80% with alcoholic liver disease over 10 years, based on studies from Japan.

4.4.2 Mechanisms of carcinogenesis

As with HBV-associated HCC, pathogenic mechanisms remain unclear. Current views for HCC implicate multiple host and environmental factors (see *Table 4.5*) that interact to favor persistence of virus and chronic inflammation in the liver. Almost all HCV-associated HCCs develop on a background of cirrhosis that may be the common denominator predisposing to malignant transformation among regenerating nodules. HCV replication persists in the liver in chronic hepatitis C and cirrhosis. HCV RNA and antigens have been detected within tumors and surrounding (non-tumorous) liver tissue.

Mechanisms of oncogenesis probably differ between HCV and HBV because their replicative cycles differ. Unlike HBV, no DNA copy of the HCV genome is generated and viral nucleic acid does not integrate into chromosomal DNA.

Studies relating to prognosis and genotype have yielded conflicting results although genotypes 1b and 1a are implicated for causing severe liver disease and HCC (see section 4.2.4.1). Genotype 1b is predominant in Japan where most HCCs are associated with HCV rather than HBV. However, a recent survey by centers involved in the Inhibition of Hepatocarcinogenesis by Interferon Study (IHIT) showed no independent association between HCC, level of viremia and genotype. Recent reports from Japan and Europe also suggest that antiviral therapy with IFNs may reduce the incidence of HCC despite cirrhosis, presumably by reducing levels of virus replication and, consequently, liver damage. In Japan, the IHIT study group carried out a retrospective, multicenter national surveillance involving 2890 patients between 1994 and 1998. Using multivariate analysis, IFN therapy (2400 patients: median total dose 480 U and 160 days duration) was associated with a reduced cumulative risk (median 4.3 years follow-up) of HCC especially among those who achieved complete virological and biochemical responses. The risk of developing HCC rose with progressive stages of fibrosis. Among the 490 untreated volunteer controls, the annual incidence of HCC, when stratified for fibrosis (stage F4), was 7.88% compared with 0.49% for sustained responders to interferon therapy.

Table 4.5. Risk factors for developing HCC with chronic hepatitis C

Host factors
- Older age
- Cirrhosis
- Male sex
- Race (reflects endemicity as well as exposure to environmental factors)
- Porphyria cutanea tarda (genetic and related to iron overload and alcohol)
- Metabolic disorders

Environmental factors
- Alcohol
- Iron overload (genetic and environmental)
- Unspecified toxins (? aflatoxin, chemicals)

These findings are intriguing. How does so short (average 6 months) a course of IFN have long-term (> 4 years) impact against hepatocarcinogenesis? Whether this impact is sustained or serves only to delay the onset of HCC, requires much longer-term follow-up. Nevertheless, such results are encouraging especially if confirmed in countries with lower rates of endemicity for HCV and HCC.

Cellular mechanisms remain to be established. Whether HCV can activate oncogenes and alter the expression of tumor suppressor genes (compare *p53* for HBV, section 3.5.5), is unclear. One hypothesis proposes that the HCV core acts to depress tumor suppressor elements such as the RB (retinoblastoma) promoter and upregulate various oncogenes, such as c-*fos*. HCV core protein is located in the cytoplasm but may interfere with growth control by binding to cellular mRNAs.

4.5 Laboratory diagnosis

Unlike for hepatitis B, conventional tests for detecting HCV antigens are not available. Instead, diagnosis relies on detecting antibodies in immunoassays and HCV RNA using RT–PCR (*Figure 4.4*). Current serological markers do not discriminate acute from chronic infection. Patterns of seroconversion are difficult to define as most patients present with acute-on-chronic, rather than *de novo* acute, infection. Detection of IgM-anti-HCV antibodies seems to be of limited diagnostic value in acute infection but may have a role in monitoring responses to antiviral therapies in chronic infection. Virus replication persists beyond 6 months with significant inflammatory activity on liver biopsy. Serial tests may be required to detect intermittent, low-level viremia. Correlation of viremia (HCV RNA) with inflammatory activity awaits development and standardization of the molecular biological techniques to quantitate reliably HCV RNA in liver tissue.

4.5.1 Immunoassays

Detection of antibodies (anti-HCV) remains the cornerstone for screening and diagnosis of HCV infection. The first-generation commercial EIA detected an antibody (anti-c100-3) related to HCV and was published simultaneously with the cloning of a complementary DNA (cDNA) representing part of the viral genome. Part of the non-structural region of the genome (NS4) was expressed in yeast fused to a portion of the human superoxide dismutase (SOD) gene product. Subsequently, detection of this antibody was found to be insensitive and non-specific in diagnosing early infection. The delay (median 22 weeks after exposure) in seroconversion to anti-NS4 and variation in the amino acid sequence of NS4 between genotypes account for most of the false negativity. Also, the anti-c100-3 assay lacks specificity for HCV infection. High rates of false positivity occurred in low-risk populations such as volunteer blood donors and people with hypergammaglobulinemia. These problems prompted the introduction of supplementary tests (see below).

Second-generation EIAs included recombinant antigens c22-3 from the nucleocapsid and c33c from another non-structural region (NS3). Inclusion of c22-3 is useful because

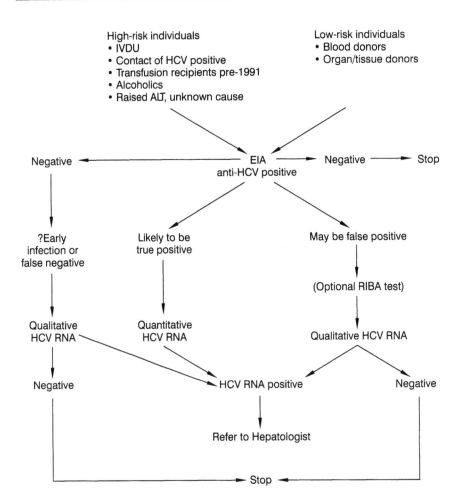

Figure 4.4. Algorithm for HCV screening. Notes: RIBA is optional to direct testing for HCV RNA; the qualitative PCR is more sensitive (~100 geq/ml) than the quantitative (bDNA) test (>200 000 geq/ml); a positive qualitative HCV RNA indicates that virus is present in serum; a quantitative HCV gives an estimate of the level of viremia (but is less sensitive than a qualitative test and may miss levels <1000 geq/ml); patients who are repeatedly reactive by EIA should be tested for viremia using RT–PCR regardless of supplemental testing using RIBA. A more cautious approach may be justified for 'low-risk' individuals.

seroconversion to anti-nucleocapsid antibodies follows rapidly acute infection and the sequence is highly conserved among HCV genotypes. Further modifications include antigens from the NS5 region and other recombinant antigens and peptides are used in assays from a variety of manufacturers. Advances in the EIA formats have been paralleled in the recombinant immunoblot assay (RIBA) and other supplementary tests. As with any screening test, the proportion of repeatedly reactive EIA results that are false-positives varies according to the prevalence of infection in the sampled population.

4.5.1.1 Supplementary tests–RIBA

The RIBA is the most commonly used supplementary test for specificity to evaluate repeatedly reactive results in the screening assays (EIAs) for anti-HCV antibodies, especially in the setting of testing low-risk individuals such as volunteer blood donors. Other supplementary tests for antibody specificity to HCV include dot-blot immunoassays and synthetic peptide assays used to quantitate antibody levels.

RIBA is a strip immunoblot assay that tests for antibodies to viral antigens immobilized on a strip of membrane (e.g. nitrocellulose) and including recombinant proteins (e.g. 5-1-1) expressed in *Escherichia coli* as well as in yeast. Various controls also are immobilized on the membrane including SOD (which is attached to the recombinant HCV proteins in the enzyme-linked immunosorbent assays) and IgG in low and high concentrations.

As the RIBA and EIA share the same antigens (such as c100-3) these assays cannot be considered as scientifically independent. They do not improve detection in early infection. In the first generation assay (RIBA 1.0), cross-reacting controls as well as the recombinant c100-3 antigens were immobilized separately. This RIBA included a related antigen (5-1-1) expressed in *E. coli* and SOD expressed in yeast to rule out cross-reactivity with yeast antigens or SOD. Less than half (45%) of initially reactive blood donations were considered specific for anti-c100-3 antibody using the RIBA 100 test.

The widely used second-generation assay, RIBA HCV 2.0, uses four antigens: 5-1-1 (*E. coli*), c100-3, c22-3 and c33c (yeast). An unequivocally positive RIBA, defined when all four bands are strongly positive (4+, maximum intensity) with respective negative and positive controls, correlates well (> 90%) with HCV RNA positivity in sera. In the immunocompetent individual, low intensity bands (< 2+) and reactions only against 5-1-1 and/or c100-3 correlate well with absence of viremia—undetectable HCV RNA levels using RT–PCR.

The third-generation assay, RIBA HCV 3.0, uses three recombinant antigens; c33c, NS5 and hSOD and two synthetic peptides c22(p) and c100(p). A positive band (>1+ intensity) to two to five of the HCV antigens is considered reactive. The increased sensitivity of the latest generation assays reflects improvements in the quality of the antigens such as c33c and c22 rather than the additional antigens such as E2 and NS5.

Problems with sensitivity and specificity of assays may be revealed, depending on the prevalence of infection in the population and the illnesses involved. In high-risk groups for HCV infection, such as IVDU and multiply transfused individuals, an unequivocal (all four antigens positive) result in the latest generation of RIBAs correlates well (> 90%) with the presence of viremia (HCV RNA). In contrast, in low-prevalence populations, such as volunteer blood donors routinely screened for parenteral risk factors, less than 50% of samples repeatedly reactive for anti-HCV antibody (by EIA) are true-positives (see *Figure 4.4*) because the repeated reactivity tends to be against individual antigens such as c33c or c22-3 and HCV RNA remains undetectable in most cases. Interpretation of these indeterminate results remains difficult in clinical practice. Conversely, these antigens may be the only ones detected in immunosuppressed individuals, such as recipients of organ grafts in whom HCV RNA can be detected in some cases.

4.5.2 Detection of HCV RNA

In the absence of serological tests for antigens, the 'gold standard' for assessing potential infectivity is direct detection of virus, by its nucleic acid. Molecular techniques that detect and quantify HCV RNA are the methods of choice in assessing viremia. Detection of HCV RNA is essential for diagnosis in the immunosuppressed, new-born and during pregnancy as anti-HCV antibodies may remain undetectable. Serial testing for HCV RNA is necessary to allow for fluctuations in levels of viremia and intermittent viremias. The minimum number, and time interval, of undetectable results that predict clearance of virus is not defined and will depend on sensitivities of the assays.

Attempts have been made to detect the negative strand of HCV RNA as a marker of replication at a particular site. These rely on a specific, positive-sense oligonucleotide to prime reverse transcription of the negative strand but are confounded with errors caused by fragments of nucleic acid priming on the positive strand. The most reliable techniques probably are those which use a 5′ tag on the RT primer as a primer binding site in the PCR.

4.5.2.1 RT–PCR

PCR is a reliable test for virus. The 5′ UTR of HCV is highly conserved among isolates and is the region of choice for diagnostic RT–PCR. Other quantitative methods include non-competitive amplification within a biotinylated colorimetric assay and using an internal standard for calibration (e.g. Amplicor monitoring). (See also section 1.4.5.)

4.5.2.2 bDNA assay

The bDNA assay also targets the 5′ UTR and provides quantitation when the viremia exceeds 2×10^5 genome equivalents (geq)/ml. (See also section 1.4.4.)

4.5.2.3 Standardization of testing for HCV RNA

A negative HCV RNA test using PCR-based techniques does not necessarily indicate clearance of virus. In theory, nested PCR can detect down to a single virus genome. However, in laboratory practice the threshold for detection (sensitivity) varies. The bDNA assay may report as negative viremias below $2.5–3.5 \times 10^5$ geq/ml. The Amplicor monitor test, which quantifies HCV RNA, may be reported as negative with less than 1000 geq/ml.

Reports should state HCV RNA as 'undetectable' rather than 'negative'. The type of assay (bDNA, etc.) and threshold of detection (cut-off) for each batch should be included with the clinical report. This assessment of sensitivity for each assay should be based on quality control evaluations in the test laboratory.

Panels of sera to be used as international standards are being evaluated through the World Health Organization and Europe (Belgian CLB labs) to standardize testing within and between laboratories and to allow meaningful comparisons of clinical trials. Ideally, samples from multicenter studies should be analyzed in a single laboratory because the threshold for detection and range of linear values varies sufficiently between laboratories to bias interpretation of data.

Commercial assays can provide internal standards for calibration. Linearity typically is between 10^3–10^6 geq/ml. Also, risk of cross-contamination with amplicons from previous reactions is reduced by including an incubation step with specific enzymes such as uracil-N-glycosylase that cleaves uracil-containing DNA at 50°C.

4.6 Histological features

4.6.1 Acute hepatitis C

The features of *de novo* acute infection are not well-defined because most patients present with an acute exacerbation of persistent infection. Damage to bile ducts can be noticeable. Most acute infections (> 80%) become chronic.

4.6.2 Chronic hepatitis C

The histology is highly suggestive with two or three features. Inflammatory activity (density of lymphocytic infiltrates) characteristically is mild and does not correlate necessarily with elevations in ALT levels.

Histological triad in chronic hepatitis C:

- Lymphocytic aggregates in portal tracts (57%)
- Bile duct damage (epithelial degeneration) (60%)
- Mild steatosis (52%)

Cytoplasmic swelling may be accompanied by eosinophilic degeneration of epithelial cells. The severity of damage may equal that seen in autoimmune diseases such as primary biliary cirrhosis and sclerosing cholangitis. Unlike these, the basement membrane remains intact and levels of canalicular enzymes, such as gammaglutamyl transpeptidase, usually remain modestly elevated. Lymphocytes tend to cluster near to portal tracts. These germinal centers resemble follicles in lymph nodes with antigen-presenting cells in the center, surrounded by B lymphocytes and an outside ring of T lymphocytes. A predominance of plasma cells may cause confusion with autoimmune hepatitis, especially since autoantibodies may be present, albeit usually in low titer. The infiltrate of lymphocytes and plasma cells may be lobular with a predominantly sinusoidal distribution, as seen with herpes infection. Acidophilic bodies may be present without associated inflammatory cells.

Fatty change is common and scattered throughout the hepatic lobule. Differentiation from alcoholic liver disease usually is not difficult. The fatty change is more severe in alcoholic liver disease and distributed predominantly around the central lobule near to the terminal hepatic venules. Any fibrosis tends to be concentrated around the central lobule in alcoholic liver disease whereas widespread distribution is more typical of chronic HCV infection. Iron deposition may occur but usually is less than, and of different distribution to, that seen in alcoholic liver disease. Iron is deposited primarily in hepatocytes rather than Kupffer cells. The elevation in hepatic iron content (> 1500 µg/g dry weight liver) following ribavirin therapy is attributed to hemolysis.

4.6.3 Hepatitis C following liver transplantation

This is covered in Chapter 6.

4.6.4 Electron microscopy

This technique has failed to delineate clearly the virus structure in clinical specimens. Descriptions of ultrastructure have added little to our knowledge based on filtration studies and immunochemical techniques.

4.6.5. Immunohistochemical features

Viral antigens can be detected in the cytoplasm of hepatocytes in chronic hepatitis C using monoclonal antibodies to HCV gene products. Sensitivity remains low; positive signals occur in only 75% of patients with viremia detected by RT–PCR of serum. Frozen cryostat sections are preferable to archival (paraffin-embedded, formaldehyde-fixed) liver tissue for optimizing detection of HCV RNA via *in situ* hybridization and results are comparable to those using immunohistochemical techniques. Detection of virus can be increased with PCR *in situ* and this technology can be applied to archival tissue. Combined *in situ* and immunological studies can characterize the inflammatory cell infiltrates, such as T-cell subsets, and their roles in causing liver cell damage.

4.7 Management

The general principles involved in caring for patients with acute and chronic hepatitis C, including complications of cirrhosis such as bleeding varices, ascites, HCC, and added risk of microbial infections, do not differ from other etiologies of acute and chronic hepatitis. (See also section 1.12.)

4.7.1 Approaches to anti-viral therapies

The aim is to clear viral nucleic acids from serum, primarily by inhibiting virus replication. Additionally, some therapies such as the α-IFNs, enhance the immune response and presumably, this favors clearance from infected hepatocytes. Recommendations from the EASL International Consensus Conference (1999) have updated the NIH Consensus Conference (1997) on treatment/management guidelines for hepatitis C. (See also section 1.13.)

4.7.1.1 Viral kinetics

A rapid rate of decline of HCV RNA in the first few days of IFN monotherapy, and with ribavirin as combination therapy, favors a sustained response.

Mathematical models of viral kinetics have produced conflicting interpretations as to whether IFN inhibits *de novo* infection of new hepatocytes or, more likely, HCV production from infected hepatocytes. Most circulating virus derives from continual rounds of *de novo* infection of cells, virus replication and cell turnover and not from cells chronically infected with HCV. Release of virus from hepatocytes ($T_{1/2}$ =1.4–3.8 days) is the rate-limiting factor rather than clearance from serum ($T_{1/2}$ = 0.35–1.25 days).

IFN-α seems to have a direct antiviral effect and probably acts by reducing the pool of circulating virus rather than by direct lysis of chronically infected hepatocytes. Serum ALT levels fall during interferon therapy (in contrast to the rise in chronic HBV infection, where IFN seems to act by stimulating the immune response). The rapid time course (< 2–4 weeks) of this fall is against immune-mediated cell lysis (ALT, $T_{1/2}$ = 3 days). Also, ALT levels remain modest despite reduction in massive viremia.

Long-term sustained response to IFN-α (section 4.7.6.1) is favored with a rapid, early decline in viremia (HCV RNA levels) during therapy. Conversely, high levels of HCV RNA are associated with sequence heterogeneity. A great diversity of quasispecies reduces the likelihood of clearing HCV and favors relapse with selection of resistant species following cessation of antiviral therapies.

> **Key Notes** Clearance of HCV RNA from the liver and serum depends on:
> - Ultimately, a multi-step process that prevents (rate-limiting) infection *de novo* of hepatocytes (naïve) from the circulating pool (reservoir) of HCV virions
> - Preventing replenishment of the pool by destroying infected hepatocytes
> - The presence of HCV genotypes 'responsive' to IFN
> - Limiting viral load

4.7.2 Interferons

The mainstay of antiviral therapy in viral hepatitis is IFN-α-2b. IFN-β has been used with limited success in small numbers of patients, including those with acute hepatitis C. Results with consensus interferon (CIFN) are similar to IFN-α-2b.

PEGylated interferon (PEG interferon alfa-2a: PEG-IFN) is a 'long-acting' IFN that has decreased systemic clearance—a 10-fold increase in serum half-life—compared with IFN alone and can be administered once weekly. PEG-IFN is chemically modified by the covalent attachment of a branched methoxy polyethylene glycol moiety to IFN. Results of trials are awaited, especially in comparison with standard combination therapy (section 4.7.1) given for 12 months. Side-effects of interferon treatment are covered in Chapter 1 (see section 1.13.2)

4.7.3 Ribavirin

Ribavirin alone in the dose range 600–1200 mg four times daily for 2–12 months can reduce serum levels of ALT but has little impact on reducing HCV RNA levels and progression of liver fibrosis. Limitations in dosing depend on the severity of hemolytic anemia. However, the minimum dose that provides maximum antiviral effect in combination with IFN is unknown and may be less than the 1000–1200 mg (depending on body weight) recommended by the manufacturer. Studies are underway using 400–800 mg/day. Hepatic iron levels rise during long-term therapy and reflect hemolysis.

Side-effects of ribavirin:

- Oral—metallic taste, dry mouth
- Skin/muscle—rashes, myalgia, muscle cramps
- Neurological—headache, irritability, mood changes, altered sleep pattern
- Gastrointestinal—nausea, dyspepsia, flatulence
- Hematological—hemolytic anemia
- Metabolic—elevated uric acid

4.7.3.1 Absolute contraindications to ribavirin

Ribavirin is contraindicated in patients with anemia and hematological conditions that favor anemia such as the hemoglobinopathies (see also section 7.11.5). Ribavirin also is contraindicated in diseases likely to be worsened by anemia such as severe heart disease and renal failure.

> **Key Notes** Ribavirin is a teratogen as shown in all animal studies
>
> Pregnancy and unreliable contraception are contraindications. Pregnancy tests should be carried out monthly during therapy. Accordingly, the manufacturers recommend a 'wash-out' period of a minimum of 6 months following discontinuation of ribavirin prior to discontinuing strict (double) measures of birth control: oral contraceptive pill with barrier methods

4.7.4 IFN-α-2b and ribavirin combination therapy

The EASL conference advocates that first line therapy is with IFN-α-2b: 3 MU subcutaneously (s.c.) three times a week (t.i.w.). However, in clinical trials, the standard combination of ribavirin (1000–1200 mg/day) with IFN-α-2b (3 MU s.c. t.i.w.) has been shown to be superior to monotherapy with IFN-α-2b or CIFN for numbers who achieve a sustained virological response. This superior outcome for combination therapy extends to all categories of patient groups—naïve (never treated) and those who failed (relapsed and non-responders) interferon monotherapy.

4.7.5 Other agents

Corticosteroids may reduce the serum ALT level in the short-term but show no benefit and may enhance virus replication. Sudden withdrawal of corticosteroids may precipitate decompensation of the liver disease. Corticosteroids increase the risk of bacterial and fungal infections in liver disease regardless of etiology.

Granulocyte–macrophage colony stimulating factor (GM-CSF) in small, uncontrolled studies reversed the leukopenia induced by IFN therapy enabling continuation of higher than usual doses in some patients. The protective efficacy of GM-CSF against episodes of microbial infection is under study.

4.7.6 Defining responses to antiviral therapies

Outcomes of antiviral therapies should be defined by clearance of viral nucleic acid (HCV RNA) using a sensitive PCR assay rather than normalization of serum levels of ALT and AST. Endpoints also should include improvements in liver histology, especially the slowing of progression of/to fibrosis.

Early publications used a reduction in serum ALT level to indicate 'response' and normalization as 'complete response' equated to clearance of viremia. Histological evaluation usually was not carried out and treatment and follow-up were relatively short-term (less than 1 year). The advent of RT–PCR to detect and quantify HCV RNA has shown a poor correlation between levels of viremia and serum elevations in ALT that can be modest despite severe liver disease. Also, many therapeutic agents (*Table 4.6*), including ribavirin, reduce the ALT level without predictable and sustained impact on lowering levels of HCV RNA.

4.7.6.1 Sustained (complete) virological response

> **Key Note**
>
> Sustained responders reach undetectable levels of HCV RNA by qualitative PCR following 6 months of therapy usually with concurrent normalization of ALT levels

Table 4.6. Other therapeutic approaches for chronic hepatitis C

Therapeutic agent	Rationale
Ursodeoxycholic acid (UDCA)	Reduces serum ALT, immunomodulatory
Pentoxyfylline	? Enhances interferon
Cytokines especially IL-2	Immunomodulatory
Alpha thymosin	Immunomodulatory
Cyclosporine	Immunosuppressant
Non-steroidal anti-inflammatory drugs (NSAIDs)	Block prostaglandins (arachidonic pathways)
N-acetyl cysteine (NAC)	Free radical scavenger
Vitamin E	Anti-oxidant
Quinolones	Interfere with viral helicase (NS3)
Venesection	Reduces hepatic iron and serum ALT level
Plasmapheresis	Reduces viremia in the short term

A sustained response is most likely in the patient with baseline (pre-treatment):

- Low levels of viremia (< 1 million geq/ml HCV RNA)
- Histological evidence of mild disease (no fibrosis, no cirrhosis)
- Genotype other than type 1

> **Key Note**
>
> Initial responders achieve undetectable HCV RNA levels within 6 months of therapy

Initial responders should be maintained on therapy for a minimum of 6 months for genotypes other than 1b and 12 months or longer for genotype 1b, to achieve a sustained virological response.

4.7.6.2 Relapse on IFN monotherapy

> **Key Note**
>
> The patient who relapses develops detectable HCV RNA after an initial complete virological response (undetectable HCV RNA) during therapy

Current recommendation is to discontinue the current regimen and retreat with IFN-α, preferably at higher dose and for longer duration and add ribavirin treatment.

4.7.6.3 Non-response

> **Key Note**
>
> The non-responder has HCV RNA detectable throughout antiviral therapy

Many experts would recommend discontinuing the IFN alone at 3 months of therapy in the non-responder and changing to combination therapy.

4.7.7 Acute hepatitis C

Antiviral therapy aims to prevent persistence of viral infection as this will lead to chronic inflammation of the liver. Presentations of *de novo* infection are uncommon in clinical practice and sources of data have dwindled following introduction of routine screening for anti-HCV antibodies of blood donors in the 1990s. Insufficient numbers of patients preclude large-scale controlled trials. Nevertheless, sustained normalization of liver enzymes with clearance of HCV RNA can be achieved in around 50%, or more, following 6 months of therapy with higher (6–10 MU) than usual doses of IFNs (α and β). Importantly, response seemed to be dose-related with no excess of significant side-effects.

This approach (see section 4.7.9) could be considered as induction therapy to achieve rapid reduction in HCV RNA. Thereafter, the dose can be titrated against side-effects to maintain undetectable viremia by sensitive PCR. Long-term outcome and safety of high-dose IFN-β have not been well studied for acute hepatitis C.

4.7.8 Chronic hepatitis C

Treatment strategies have been those adapted from chronic hepatitis B (see section 3.8.2), including use and dosing schedules of IFN, monitoring ALT and aspartate aminotransferase (AST) levels to indicate 'response' to therapy and anticipating side-effects and outcomes.

Simply, there is no consensus on who to treat and which protocol is optimal for antiviral therapies. Controversy remains over selection of many patients who stand to benefit such as those who seem to maintain their liver enzymes within the normal range, patients with untypical serological profiles, concurrent autoimmune features, the immunosuppressed, and children.

4.7.8.1 Who should be followed?

The patient with persistently normal serum ALT levels (< 50 IU/l on serial determinations: around three times each year). Decision not to treat is based on likely good prognosis and poor response to IFN monotherapy and standard combination therapy (see

Key Notes Management issues in chronic hepatitis C

Who to treat?

- Progression of liver disease is unpredictable
- Mild liver disease may persist for many years

The paradox of antiviral therapy

- The majority of patients fail to clear HCV RNA with antiviral therapy
- HCV RNA remains detectable in 10–40% despite normalization of ALT
- Most efficacious in mild liver disease
- Less efficacious in advanced liver disease with cirrhosis
- Benefits of preventing cirrhosis and HCC likely but unknown

section 4.7.1). Current recommendations are to monitor closely and regularly every 3–4 months for elevations in liver enzymes and withhold treatment until elevations in ALT are detected. This rationale also presumes that 'healthy carriers' exist and that although their natural history is unpredictable progression to fibrosis and cirrhosis over time is unlikely. A few exceptions will have cirrhosis already, others may progress silently and elevations in enzymes can be elusive. More patients will be considered for antiviral therapies as success rates improve and outcomes become more predictable based on serological profiles.

Key Notes A normal serum ALT level in chronic hepatitis C

- Found in up to 40% of individuals with anti-HCV antibodies
- Severe liver disease uncommon but may be present
- Cirrhosis can be found in symptomless individuals
- HCV RNA may be present, including after antiviral therapy
- Monitor with HCV RNA levels
- May occur with other therapies; ursodeoxycholic acid, ribavirin, venesection etc.
- Can occur independently of HCV RNA levels

4.7.8.2 Who should be treated?

The patient with significant, but compensated, chronic liver disease: persistently elevated serum ALT level (> 50 IU/l) for more than 6 months on three separate readings, detectable HCV RNA and evidence of progressive liver disease on biopsy: septal fibrosis and significant necroinflammatory activity.

Key Notes Antiviral therapies: current recommendations for chronic hepatitis C

- HCV RNA positive
- Well compensated liver disease
- No previous history of ascites, encephalopathy, other complications
- Histological evidence of chronic viral hepatitis
- Elevated serum ALT (or AST) levels (serial estimations necessary)

The decision to treat should not be influenced by mode of acquisition, HCV RNA level, genotype and high-risk behavior, although these and poor compliance will limit success.

4.7.8.3 Treatment contraindications

Antiviral therapies can make worse the patient with decompensated chronic liver disease (ascites, encephalopathy), severe depressive illness, low complete blood count (CBC; cytopenia), thyroid disorders and autoimmune disease, and rejection in a renal transplant. Therapy is unlikely to be effective with active alcohol use and/or active illicit IVDU, probably also due to poor compliance by the patient. (See also section 1.13.2.)

4.7.9 Optimizing therapy

The consensus view from the National Institutes of Health conference held in the US in 1997 was to offer IFN-α-2b monotherapy at 3 MIU t.i.w. for 6 months and to reserve combination therapy with added ribavirin for those who relapse and for non-responders. The EASL consensus statement in 1999 updated these recommendations to include combination therapy for naïve as well as relapsers on previous IFN-α-2b monotherapy.

Ribavirin was introduced at daily doses of between 1000–1200 mg total dose per day, given orally, based on body weight. Studies are underway to assess the impact of lower doses (400–600 mg total daily dose, t.i.w.) in combination with long-acting IFNs or IFN-α-2b.

Key Notes Clearance of HCV RNA with IFN and ribavirin is most likely with:

- A low serum HCV RNA: $<10^6$ geq/ml
- Duration > 6 months of therapy
- Age < 50 years when infected
- Short duration of disease
- Genotypes other than type 1[a]
- No cirrhosis
- No concomitant liver disease (HBV, alcohol)
- Normal iron loads in the liver

[a] May reflect tendency to lower levels of viremia than in some other genotypes and recent infection in young patients (cirrhosis less likely) rather than true resistance of the virus

4.7.10 Induction therapies, daily IFN and escalating dose schedules

The scientific rationale to minimize the risk of relapse and emergence of resistant variants favors use from the outset of as high a dose of antiviral therapy as tolerated (high induction therapy) and discourages escalating regimes. Unfortunately, guidelines are not available and clinical trials have lacked the design and power to address these concerns, or have given conflicting results. Interest in resolving these issues has re-emerged with the introduction of long-acting IFNs and early data reporting comparable response rates to IFN/ribavirin combination therapy t.i.w.

HCV RNA levels may become undetectable in some non-responders to the conventional 3 MU t.i.w. regimes when retreated with 5–10 MU t.i.w., at least as induction therapy, because most patients cannot tolerate such high doses for the duration of therapy. These observations suggest inadequacy of low, and escalating, doses (1–3 MU t.i.w.) as initial therapy. As almost all patients relapse on cessation of low-dose therapy, several authorities favor maintaining these patients long-term on IFN with doses titrated to sustain undetectable HCV RNA levels as well as normal levels of ALT. The duration of maintenance therapy remains ill-defined.

4.7.11. Naïve patients

The combination of IFN-α-2b (3 MU t.i.w) with ribavirin (1000–1200 mg/day, depending on body weight) should be offered if there are no contraindications. Many authorities limit duration of therapy to 6 months for genotypes 2 or 3 regardless of the level of viremia and for genotype 1 with baseline HCV RNA levels below 2 000 000 geq/ml. Treatment for 12 months improves by 10–15 % the rate of sustained response with genotype 1 with viremia levels above 2×10^6 geq/ml (*Table 4.7*).

Table 4.7. Virological responses in naïve patients with chronic hepatitis C (after McHutchinson et al., 1998)

	IFN-α-2b monotherapy 3 MU t.i.w.	INF-α-2b+ribavirin combination 3 MU t.i.w. + 1000–1200 mg/day
Duration of therapy	48 weeks	48 weeks
Patients (*n*)	225	228
Female (*n*)	75	76
Genotype 1	72%	73%
Serum HCV RNA[a] (geq/ml)	4.8×10^6	5.7×10^6
Patients with $> 2 \times 10^6$ geq/ml	72%	67%
48 weeks	54/225 (24%)	115/228 (50.4%)
week 72 (24 weeks after ETR)	29/225 (12.9%)	87/228 (38%)
SVR, genotype 1	7%	28%
SVR, genotype (non-1)	29%	66%

[a]HCV RNA sensitivity: detection down to 10 geq/ml. ETR, End of treatment response; SVR: sustained virological response.

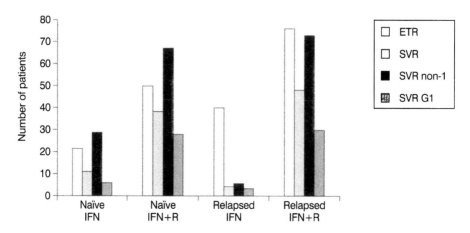

Figure 4.5. Sustained virological response to IFN-α-2b monotherapy and combination therapy with ribavirin (after McHutchison *et al.*, 1998 and Davis *et al.*, 1998). SVR, Sustained virological response at 72 weeks; SVR non-1, for genotype other than 1; SVR G1, for genotype 1; IFN, IFN-α-2b; R, Ribavirin.

IFN monotherapy (3 MU or 9 μg of synthetic IFN t.i.w.) should be reserved for intolerance/contraindications to ribavirin and maintained for 12 (not 6) months.

4.7.12 Relapsers and non-responders to IFN monotherapy after 6 months

The option is to treat with combination (IFN 3 MU t.i.w. and ribavirin 1000–1200 mg/day) therapy or retreat with monotherapy for 12 months. Therapy can be discontinued after 3–6 months on retreatment for persistently detectable HCV RNA levels (*Table 4.8*).

Table 4.8. Virological responses in patients with chronic hepatitis C who had relapsed with prior IFN-α-2b monotherapy (after Davis *et al.*, 1998)

	IFN-α-2b monotherapy 3 MU t.i.w.	INF-α-2b + ribavirin 3 MU t.i.w.+ 1000–1200 mg/day
Duration of therapy	48 weeks	48 weeks
Patients (*n*)	172	173
Female (*n*)	60	61
Genotype 1	55%	57%
Serum HCV RNA[a] (geq/ml)	5.2×10^6	4.8×10^6
Parents with $> 2 \times 10^6$ geq/ml	76%	74%
ETR at 48 weeks	69/172 (40%)	133/173 (76.9%)
SVR at week 72 (24 weeks after ETR)	8/172 (4.6%)	84/173 (48.5%)
SVR, genotype 1	3%	30%
SVR, genotype non-1	6%	73%

[a]HCV RNA sensitivity: detection down to 10 geq/ml. ETR, End of treatment response; SVR, sustained virological response.

4.7.13 Relapsers and non-responders to combination therapy

Retreatment has low success in clearing HCV RNA in such patients. However, sustained virological response can be achieved in a few previous relapsers, the alcoholic who becomes abstinent and following reduction in iron overload by venesection. Improvement in histological activity index can occur despite failure to clear HCV RNA.

> **Key Notes** Histology improves in 60% of non-responders
> * Reduced necroinflammatory activity
> * Slowed progression of fibrosis
> * Likely reduced risk of HCC

4.7.14 Do cirrhotic patients benefit from antiviral therapy?

High-dose antiviral therapy is poorly tolerated in the cirrhotic but lower than usual doses (1–3 MU t.i.w) of IFN-α-2b have been used with some success although virological relapse remains a risk. Long-term follow-up studies are underway in some responders maintained on low-dose interferon (1–3 MU t.i.w.) who show significant improvement in clinical well-being as well as histological inflammatory activity. Whether such maintenance therapy improves long-term prognosis and reduces the risk of liver cancer, is under study.

4.7.15 Treating extrahepatic complications of HCV

Interferon monotherapy and/or ribavirin can improve the vasculitis, nephropathy and neuropathies in around 30–60% of cases; improvement is associated with reduction in cryoglobulinemia. Unfortunately symptoms and signs tend to recur on cessation of therapy. Such patients require long-term maintenance therapy. The duration and optimal dosages are under evaluation.

4.7.16 Treating pediatric patients

Reluctance to treat children is based on their predilection for having only mild disease on liver histology, the uncertainties of balancing benefits against long-term risks, imperfect monitoring to limit any adverse impact on growth and development and practical difficulties of administering medications, especially by injection. Dosages of IFNs are based on surface area (typically 3 MU/m², t.i.w. for 6 months). Responses and side-effects appear to be similar to those seen in adults. Trials are underway using combination therapy with IFN-α-2b but concerns remain over the teratogenic potential of ribavirin (see section 7.11.6 and section 7.12.2).

4.7.17 Treating hemophiliacs

Large-scale trials are underway. The understandable reluctance to carry out liver biopsy has restricted information on disease severity and, consequently, immediate need for treatment.

4.7.18 Treating HCV infection in the grafted liver

Interferons should be used with caution in patients who have a liver graft due to the increased risk of precipitating rejection. Also, a sustained virological response is unlikely with the majority due to the ensuing very high viremias and immunosuppression. In this setting, low dose therapy favors emergence of resistant variants. (See also section 6.7.)

4.7.19 What about HIV and HBV co-infections?

As the prognosis of HIV infection continues to improve, antiviral therapy should be considered for hepatitis C with concurrent early HIV infection (see Chapter 8). Some patients seem to be at especial risk of rapid progression of their hepatitis C due to immunosuppressive effects of HIV. Similar considerations prevail for concurrent hepatitis B, although clearance of both viruses is less likely than for HCV alone. Occult HBV infection (HBV DNA in liver biopsy tissue despite undetectable HBsAg in serum) was found in 66 (33%) of 200 Italian patients with hepatitis C, including 20 without anti-HBc. Importantly, detection of HBV DNA in liver tissue was twice as frequent among those who failed IFN-α-2b monotherapy compared with responders.

Key Notes Individual cases for referral to specific trials should include

- Patients with mild hepatitis
- Persistently normal ALT levels
- Compensated cirrhosis
- Age < 18 years
- Age > 60 years
- Co-infected patients (HIV, hepatitis B)
- Hemophiliacs

4.7.20 Monitoring antiviral therapy

Individual laboratory tests show poor sensitivity and specificity for detecting progression to significant liver disease and for responses to antiviral therapy. These can be improved with serial profiles of CBCs, levels of liver enzymes (AST and ALT) as well as HCV RNA. CBCs should be performed weekly during the first 4 weeks of therapy and every 1–3 months during therapy, and 3–6 months afterwards. In addition, levels of hemoglobin and reticulocyte counts should be monitored closely for any fall especially from hemolysis due to ribavirin. A decrease in the level of hemoglobin by 30–40 g/l may occur and may suddenly cause symptoms. Thyroid function (thyroid stimulating hormone as well as T3, T4 and free thyroxine index (FTI)) should be tested every 3 months.

4.7.20.1 When should liver histology be obtained?

Liver biopsy is not recommended routinely for patients with persistently normal liver enzymes (AST and ALT levels), although up to 20% could have significant and progressive disease. A baseline (pre-treatment) liver biopsy is not essential prior to instigating antiviral therapy unless there is concern over possible decompensation with undisclosed advanced disease or suspected additional disease, such as alcoholic hepatitis, that will reduce the efficacy of therapy. Some clinicians are prepared to treat patients without knowledge of prior liver histology, reserving this for future decision-making in the individual incomplete responder.

A liver biopsy is usually recommended before selection for antiviral therapy to assess outcome, especially in the context of a clinical trial. Histological evidence of significant disease (beyond stage 1) activity and evidence of fibrosis (see section 1.9) would favor early therapy. Minimal hepatitis without fibrosis affords the option to delay therapy especially with ribavirin because of concerns about side-effects and teratogenic issues for patients wishing to complete a family.

Assessment of liver histology subsequent to antiviral therapy may be useful in non-responders with suspected (e.g. alcoholic) underlying liver disease or tumor. In the few studies with objective assessment of histological changes, around 70% of patients, including virological non-responders to antiviral therapy, showed some improvement, notably in necroinflammatory activity rather than degree of fibrosis. Evidence is accumulating that antiviral therapy may slow the rate of progression of fibrosis, even in non-responders, and consequently, might reduce the risk of development of HCC.

4.7.20.2 Atypical serological profiles

There is no consensus on how to manage patients with limited antibody profiles such as anti-core only; anti-c33, anti-22c and those with detectable HCV RNA despite undetectable antibodies by RIBA 2.0 and sensitive EIAs. Serial monitoring for anti-HCV antibodies and HCV RNA is recommended because some of these patients may have acute hepatitis C and were tested before seroconversion. In the immunosuppressed, as with HIV and following organ transplantation, monitoring must rely on HCV RNA by sensitive RT–PCR because tests based on antibodies may remain falsely negative due to delayed or suppressed seroconversion.

4.7.20.3 What about iron overload?

Successful clearance of HCV RNA is less likely in patients with high levels of hepatic iron. Also, an inverse relationship exists between the amount of iron deposited in the liver and reduction in serum ALT levels following IFN therapy. Some authorities recommend venesection prior to antiviral therapy in such cases.

4.8 Future approaches to antiviral therapies

Many of the principles of antiviral therapy have been adopted from studies in chronic hepatitis B. The disappointing results from clinical trials have led to a critical re-evaluation of the feasibility and validity of such approaches for chronic hepatitis C.

Variation of response to antiviral therapies between patients may reflect effective dose as well as viral load. Current protocols do not take into account the impact of body weight, obesity etc. on the distribution of drug. Whether doses should be standardized by surface area (per m²) as in pediatric practice, requires consideration.

Novel therapies such as anti-sense oligonucleotides (which prevent translation of viral RNA) and ribozymes (which cleave RNA targets) show promise in experimental systems, such as cell culture. There are major problems, however, in delivering these molecules to the target organ (liver) *in vivo*. The short half-life of some of these molecules also may be problematic. (See also section 1.13.)

Therapies on the horizon:

- Combinations > monotherapies
- Protease inhibitors (*trans*-acting NS3/NS4a dimer)
- Helicase (NS3) inhibitors
- Amantidine
- Immunosuppressive drugs

4.8.1 Cirrhosis and HCC

All patients with cirrhosis (see section 1.8.5) should be screened twice yearly using ultrasound imaging to detect small, resectable (< 3 cm diameter) tumorous nodules. Serial analyses of alpha-fetoprotein levels improve sensitivity but remain in the normal range (< 25 U) in 20–40% of cases of proven HCCs associated with cirrhosis and hepatitis C. Encouraging data from Japan suggest that IFN therapy may reduce the annual incidence of HCC in patients with significant fibrosis (section 4.4.2).

4.8.2 Liver transplantation

All aspects of liver transplantation with respect to HCV are covered in Chapter 6.

4.9 Prevention

The reservoir of infection can be sustained through high-risk behaviors, despite significant progress with screening donated blood and organs and universal precautions against parenteral spread. Also, significant numbers of patients deny any identifiable risk factors; this limits the impact of counseling and other general preventative measures. Immune globulin is not available to prevent HCV infection and there are no immediate prospects of a vaccine (see section 4.10.1).

4.10 Future issues

Improvements in long-term sustained virological response are likely with the long-acting IFNs under trial as monotherapy and in combination with ribavirin. As for HBV, evidence is increasing for the immune response to take an adverse role in liver injury. Consequently, the role of immunosuppression (cyclosporine) and other immunomodulatory drugs (including ribavirin) are being reviewed with and without liver transplantation. In addition, the search continues to find diverse antiviral agents such as amantidine and protease inhibitors and combinations of drugs to maintain a high barrier to antiviral resistance and prevent the emergence of resistant quasispecies.

4.10.1 Prospects for hepatitis C vaccines

Repeated acute infections may occur in one individual (in experimental infections of chimpanzees) suggesting limited duration of any protective immunity. As with other rapidly replicating viruses, the significant and changing genetic (sequence) diversity (section 1.3.1) in the most obvious target antigens, the surface glycoproteins, remains a challenge for developing a vaccine against HCV. The challenge is to identify target antigenic domains that are genetically stable and elicit a neutralizing antibody response. Identification of specific domains would make feasible the development of multivalent vaccines to accommodate the various genotypes.

As with HBV, chronic carriers of HCV make antibodies to a variety of virus-specified proteins including the structural nucleocapsid protein and various non-structural proteins. There are no data on how pre-existing antibodies to the nucleocapsid protein might ameliorate an acute HCV infection and abrogate persistent infection.

Prospects for developing live, attenuated and killed vaccines are hampered by the scarcity of suitable primate hosts, lack of markers of virulence and attenuation and difficulties with developing an efficient and reliable cell culture system for virus growth.

Problems in vaccine design:
- Diversity (quasispecies)
- Search for neutralizing epitope/s
- Universal epitope/s seem elusive
- Escape variants
- Lack of suitable animal models

4.10.1.1 Protein and peptide-based HCV vaccines

The most direct approach seems to be to develop a vaccine based on HCV proteins expressed by recombinant DNA technology. Constructs that express the nucleocapsid protein, but not E1 and E2, can induce CTL responses in mice. The surface glycoproteins (E1/E2) of the prototype strain of HCV-1 have been expressed in cultured human (HeLa) cells using recombinant vaccinia virus, purified and used with adjuvant to immunize seven chimpanzees. Five animals were protected against challenge with live virus, the remaining two, with low antibody levels, had short, resolving acute infections. However, several criticisms are leveled at this study. The challenge dose of 10 CID_{50} (chimpanzee infectious dose) was low, of the same strain as the immunizing innoculum and given when the antibody response was high, only 2 weeks after the final immunizing dose. Antibody levels fell within weeks of immunization, and later challenge with heterologous genotypes resulted in infection. Thus, the major potential problem of variability of the surface glycoproteins was circumvented. Whether a cocktail of antigens might protect against a variety of HCV strains warrants investigation.

An alternative approach might be to immunize with a protein with less sequence variability, such as the nucleocapsid protein. Chronic infection with HCV is associated with antibodies against this protein but an immune response prior to infection might be protective, especially if cellular as well as antibody responses were stimulated. This approach has been attempted with HBV in chimpanzees challenged following immunization with the core antigen. The subsequent detection of antibodies to the surface protein, HBsAg, indicated recovery from a short and mild acute infection.

4.10.1.2 HCV vaccine candidates

Strategies that target only groups with risk factors are unlikely to eliminate HCV from the community because up to 30% of infected individuals detected on routine screening deny any risk factors. Similar strategies and high risk groups failed to reduce the incidence of hepatitis B in countries such as the USA.

Further reading

Alter, M.J., Kruszon-Moran, D., Nainan, O.V., et al. (1999) The prevalence of hepatitis C virus infection in the United States, 1988 through 1994. New Engl. J. Med. **341**: 556–562. Centers for Disease Control—Sentinel Counties Studies NHANES III. The natural history of community-acquired hepatitis C in the United States. Follow-up of patients with non-A, non-B hepatitis in Sentinel Counties studies, risks of persistent viremia and chronic hepatitis.

Andreone, P., Gramenzi, A., Cursaro, C., et al. (1999) Interferon-α plus ribavirin in chronic hepatitis C resistant to previous interferon-α course: results of a randomized multicenter trial. J. Hepatol. **30**; 788–793.
Disappointing lack of sustained virologic response in previous relapsers to IFN-α monotherapy.

Bonkovsky, H.L., Woolley, J.M. and the consensus interferon Study Group. (1999) Reduction of health-related quality of life in chronic hepatitis C and improvement with interferon therapy. Hepatology **29**: 264–270.

Cacciola, I., Pollicino, T., Squadrito, G., et al. (1999) Occult hepatitis B virus infection in patients with chronic hepatitis C liver disease. New Engl. J. Med. **341**: 22–26.
Study of 200 Italian patients with HCV and liver disease; 33% had HBV DNA detectable in liver compared with 14% controls without HCV. HBV DNA was detected twice as frequently among those who had failed IFN-α-2b monotherapy and more often associated with cirrhosis than milder liver disease. No specific gene rearrangements were noted on sequencing but the low level viremias and undetectable antigens suggest possible suppression of gene expression by HCV.

Centers for Disease Control. (1999) Public Health inter-agency guidelines for screening donors of blood, plasma, organs, tissue and semen for evidence of hepatitis B and hepatitis C. *MMWR* **40**: (RR-4): 1–17.
US Public Health Service recommendations for counseling and medical management of persons with anti-HCV antibodies.

Consensus Statement. (1999) EASL International consensus conference on hepatitis C. *J. Hepatol* **30**: 956–961.
Updates the NIH consensus conference of 1997 and discusses combination therapy of IFN-α with ribavirin.

Davis, G.L. (1999) Hepatitis C. In: *Schiff's Diseases of the Liver.* (eds E.R. Schiff, M.F. Sorrell, W.C. Maddrey). Lippincott-Raven, Philadelphia, pp. 793–836.

Davis, G.L., Esteban-Mur, R., Rustgi, V., *et al.* (1998) Interferon alfa-2b alone or in combination with ribavirin for the treatment of relapse of chronic hepatitis C. *New Engl. J. Med.* **339**: 1493–1499.
Of 345 patients with previous relapse on IFN-α monotherapy, showing sustained response of 49% with combination therapy compared with 5% with repeat IFN-α monotherapy. A back-to-back paper with McHutchinson et al. who treated naïve patients.

Di Bisceglie, A.M. (1995) Hepatitis C and hepatocellular carcinoma. *Sem. Liver Dis.* **15**: 64–69.
Review with 72 references on clinical and epidemiological data.

Di Bisceglie, A.M., Conjeevaram, H.S., Fried, M.W., *et al.* (1995). Ribavirin as therapy for chronic hepatitis C—A randomized, double-blind, placebo-controlled trial. *Ann. Intern. Med.* **123**: 897–903.

Enomoto, N., Sakuma, I., Asahina, Y.M., *et al.* (1996). Mutations in the nonstructural protein 5A gene and response to interferon in patients with chronic hepatitis C virus 1b infection. *N. Engl. J. Med.* **334**: 77–81.
Identification of an IFN-sensitive region in a non-structural region of the HCV genome that determined complete response to IFN therapy in patients with this variant.

Farci, P., Alter, H.J., Govindarajan, S., *et al.* (1992) Lack of protective immunity against reinfection with hepatitis C virus. *Science* **258**: 135–140.
Markers of viral replication and host immunity were studied in five chimpanzees sequentially inoculated over a period of 3 years with different HCV strains of proven infectivity. Each rechallenge of a convalescent chimpanzee with the same or a different HCV strain resulted in the reappearance of viremia, suggesting that HCV infection does not elicit protective immunity against reinfection.

Farci, P., Shimoda, A., Wong, D., *et al.* (1996) Prevention of hepatitis C virus infection in chimpanzees by hyperimmune serum against the hypervariable region 1 of the envelope 2 protein. *Proc. Natl. Acad. Sci. USA* **93**: 15394–15399.
Emergence of quaisispecies in transmission studies

Fattovich, G., Giustina, G., Degos, F., et al. (1997) Morbidity and mortality in compensated cirrhosis type C: a retrospective follow-up of 384 patients. *Gastroenterology* **112**: 463–472.
Multi-center study in Europe with 5-year follow-up showing only slow progression of disease with relatively long life expectancy.

Feray, C., Samuel, D., Gigou, M., et al. (1995) An open trial of interferon alfa recombinant for hepatitis C after liver transplantation: antiviral effects and risk of rejection. *Hepatology* **22**: 1084–1089.
Open study of 14 (out of 46) graft recipients with HCV-related chronic active hepatitis treated with IFN-α-2b 3 MIU thrice weekly for 6 months. Treatment had little effect on HCV RNA and histological activity and significantly increased the episodes of chronic rejection.

Gretch, D.R. (1997) Diagnostic tests for hepatitis C. *Hepatology* **26** (Suppl 1): 43S–47S.
One of a series of articles in this supplement on Hepatitis C—good overview.

Imai, Y., Kawata, S., Tamura, S., et al. (1998) Relation of interferon therapy and hepatocellular carcinoma in patients with chronic hepatitis C. *Ann. Intern. Med.* **129**: 94–99.
A retrospective study of 419 consecutive patients previously treated with IFN-α-2b for 6 months and compared with 144 non-treated controls. During follow-up of around 47 months, HCC was detected in 28 previously treated patients and 19 controls. The risk of HCC rose between groups from sustained response to relapse to non-responders.

Kakimi, K., Kuribayashi, K., Iwashiro, M., et al. (1995) Hepatitis C virus core region: T helper cell epitopes recognized by BALB/c and C57BL/6 mice. *J. Gen. Virol.* **76**: 1205–1214.
Non-overlapping T-cell antigenic determinants of HCV core proteins are recognized differently between mice of specific haplotypes. Any extrapolation to humans would indicate that multivalent vaccines will have to be developed to cover the range of HLA haplotypes.

Kelen, G.D., Green, G.B., Purcell, R.H., et al. (1992) Hepatitis B and C in emergency department patients. *New Engl. J. Med.* **326**: 1399–1404.
High seroprevalence (18%) of hepatitis C in patients attending Casualty (ER) Departments in the USA before the 1990s. The seroprevalence in IVDU (80%) is falling.

Kenny-Walsh, E. (1999) Clinical outcomes after hepatitis C infection from contaminated anti-D immune globulin. Irish Hepatology Research Group. *New Engl. J. Med.* **340**: 1228–1233.
Follow-up of over 17 years of a cohort of 704 women who were HCV RNA positive; only 2% cirrhosis—although 51% had fibrosis; normal ALTs in 45%. Study cited as evidence of good long-term prognosis for HCV infection in young women.

Lampertico, P., Rumi, M., Romeo, R., et al. (1994) A multicenter randomized controlled trial of recombinant interferon alpha 2b in patients with acute transfusion associated hepatitis C. *Hepatology* **19**: 19–22.
One of the few controlled trials on acute hepatitis C. Undetectable HCV RNA in 39% of patients with acute hepatitis C 6 months after treatment with IFN-α-2b. All untreated controls developed persistent viremia.

Lau, D. T.-Y., Kleiner, D.E., Ghany, M.G., *et al.* (1998) 10 year follow-up after interferon-α therapy for chronic hepatitis C. *Hepatology* **28**: 1121–1127.
Encouraging long-term (up to 13 years) histological follow-up in 10 patients with sustained virological response to IFN-α monotherapy (6 months treatment) showing lack of progression of fibrosis in all but one patient.

Lerouxroels, G., Esquivel, C. A., Deleys, R., *et al.* (1996) Lymphoproliferative responses to hepatitis C virus core, E1, E2, and NS3 in patients with chronic hepatitis C infection treated with interferon alfa. *Hepatology* **23**: 8–16.
Peripheral blood mononuclear cells recognize consistently the carboxyl terminal portion of the core protein of HCV encompassing specific amino acid residues. Specific T-cell responses in chronic hepatitis C may contribute to clearance of virus, including following antiviral therapy.

Major, M.E., Feinstone, S.M. (1997) The molecular virology of hepatitis C. *Hepatology* **25**: 1527–1538.
Comprehensive review with 163 references.

Marusawa, H., Osaki, Y., Kimura, T., *et al.* (1999) High prevalence of anti-hepatitis B virus serological markers in patients with hepatitis C virus related chronic liver disease in Japan. *Gut* **45**: 284–288.
High prevalence of anti-HBc antibodies among 2014 Japanese patients with chronic hepatitis C; 49.8% with cirrhosis and 59.4% with HCC.

Mathurin, P., Moussalli, J., Cadranel, J.-F., *et al.* (1998) Slow progression rate of fibrosis in hepatitis C virus patients with persistently normal alanine transaminase activity. *Hepatology* **27**: 868–872.
Encouraging report of French study (by Poynard and Opolon) comparing progression of fibrosis in 102 patients with normal range ALT (twice as slow) compared with 102 with elevated ALTs.

McHutchinson, J.G., Gordon, S.C., Schiff, E.R., *et al.* (1998) Interferon alfa-2b alone or in combination with ribavirin as initial treatment for chronic hepatitis C. *New Engl. J. Med.* **339**: 1485–1492.
In a double-blind and controlled study of 912 patients, the combination of IFN-α (3 MU t.i.w.) with ribavirin (1000–1200g daily for 24 or 48 weeks) was superior (31–38% sustained virological response) compared with IFN-α alone with placebo for ribavirin (6–13% sustained virological response. (Compare Davis et al.—back-to-back paper on previously relapsed patients).

National Institutes of Health (1997) consensus development conference panel statement. Management of hepatitis C. *Hepatology* **26** (Suppl 1): 2S-10S.
Clear and concise—based on interferon monotherapy prior to licensing of ribavirin in the USA. Updated by EASL consensus (European) conference in 1999.

Neumann, A.U., Lam, N.P., Dahari, H., *et al.* (1998) Hepatitis C viral dynamics in vivo and the antiviral efficacy of IFN α therapy. *Science* **282**: 103–107.
The major initial effect of IFN-α is to block production and release of virions. Contrast the papers by Zeuzem et al.

Okamoto, H., Okada, S., Sugiyama, Y., *et al.* (1991) Nucleotide sequence of the genomic RNA of hepatitis C virus isolated from a human carrier: comparison with reported isolates for conserved and divergent regions. *J. Gen. Virol.* **72**: 2697–2704. 1994.
Early classification of HCV into three major genotypes (I-III) and subtypes (a,b,c), superceded by that of Simmonds et al. *– 1994.*

Pileri, P., Uematsu, Y., Campagnoli, S., *et al.* (1998) Binding of hepatitis C virus to CD81. *Science* **282**: 938–941.
One possible putative receptor for binding of HCV. Using in vitro *assays, the HCV envelope protein E2 was shown to bind to CD81 expressed as a recombinant fusion protein with species (human, not murine) specificity. As CD81 is found on most cells, if confirmed, these observations indicate that HCV can bind to most cells.*

Poynard, T., Bedossa, P., Chevalier, M., *et al.* (1995) A comparison of three interferon alpha-2b regimens for the long-term treatment of chronic non-A, non-B hepatitis. *New Engl. J. Med.* **332**: 1457–1462.
Randomized studies on 303 patients from France following standard 6 months treatment with IFN-α-2b at 3 MU t.i.w. Normalization of liver enzymes (ALT levels) was maintained at 3.5 years (2 years following cessation of 18 months therapy) in 22% of patients given the same dose for a further 12 months compared with 8% who received only 6 months of therapy. Normalization of ALT levels was maintained in only 10% of patients given a lower dose (1 MU t.i.w.) for the additional 12 months of treatment. A large study from Europe showing the limitations of trials and outcomes. IFN-α-2b is ineffective in the majority of patients with chronic hepatitis C. Improvements in histological features were not assessed after 18 months and normalization of liver enzymes did not predict reliably clearance of HCV RNA.

Poynard, T., Bedossa, P., Opolon, P., *et al.* (1997) Natural history of liver fibrosis progression in patients with chronic hepatitis C. *Lancet* **349**: 825–832.
French study of 2235 patients showing that host factors (older age, alcohol intake and male sex) rather than virus genotype are associated with progression of fibrosis.

Rothschild, M.A., Berk, P.D., Seeff, L.B. (1995) Hepatitis C. *Sem. Liver Dis.* **15**: 1–122.
Review articles by mostly American authorities on hepatitis C including epidemiology, genetic heterogeneity, therapy, transplantation and non-A, non-B, non-C hepatitis.

Sangiovanni, A., Morlaes, R., Spinzi, G., *et al.* (1998) Interferon alfa treatment of HCV RNA carriers with persistently normal transaminase levels: a pilot randomized controlled study. *Hepatology* **27**: 853–856.
Italian study of 31 patients with persistently normal ALT levels with 16 showing no response to interferon α monotherapy (3 MU t.i.w. for 6 months) compared to no treatment (15 patients). Authors warn against treating such patients because ALT levels flared in 10 of the 16 treated. Whether the rise in ALT represented an attempt to clear virus, remains unknown.

Shimizu, Y.K., Feinstone, S.M., Kohara, M., Purcell, R.H., Yoshikura, H. (1995) Hepatitis C virus: detection of intracellular particles by electron microscopy. *Hepatology* **23**: 205–209.
Ultrastructural identification and characterization of HCV in cell culture and liver tissue from an infected chimpanzee. Intracellular enveloped virus-like particles around 50nm in T- and B-cell lines following infection with virus.

Simmonds, P. (1995) Variability of hepatitis C virus. *Hepatology* **21**: 570–583.
Review of HCV genotypes.

Simmonds, P., Alberti, A., Alter, H.J., *et al.* (1994) A proposed system for the nomenclature of hepatitis C viral genotypes. *Hepatology* **19**: 1321–1324.
The basis of the most widely accepted system of classifying HCV genotypes.

Thomas, S.L., Newell, M.L., Peckham, C.S., Ades, A.E., Hall, A.J. (1998) A review of hepatitis C virus (HCV) vertical transmission: risk of transmission to infants born of mothers with and without HCV viraemia or human immunodeficiency virus infection. *Int. J. Epidemiol.* **27**: 108–117.
Of 976 eligible infants from 28 studies, <10% (0–11%) showed vertical transmission of HCV, especially from mothers with HCV RNA levels below 10^6 copies/ml.

Tong, M.J., El-Farra, N.S., Reikes, A.R., *et al.* (1995) Clinical outcomes after transfusion associated hepatitis C. *New Engl. J. Med.* **332**: 1463–1466.
Serious complications of hepatitis C reported from a specialized referral center in the USA; 51% cirrhosis and 15% mortality within 1–15 years of follow-up. This study contrasts with surveys from non-specialized centers that show little or no excess mortality above controls.

Von Weizsacker, F., Wieland, S., Kock, J., *et al.* (1997) Gene therapy for chronic viral hepatitis: ribozymes, antisense oligonucleotides and dominant negative mutants. *Hepatology* **26**: 251–255.
Elegant discussion of futuristic concepts.

Wyatt, C.A., Andrus, L., Brotman, B., *et al.* (1998) Immunity in chimpanzees chronically infected with hepatitis C virus: role of minor quasispecies in reinfection. *J. Virol.* **72**: 1725–1730.
HCV sequences following reinfection were detectable as minority species in the original inoculum and, presumably, selected by the immune system of the chimpanzee.

Yoshida, H., Shiratori, Y., Moriyama, M., *et al.* (1999) Interferon therapy reduces the risk of hepatocellular carcinoma: national surveillance program of cirrhotic and non-cirrhotic patients with chronic hepatitis C in Japan. *Ann. Intern. Med.* **131**: 174–181.
A retrospective national surveillance study (2400 treated, 490 volunteer controls) by centrers involved in the Inhibition of Hepatocarcinogenesis by Interferon Therapy (IHIT); reduction in HCC over 4 years follow-up following approximately 6 months of IFN therapy. Risk of HCC was associated with male sex and increasing age. Annual incidence of HCC increased with degree of fibrosis but was independent of HCV RNA level and genotype.

Zeuzem, S., Schmidt, J.M., Lee, J.-H., Ruster, B., Roth, W.K. (1996) Effect of interferon alfa on the dynamics of hepatitis C virus turnover *in vivo*. *Hepatology* **23**: 366–371.
Pharmacokinetics of clearance of HCV.

Zignego, A.L., Brechot, C. (1999) Extrahepatic manifestations of HCV infection: facts and controversies. *J. Hepatol.* **31**: 369–376.
Comprehensive review with 99 references giving critical insight into mixed cryoglobulinemia, lymphoproliferative disorders and various extrahepatic diseases reported to be associated with HCV infection.

Chapter 5

Acute Liver Failure

5.0 Issues

Acute liver failure (ALF, also known as fulminant hepatitis) develops as an uncommon complication of acute viral infection. ALF describes a spectrum of acute liver disorders (see section 5.1) that are complex, require specialist management and carry a high mortality. Prompt diagnosis of viral from non-viral etiology delineates likely prognosis without liver transplantation. In expert hands, with medical management alone, survival rates of 40–60% can be achieved for fulminant hepatitis A and B if referred before development of cerebral edema. In contrast, the uniformly poor prognosis of most other causes of ALF requires early diagnosis and assessment for suitability for liver transplantation.

The true incidence of ALF is unknown. While all hepatotropic viruses can cause ALF, their relative contributions vary between developed and developing countries. Published data reflect the bias inevitable when reporting any uncommon condition managed in specialized centers.

Key Notes Epidemiology of ALF with viral hepatitis

- All hepatotropic viruses can cause ALF
- Epidemiology varies between developed and developing countries
- Rare despite common exposure in high-risk groups
  - Suggests unusual virus–host interaction
  - High dose of virus not essential
  - Detection of viral nucleic acids and serological markers does not prove pathogenesis

This chapter will focus only on those issues relating to viral hepatitis. The reader should consult other texts, including by Lee and Williams, for general issues relating to ALF.

5.1 Definitions

ALF replaces the popular term 'fulminant hepatitis'. All definitions reflect clinical observations that serve to predict prognosis based on time intervals between first recognition of illness and onset of hepatic encephalopathy (*Table 5.1*). ALF embraces the original terms, fulminant hepatic failure and late onset hepatic failure, that stratify patients before, and after, an arbitrary 8-week time interval between the onset of first symptoms and onset of hepatic encephalopathy, respectively.

There is no consensus on definition, especially on exclusion criteria such as underlying chronic liver disease. The early definitions by Trey and Davidson and Gimson *et al.* are most widely accepted and exclude patients with known underlying chronic liver disease. In a separate classification (O'Grady *et al.*, 1993), ALF embraces the terms hyperacute, acute and subacute liver failure, according to shorter time intervals and includes symptomless chronic liver disease (Bernuau *et al.*, 1986; O'Grady *et al.*, 1993).

Hepatic encephalopathy (*Table 5.2*) is integral to the definition of ALF, although some European groups include patients without hepatic encephalopathy (grade zero) but with abnormal synthetic function, such as low levels of Factor V.

5.1.1 Distinguishing acute-on-chronic hepatitis

Hepatic encephalopathy can be a presenting feature of a flare of acute-on-chronic hepatitis B or C, such as after instigation or withdrawal of immunosuppressive drugs. Most authorities prefer to distinguish from ALF such cases with known pre-existing liver disease (excepting Wilson's disease) as the pathogenic mechanisms probably differ, at least for the hepatic encephalopathy.

Table 5.1. Definitions of ALF

Authors	Designation	Definition
Trey and Davidson 1970	Fulminant hepatic failure	Onset of hepatic encephalopathy within 8 weeks of first symptoms in the absence of pre-existing liver disease
Gimson *et al.*, 1986	Late onset hepatic failure	Onset of hepatic encephalopathy > 8 and < 26 weeks after first symptoms in the absence of chronic liver disease
Bernuau *et al.*, 1986[a]		Jaundice to encephalopathy
	Fulminant hepatic failure	0–14 days
	Subfulminant hepatic failure	14–84 days
O'Grady *et al.*, 1993[a]		Jaundice to encephalopathy in the absence of symptomatic liver disease
	Hyperacute liver failure	0–7 days
	Acute liver failure	8–28 days
	Subacute liver failure	29–84 days

[a]Definitions include symptomless chronic liver disease.

Table 5.2. Grades of hepatic encephalopathy

Grade[a]	Clinical features
0	Alert, oriented
I	Confused, altered mood, psychometric defects
II	Drowsy, inappropriate behavior
III	Stuporous, marked confusion but speaking and obeying simple commands, inarticulate speech
IV	Coma, unrousable

[a]Excludes influence of sedatives, alcohol and central nervous system drugs.

5.1.2 Distinguishing superinfection

ALF should be separated from *de novo* acute viral hepatitis (e.g. hepatitis A) that can pursue a severe, complicated course when superimposed on underlying chronic liver disease (e.g. from alcoholic hepatitis and chronic hepatitis B and C).

5.2 Epidemiology

The exact epidemiology is unknown. Worldwide, indeterminate etiology (non-A–E) competes with hepatitis B as the most common, presumed viral, cause of ALF (*Table 5.3*). ALF from viral hepatitis remains uncommon despite widespread exposure to the major hepatotropic agents. This paradox is central to any hypotheses on the pathogenetic mechanisms underlying ALF (see section 5.3). Outbreaks and clustering of cases transmitted from a common source are extremely uncommon in the absence of other confounding factors. Severe acute hepatitis and complications are more likely with pre-existing alcoholic liver disease and hepatitis C as well as older age (for hepatitis A), intravenous drug use (IVDU; for hepatitis B and B + D) and pregnancy in regions endemic for hepatitis E. Successive cases of ALF have been reported for fulminant hepatitis B transmitted over a time interval of years from a symptomless contact.

As with other uncommon sporadic conditions requiring expert management, reporting of data are biased and differ between developed and developing regions of the world.

5.2.1 Developed regions

In Europe most data are generated from select transplant centers with especial expertise in the field. In the USA, cases tend to be scattered across the country. Also, reports of the proportion of viral to non-viral etiologies reflect the contribution of acetaminophen (paracetamol) when taken in excess.

Crude estimates of ALF from acute viral hepatitis have been extrapolated from case fatality rates (*Table 5.4*) and suggest around 1000–2000 cases per year in the USA. In the USA (*Table 5.5*) and UK, ALF of unknown (indeterminate) etiology probably accounts

Table 5.3. Acute liver failure of viral etiology

Major viruses	Region	Features
Hepatitis A virus	Worldwide	Very rare
Hepatitis B virus	Worldwide	Wild-type and variants
Hepatitis C virus	Solitary case reports	Very rare
Hepatitis D virus	Mediterranean, South America	Co-infection with HBV
Hepatitis E virus	Developing countries	Young adults: high mortality in pregnancy
GBV-C/hepatitis G virus	Coincidental	Significance in pathogenesis unlikely
Herpesviruses: herpes simplex viruses 1 and 2, varicella zoster virus, Epstein–Barr virus, cytomegalovirus, human herpes virus-6	Individual reports	Pregnancy, immunosuppressed, children
Parvovirus	Individual series	Children, immunosuppressed
Adenoviruses	Individual reports	Children, immunosuppressed
Exotic agents (see Chapter 8)	Tropical regions	Children
Yellow fever virus	West Africa, South America	Children
Dengue viruses	China, South East Asia, South and Central America	Children
Ebola and Marburg viruses	West and Central Africa, Nigeria, Sierra Leone	Outbreaks
Rift Valley fever virus	Sub-Saharan Africa, Sudan, Egypt, Senegal	Forest workers, animal handlers

for the largest proportion of cases after exclusion of drug-related hepatotoxicity, particularly from excess acetaminophen (paracetamol).

5.2.2 Developing nations

In contrast, in developing countries and the East, viral hepatitis remains the most common cause—despite widespread exposure to hepatotropic viruses—beginning from early childhood. Differences in speed of onset of hepatic encephalopathy also vary between developed and developing nations as these reflect viral etiologies. In the study from India (*Table 5.6*), with a predominance of acute viral hepatitis, all patients developed hepatic encephalopathy within 4 weeks of clinical illness. The relatively high percentage of survivors of ALF from non-A, non-B etiology in India reflects the inclusion of a high proportion with hepatitis E [spontaneous survival is < 20% for indeterminate (non-A–E) cases in the West].

Epidemiologies also vary with inclusion of pediatric cases. Babies and children can develop ALF from all agents implicated in adults. Indeterminate etiology accounts for a significant proportion of cases in pediatric practice. In addition, ALF has been

Table 5.4. Deaths from acute viral hepatitis in the United States (1983–1995). Viral Hepatitis Surveillance Program at the Centers for Disease Control and Prevention

	Hepatitis A	Hepatitis B	Non-A, non-B
Total	341 414	267 620	49 291
Deaths (%)	1366 (0.4)	3211 (1.2)	986 (2.0)

reported, albeit rarely, with common childhood infections such as echoviruses and adenoviruses, as well as the herpesviruses (see section 7.8), including in the immuno-suppressed individual (*Table 5.3*).

5.2.3 Hepatitis A

HAV rarely causes ALF, including during epidemics. Sporadic cases of ALF have been reported in children and adults. In the West, severity of clinical illness rises with age. Excess mortality rates occur in adults over 40 years of age and those with underlying liver disease. More overt illness is anticipated with the shift in herd immunity (see section 1.2) towards older adults. During an epidemic of hepatitis A in Tennessee during 1994–1995, 256 of 1700 diagnosed cases of hepatitis A were hospitalized. Among those over 40 years, extrahepatic and biliary complications were common (25%) and three of the five cases of ALF were in this older age group.

Severe hepatitis A:

- Age > 40 years
- Underlying chronic liver disease (alcoholic hepatitis, hepatitis C)
- No excess mortality in pregnancy
- Survivors show complete recovery and immunity from future infection

In developing countries, where more than 90% of children are exposed to HAV before their fifth year and almost all adults are immune, fulminant hepatitis A is very rare in adults. Conversely, severe hepatitis A is most likely in the presence of other infections that can target the liver—such as dengue, schistosomiasis and chronic hepatitis B and/or C—as well as alcoholic liver disease. These should be distinguished from ALF.

Table 5.5. Etiology of ALF in the USA, 1994–1996 (295 patients) (after Schiodt *et al.*, 1999)

Etiology	No (%)	Transplanted	Spontaneous recovery
		121 (41%)	74 (25%)
Acetaminophen overdose	60 (20)	7 (12)	34 (57)
Hepatitis A virus	21 (7)	7 (35)	8 (40)
Hepatitis B virus	30 (10)	19 (63)	4 (13)
Unknown	44 (15)	26 (60)	11 (26)
Drug-related	34 (12)	17 (50)	7 (21)
Miscellaneous	108 (37)	45 (43)	10 (9)

Table 5.6. Etiologies of ALF in India (after Acharya *et al.*, 1996)

Etiology	% Survivors (*n* = 143)	% Non-survivors (*n* = 280)	Total (*n* = 423)
Hepatitis A virus	2	1.4	1.7
Hepatitis B virus	27	27.8	27.6
Hepatitis C virus	3.5	3.9	3.8
Non-A, non-B virus	62.2	63	62.4[a]
Anti-tuberculosis drugs	5	4.3	4.5

[a]Including 8.7% HBsAg seropositive. Of 50 consecutive cases diagnosed as non-A, non-B, 40% had detectable HEV RNA or HEV RNA + HCV RNA on RT–PCR; none had HBV DNA.

As most patients with fulminant hepatitis A develop hepatic encephalopathy within 8 weeks of first symptoms, spontaneous recovery (prognosis without liver transplantation) exceeds 50%. There are individual reports of severe, including fulminant, hepatitis A in children and adults following relapse after a biphasic course. Hepatitis A virus (HAV) and viral antigen (HAAg) have been detected in stool and liver, suggesting persistence of virus in these patients (see section 2.5).

5.2.4 Hepatitis E

The few reports of hepatitis E virus (HEV) infection in the West seem confined mostly to travelers returning from high-risk areas. Severe hepatitis E, including ALF, occurs in major epidemics typically in developing countries with poor standards of sanitation and hygiene.

Most cases are young adults, despite probable exposure previously in childhood. Excess mortality (around 20%) has been reported in pregnancy. Of the females with ALF in India, 25% were pregnant (compared with 3% in the general Indian population). However, mortality rates for sporadic ALF (contrast epidemic hepatitis E, section 2.1.4) were comparable to non-pregnant females and males in the same series and for the differing viral etiologies. Whether this especially severe outcome for developing countries reflects virus virulence, host factors, including the peculiar state of pregnancy and possible pre-existing liver disease, and suboptimal facilities, is unclear.

Severe hepatitis E in developing countries:

- Significant proportion of non-A, non-B hepatitis
- Young adults
- Pregnant women (some regions only)
- Survivors show complete recovery
- Survivors may not be protected from future infection by HEV

5.2.5 Hepatitis B and B+D

Hepatitis B virus (HBV) and HBV with hepatitis D virus (HDV) co-infection are common causes of ALF in France and some other European countries but not in the UK and USA. The patient typically is a young adult who presents after sexual contact with a symptomless partner, discovered afterwards to be seropositive for HBsAg and anti-HBe. No specific HLA haplotype has been identified that predisposes to ALF but this area requires further study.

Outbreaks of severe hepatitis B, and B+D co-infection, have been reported among drug-users. Early reports did not exclude underlying chronic liver disease, especially from coexistent hepatitis C and alcohol abuse. Rarely, outbreaks of severe acute hepatitis B, including an excess number that progressed to ALF, have been linked to common source, such as through nosocomial spread.

Fulminant hepatitis B is a heterogeneous disorder. As most patients present within 8 weeks of first symptoms, spontaneous survival is better than with indeterminate etiology (*Table 5.4*) and those who pursue a subfulminant course beyond 8 weeks.

A prospectively studied panel of patients from the UK, who did not proceed to liver transplantation, could be divided into those with rapid (< 1–2 weeks), and slow (> 3 weeks), progression according to rates of clearance of HBsAg, HBV DNA, pre-S2 antigen, development of anti-pre-S2 antibodies, intervals between onset of clinical illness and hepatic encephalopathy, and prognosis for recovery without liver transplantation.

Severe hepatitis B and B+D:

- Sporadic despite widespread exposure and transmission between high-risk groups
- Typical transmission from symptomless, anti-HBe seropositive carrier
- Vertical transmission from HBeAg seronegative mother
- Hepatitis B+D virus co-infection
- Rapid clearance of viral antigens favors spontaneous recovery
- Survivors develop immunity against future infection
- Survivors rarely develop chronic liver disease from residual HBV infection

5.2.5.1 Vertical transmission and ALF

ALF is rare in babies but favored following transmission of HBV from an HBeAg seronegative mother with chronic hepatitis B. Maternal levels of HBV DNA can be low but rise around the time of delivery. Vertical transmission of HBV variants that fail to synthesize HBeAg (see sections 3.2.4.1 and 7.5.2.1) may be involved in some cases. Survivors show serological evidence of clearance of virus and subsequent immunity (anti-HBs) to future infection. Chronic hepatitis B is rare in survivors.

5.2.6 Hepatitis C

Hepatitis C virus (HCV) rarely causes ALF, including in high-risk groups (*Table 4.1*). The few cases described tend to be immunosuppressed and/or have hematological disorders and/or received chemotherapy with recent withdrawal. Some have concurrent HBV infection. One case report documents seroconversion to anti-HCV antibodies (acute hepatitis C) and development of hepatic encephalopathy, coagulopathy and massive hepatic necrosis within 5 weeks of blood transfusion. Serum levels of HCV RNA ($>10^8$ genome equivalents/ml) peaked with high transaminase levels prior to falling to low levels [detectable only by sensitive polymerase chain reaction (PCR)] beyond 6 weeks after blood transfusion. Whether hepatitis C contributes significantly to the pool of ALF without pre-existing liver disease, concurrent HBV infection and immunosuppression, remains controversial. Serological tests do not discriminate acute from chronic infection. Also, immunosuppressed individuals may not seroconvert to anti-HCV antibodies and the recombinant immunoblot assay (RIBA) may be falsely negative or indeterminate, with detection of less than four antibodies (see section 4.5.1). Withdrawal of chemotherapy has been associated with liver failure in chronic carriers of HCV, as well as HBV, but reports are too few to compare clinical and pathological findings with fulminant hepatitis B.

5.2.7 Indeterminate etiologies

The term non-A–E hepatitis is misleading because this implies a single and novel cause. Multiple viruses, non-viral agents and predisposing host factors probably contribute to this group. In the West, this category accounts for a significant proportion of assumed viral etiologies especially in children and young adults. Some cases of autoimmune hepatitis can masquerade as fulminant hepatitis. In contrast, in developing regions of the world, hepatitis E should be excluded. In specific cases of ALF of unknown etiology, after rigorous exclusion of the major hepatotropic agents, evidence is accumulating in favor of at least one potentially novel agent (candidate hepatitis F), based on recurrence of ALF following liver transplantation (see section 5.6.5).

5.2.7.1 Aplastic anemia

This is a well-recognized, albeit rare, association with ALF of unknown cause, especially in children. In the West, aplastic anemia complicates 2–5% of such cases. This complication is more common (4–25%) in the Far East. Most cases are young boys who present with a severe pancytopenia within 3 months of an attack of hepatitis. Failure of the bone marrow can be precipitous and usually is fatal if left untreated.

Aplastic anemia also is a recognized complication following liver transplantation for fulminant non-A–E hepatitis. Bone marrow failure developed in 28% and 33% of patients following liver transplantation in two studies from the USA. 'Toxic' factors likely contribute as recovery of bone marrow has occurred with use of immunosuppression and anti-thymocyte globulin. Parvovirus B19 and Epstein–Barr virus (EBV) serological markers and nucleic acid sequences have been detected in the serum of a few patients, mostly children, with ALF of uncertain cause. Their clinical significance remains unclear.

5.2.7.2 Cryptic infections

Some cases of unknown etiology with undetectable serological markers have been attributed to HBV and/or HCV based on detection of nucleic acids in serum, and liver, by the PCR. Disclosure of infection can occur following grafting (and immunosuppression) with detection of HBsAg, HBV DNA by dot-blot hybridization, and HBeAg. The clinical outcomes of these infections tend to be mild. The relative contribution of these untypical cases to the pool of ALF is unknown. Caution is warranted in attributing ALF to any cryptic infection based on PCR detection alone (see section 5.2.7.3). Serial analyses to profile changes in viremia are required before incriminating an infectious agent in the pathogenesis of any disease, especially in patients who have received multiple transfusions.

5.2.7.3 GBV-C/hepatitis G virus, TT virus and SEN virus

GBV-C/hepatitis G virus (HGV) and TT virus sequences have been detected in liver and serum in some cases of ALF of unknown cause. Their significance is unclear in these patients given their relatively common exposure to blood products and ubiquitous nature of these agents. SEN virus, named after the first patient, has been detected in some patients who developed acute hepatitis following blood transfusions. Studies are in progress to determine the role of this virus in ALF. (See also section 8.9).

5.2.8 Herpesviruses

Herpes simplex viruses, varicella zoster virus (VZV), cytomegalovirus (CMV), human herpes virus type 6 and EBV are rare causes of ALF but are disproportionately represented in the immunosuppressed, children, neonates and pregnant women (see sections 7.2.4 and 7.8). Around 20 cases of ALF attributed to EBV have been published. The majority are in children and young patients with overt immunodeficiency, especially X-linked lymphoproliferative disorders and deficiencies in the complement pathways and human immunodeficiency virus (HIV) infection. Boys with X-linked lymphoproliferative disease develop hypogammaglobulinemia, B-cell lymphoma and aplastic anemia, as well as liver failure. Clinical features are not diagnostic; skin rash and sore throat are not invariable and the atypical mononucleosis is seen with other viral infections, including CMV (concurrent CMV infection is common) and early HIV infection. Hemorrhagic manifestations can be striking and a Reye's-like syndrome and erythrophagocytosis have been described. Almost all cases are fatal but individual publications document spontaneous recovery with antiviral therapies (e.g. ganciclovir) and successful outcome following liver transplantation, including without infection of the graft.

5.2.9 Exotic agents

Yellow Fever virus, Rift Valley Fever virus, Marburg and Lassa viruses are among several viruses causing hemorrhagic fever with severe acute hepatitis that can evolve into ALF (*Table 8.5*). These agents should be considered especially in travelers bitten by insects in endemic areas such as the African countries. Early liaison with a reference center is necessary to arrange the most appropriate tests for diagnosis, optimize management and minimize exposure and risks of transmission to attendant healthcare and laboratory personnel.

5.2.10 Other causes

Rare cases of ALF occur in association with common infections, especially in neonates, children, pregnant women and the immunosuppressed host (see Chapter 7).

Risk factors for neonatal echovirus infection in neonates and children:

- Prematurity
- Male sex
- Maternal infection (vertical/perinatal transmission)
- Cesarian section for abdominal illness in the mother

5.3 Pathogenesis

The key to the pathogenesis of ALF lies in explaining its rarity despite widespread exposure to the major hepatotropic viruses, especially among high-risk groups such as IVDU. Accordingly, virus virulence *per se* (pathogenic phenotypes) is insufficient to explain most cases of fulminant viral hepatitis especially because transmission typically is from a symptomless source. Further, ALF is sporadic in nature; any clustering of cases tends to occur in high risk groups but with additional confounding factors such as underlying chronic liver disease from alcohol abuse and hepatitis C and from multiple co-infections (e.g. HBV + HDV). These observations have led most authorities to implicate primarily host, rather than viral, factors in the pathogenesis of ALF. An exaggerated immune response is involved in ALF due to HBV and, possibly, HAV and HEV. Other host factors that predispose to ALF from various viruses include older age, immunodeficiency, pregnancy, malnutrition and in neonates and young children. Data are insufficient to implicate HLA haplotypes.

Evidence is accumulating, however, to implicate viral factors in fulminant hepatitis, based mostly on studies of hepatitis B. Host and viral factors in fulminant hepatitis B interplay to account for dynamic changes in virus quasispecies at the time of transmission and early clearance of selected (minority population) variants prior to clinical presentation.

5.3.1 Host factors in fulminant hepatitis B

In ALF due to hepatitis B, the traditional view is that HBV is not cytopathic. Instead, severe liver injury is caused by an exaggerated cellular immune response. Immunological data are convincing for an exaggerated immune attack by cytotoxic T cells, targeted against HBcAg and, possibly, also HBeAg. A transgenic mouse model of fatal necroinflammatory liver disease implicated MHC-class I-restricted, CD8+ T cells in causing the massive necrosis of HBsAg positive hepatocytes. Class II MHC also seems to play a role. In a single case report by Missale *et al.*, it was found that proliferative responses of peripheral blood mononuclear cells to HBV nucleocapsid and envelope antigens were sensitized to CD4+, HLA class II-restricted T cells. Specificity to immunodominant epitopes increased following transplantation, when the patient developed fibrosing cholestatic hepatitis (FCH). Following transplantation across an

HLA mismatch, macrophages process viral peptides and present derivatives to CD4+ T cells in an MHC-restricted (class II) manner that relies only on host-derived elements for T-cell activation.

Massive necrosis of hepatocytes results from the seemingly inappropriate response of the host to clear virus-infected hepatocytes. Low levels of HBV replication and early clearance of antigens (HBsAg, HBeAg) support this view. Also, classical ALF does not recur following liver transplantation (see section 6.3) despite immunosuppressive therapy that enhances virus replication. Close homology of viral sequences between transmitter and index case has been cited as evidence in favor of host factors in the pathogenesis of fulminant hepatitis B. However, these observations require appraisal as evidence accumulates to implicate viral, as well as host, factors.

Host factors in fulminant hepatitis B:

- Rare despite common exposure
- Sporadic cases (clustering, very rare)
- Transmission from symptomless carrier
- Exaggerated immune response
- Selective clearance of virus variants

5.3.2 Virus variants in fulminant hepatitis B

Variant forms of HBV (core promoter mutations, HBeAg minus, precore, and pre-S2 defective, variants) have been associated with outbreaks of severe hepatitis B from a single source, albeit as rare events. Individual reports of successive cases of ALF among unrelated (HLA mismatched) index cases have occurred following transmission of HBV from a symptomless chronic carrier. Although no specific 'fulminant genotype' has been identified, higher numbers of mutations throughout the HBV genome have been reported in fulminant hepatitis B compared with chronic hepatitis B. Several publications reported a significant association with infection by HBV variants unable to produce HBeAg as a result of a translational stop codon in the precore region of the genome (G1896A) and point mutations in sequences in other regions of the genome, such as in the X gene. In particular, mutations have been found in the core promoter/enhancer II motif that may act to increase levels of transcription of the pregenomic RNA (see section 3.2.4.1). Any association of ALF with the pre-core stop mutation may reflect co-segregation with mutations and variations elsewhere in the genome.

Virus factors (variants) in fulminant hepatitis B:

- HBeAg-negative variants
- Selective transmission of variants from mother to child
- Early clearance of minority variants
- Successive cases of ALF reported
- Outbreaks (clustering) of ALF from a common source

These associations remain controversial because they are not unique to fulminant hepatitis B. Similar mutations have been found in acute, uncomplicated hepatitis as well as chronic hepatitis B, especially in anti-HBe, rather than HBeAg, seropositive patients. Most studies on sequence variations in the HBV genome have been limited to selected regions (especially the precore region) and analysis was based on serum sampled after presentation when the viremia was very low. Whether these mutations are cause or effect, also remains to be resolved. Immune pressure is extreme in fulminant hepatitis B; surviving virus may represent escape variants rather than especially pathogenic varieties. That the same mutations may be found during exacerbations of chronic hepatitis, would support this view.

Using a phylogenetic approach, one group from the UK found strains of HBV with particular combinations in these regions that were unique, or nearly so, to fulminant hepatitis B when associated with aberrant cysteine and methionine residues in the X gene. Further, cluster identity was closely related to mortality and rapidity of progression of disease regardless of geographic origin. A1896 variants (typically of genotype D) were associated with rapid clearance of HBV DNA and spontaneous recovery, whereas C1858 (typically genotype A) with G1896 was associated with progressive deterioration and death from relentless disease.

5.3.2.1 Viral dynamics in fulminant hepatitis B

Fulminant hepatitis B seems to be an ideal model for studying virus dynamics under extreme immune pressure. Evidence implicates dynamic changes in virus quasispecies in the contact as well as the index case. Some of the strongest evidence for virus heterogeneity and dynamics comes from studies on vertical transmission. Babies born of HBeAg seronegative mothers are at especial risk of developing severe hepatitis B (see section 7.5.2.1). HBeAg seronegative mothers typically have a heterogeneous population of HBV genomes compared with HBeAg seropositive mothers who are chronically infected with HBV. The predilection for developing fulminant hepatitis B in the neonate may result from vertical transmission of selected minority variants that represent a small proportion (< 5%) of the maternal population. Sampling the contact (mother) at the time of presentation may not reflect viral events at the time of transmission. Also, most sequence analyses of individual virus variants have been limited to very few cases in ALF and focused on select regions of the virus genome.

Following transmission, evolution of HBV species is especially rapid in fulminant hepatitis B compared with acute uncomplicated infection and chronic hepatitis B. Accordingly, by the time the typical patient presents with ALF, the species detected may not be representative of the transmitted population. Those minority variants implicated in the onset of ALF will probably have been targeted by the immune system for early clearance.

Fulminant hepatitis B is a heterogeneous disorder probably due to the variation in response of the immune system of the host to specific virus variants. There is strong evidence that T- and B-cell responses are specific to genotype and subtype of HBV. In the prospective study from the UK, those patients with the characteristically rapid clinical and serological evolution ('rapid ALF') survived probably because clearance of virus occurred relatively early, allowing liver regeneration to take place. All survivors had variants typical of genotype D: U1858/A1896 or U1858/G1896, while deaths consequent to a slower clinical and serological evolution (> 21 days to coma and seroconversion) were associated with C1858/G1896 infections (typical of genotype A). These findings raise the possibility that variants of genotype A may be the primary target of the immune response in fulminant hepatitis B. Variants with high replication efficiency may be the residual population following seroconversion and recovery as similar changes in genotypes have been noted following seroconversion in acute, uncomplicated hepatitis B.

Fulminant hepatitis B is associated with specific HBV variants:

- T1762, A1764 and T1653 in the basal core promoter
- Impaired nuclear factor binding activities within the basal core promoter
- A1896 in association with mutations elsewhere in the genome
- Genotype D and high replication efficiencies in fulminant hepatitis B linked to good prognosis
- Genotype A and low replication efficiencies in fulminant hepatitis B linked to poor prognosis
- Mutations leading to aberrant expression of X gene proteins
- Mutations leading to aberrant cysteine and methionine residues in the X gene

5.3.2.2 Replication competence and pathogenicity

Assessing virus virulence is difficult and data are conflicting in fulminant hepatitis B. Interpretation of replication efficiency is limited mostly to results from cultured cells. An efficient *in vitro* system is lacking and HBV has not been grown successfully long-term in tissue culture. Caution is required before equating increased replication fitness with pathogenic potential, especially with experiments that exclude the immune axis in fulminant hepatitis B.

Upregulation of pregenomic mRNA has been demonstrated *in vitro* using HBV DNA from an infectious clone, isolated from a common source outbreak of fulminant hepatitis B. However, ALF was not invariable and some patients developed uncomplicated hepatitis B. Also, in a larger transfection study of HBV genomes from seven patients with fulminant hepatitis B, no differences were found in replication competence compared to wild-type and with any defect in expression of HBeAg.

Attempts to interpret the pathogenetic mechanisms in fulminant hepatitis B have been confused by extrapolating results from FCH (see sections 3.2.4.1 and 6.3.3). Transgenic animal models of severe hepatitis B resemble more closely FCH than classical fulminant hepatitis B. Also, only HBV genomes from the patients with FCH show consistently high replication competence *in vitro*. Individual reports have described novel, cytopathic HBV variants with high replication efficiencies, attributed to enhanced binding of nuclear factors to a novel site created by an 11 nt insertion in the basal core promotor (BCP), or to mutations in the BCP (C1766G). Such changes enhance transcription of pregenomic mRNA and increase the rate of synthesis of HBV DNA and virions.

Increased replication efficiency does not necessarily correlate with increased pathogenicity. Proof of pathogenic potential requires development of fulminant disease following transmission. Unfortunately for research purposes, fulminant hepatitis is extremely difficult to reproduce in animal models.

5.3.2.3 Hepatitis A and E

Studies are too few to assess the relative importance of host and viral factors. Appreciation is growing of the considerable genetic diversity among isolates of HAV and HEV from various regions of the world but there is no evidence of especially pathogenic isolates.

Limited sequence analyses, mostly of the 5′ untranslated region (UTR), have not revealed any unusual changes in HAV RNA in fulminant hepatitis A (mutations in the 5′ UTR of the vaccine strains of polioviruses are associated with decreased pathogenicity and attenuated strains of HAV also have mutations in the 5′ UTR). Individual reports suggest that virus heterogeneity may play a role in determining severity of hepatitis A. An HAV variant originating in India was associated with severe, protracted hepatitis A in a pregnant mother and, subsequently, her baby born 4 months later. Sequence analysis of the 5′ UTR showed various deletions that might affect the efficiency of internal ribsome entry site-dependent translation (see section 2.2.3).

The complete genome sequence was determined from HEV RNA amplified directly from explanted liver tissue of a male who developed fulminant hepatitis E after returning to England from the Indian subcontinent. There was no conclusive evidence that variations in sequence (other than of a degree typical within the Burmese genotype, see section 2.3.4) contributed to a pathogenic phenotype.

5.3.2.4 Hepatitis C

Data are insufficient to draw any conclusions on possible pathogenic mechanisms. A single case report documents low genetic diversity for genotype 1b between clones isolated from two samples in a patient who developed acute liver failure 4 weeks after a blood transfusion.

5.4 Clinical features

Viral hepatitis cannot be distinguished clinically from non-viral causes of ALF, such as drug-related hepatotoxicity, particularly from acetaminophen (paracetamol) overdose, anti-epileptics, non-steroidal anti-inflammatory agents and anti-tuberculous therapies, carbon tetrachloride, halothane and poisonous mushrooms (*Amanita phalloides*), acute fatty liver of pregnancy and Reye's syndrome (see also section 7.9). Furthermore, chronic liver diseases such as Wilson's disease and autoimmune hepatitis, as well as malignant infiltration and Budd–Chiari syndrome, can present with clinical features indistinguishable from those due to a viral infection. Other microbial infections, such as tuberculosis, leishmaniasis, leptospirosis and severe sepsis, can mimic ALF.

The size of the liver varies according to the stage of illness rather than with etiology. The liver usually is small in severe ALF; hepatomegaly suggests infiltration, typically from malignancy, fat or hemorrhage. Splenomegaly is uncommon and should alert the possibility of underlying chronic liver disease, especially Wilson's disease, and infiltration, including lymphoproliferative disorders. Any skin rashes are non-specific; cutaneous stigmata, including genital ulceration, may be absent in herpesvirus infections.

ALF may present with bone marrow aplasia in children (see section 7.9.1).

5.5 Laboratory diagnosis

Prognosis without grafting relates directly to etiology. In clinical practice, patients with fulminant hepatitis A or fulminant hepatitis B have more than a 50% chance of spontaneous recovery (without the need for liver transplantation), whereas those with non-A, non-B have less than 20% chance of spontaneous survival.

5.5.1 Serological markers

Rapid diagnosis relies on conventional serological markers that remain the cornerstone for diagnosing viral hepatitis. While general principles of clinical management are regardless of etiology, accuracy of virological diagnosis is essential for predicting prognosis if transplantation is not an option, and for identifying those cases in which antiviral therapy can make a difference to outcome (see sections 6.3 and 6.4) before and after transplantation.

Diagnosis of fulminant viral hepatitis requires a high index of suspicion and skill in interpreting limited availability of tests in an emergency situation. Many of the pitfalls in their interpretation are exemplified by ALF (*Table 5.7*). Serial analyses to profile changes in viremia are required before incriminating an infectious agent in the pathogenesis of any disease, especially in patients who belong to high-risk groups for viral infections. Multiple viral etiologies should be sought using panels of tests and repeated in doubtful cases. Serological markers for HBV and HCV can coexist in patients presenting with liver failure. Whether these represent acute-on-chronic, or simultaneous co-infections, remains unclear.

A negative test for HBsAg does not exclude hepatitis B because this antigen is cleared rapidly in fulminant hepatitis B. Conversely, persistence of HBsAg, especially at high-titer, may indicate a chronic carrier with an additional etiology for the liver failure. Immunoglobulin (Ig)M anti-HDV may be detectable only transiently in co-infections with HBV. Conventional enzyme immunoassays for anti-HCV do not discriminate acute from chronic infection. ALF due to HCV appears to be very rare. However, seronegativity for antibodies to HCV does not exclude the diagnosis because seroconversion can be delayed in acute infection. Conversely, a positive anti-HCV antibody test more often indicates underlying chronic infection; another cause should be sought for the fulminant presentation. IgM antbodies to HEV and herpesviruses also may be detected only transiently. Instead, seroconversion to IgG antibodies (IgG anti-HEV, IgG anti-HSV, IgG anti-EBV, IgG anti-CMV) may assist in the diagnosis of these viral infections.

5.5.2 Polymerase chain reaction

Nucleic acids of HBV, HCV and GBV-C/HGV have been detected using the PCR in some patients with ALF attributed initially to non-A–E hepatitis. In the immunosuppressed patient, detection of viral nucleic acids is more reliable than antibody tests that may be falsely negative. However, the significance of a positive PCR remains difficult because detection does not prove etiology. Importantly, levels of nucleic acid may fall below the detection limits of some branched chain DNA assays. Serial testing by quantitative

Table 5.7. Pitfalls in serological tests for diagnosing viral hepatitis and acute liver failure. (Note that antibody tests may be falsely negative in the immunosuppressed host)

Virus	Marker	Comment
HAV	IgM anti-HAV	Reliably high
HBV	HBsAg HBeAg HBV DNA	Rapid clearance, may be negative
HBV	IgM anti-HBc	Reliably high
HBV	anti-HBc anti-HBs anti-HBe	Rapid clearance of antigens; may be positive
HCV	NK	Delayed seroconversion Difficult to discriminate from acute-on-chronic infection
HDV	IgM anti-HDV HDV RNA	Short-lived, may be negative HDV may interfere with HBV markers
HEV	IgM anti-HEV HEV RNA	Short-lived, may be negative
	IgG anti-HEV	May be positive early in acute infection
GBV-C/HGV	GBV-C/HGV RNA	Significance unknown; most patients have received blood transfusions and products
Herpesviruses	Antibodies	Delayed seroconversion; paired titers required
Parvovirus	Antibody tests against B19	Positive in some cases of Non-A–E, especially children
Adenoviruses	Panels of antibodies	Paired titers required

NK, not known.

assays using sensitive PCR may be necessary to detect low levels of viremia that are changing in temporal relation to the illness. A decline is more in favor of a viral etiology, whereas persistent viremia may indicate underlying chronic infection.

5.5.3 Liver histology and immunohistochemistry

Conventional percutaneous liver biopsy is contraindicated in ALF. The risk of bleeding is high with prolongation of prothrombin time (PT) of more than 2 s, especially with concurrent thrombocytopenia and renal failure. Transjugular liver biopsy is safe in expert hands but should be reserved for the individual patient suspected of having other, non-viral disease(s) that impact on management, such as Wilson's disease or acute fatty liver of pregnancy.

Histological examination of liver tissue is of limited value in diagnosing the etiologies of ALF. Viral nucleic acids detected by *in situ* hybridization in liver tissue can provide a rapid diagnosis in selected cases if sampling is feasible. Results may become available only after the window of opportunity closes on management options such as liver transplantation.

Patterns of hepatic necrosis show some correlation with etiology. Liver histology due to fulminant viral hepatitis typically shows zonal or massive cell drop-out rather than coagulative necrosis attributed to 'single hit' agents such as acetaminophen. Immuno-histochemical stains are of limited value in fulminant hepatitis due to the widespread destruction of the liver.

5.5.3.1 Fulminant hepatitis A

Liver histology is characterized by preservation of hepatocytes around the hepatic venules (Zone III of Rappaport); major destruction is focused around the periportal areas that are rich in plasma cells. Microvesicular fat can be a prominent feature. Autoimmune features, including plasma cells, rosetting and interface hepatitis (piece-meal necrosis) can be prominent in some cases of severe hepatitis A.

5.5.3.2 Fulminant hepatitis B

Submassive necrosis (cell drop-out) is characteristic but sparing of hepatocytes is more random in zonal distribution than for fulminant hepatitis A. Most cases of fulminant hepatitis B show only minimal staining for HBsAg and HBcAg but the distribution does not correlate with severity of liver disease. In contrast, in FCH (see section 3.2.4.1), the liver cells show ballooning with massive accumulation of nuclear and cytoplasmic HBsAg and HBcAg. Periportal fibrosis with proliferation of ductal plates can be promi-nent in FCH whereas fibrosis is absent in fulminant viral hepatitis.

5.5.3.3 Fulminant hepatitis C

Clinical and histological descriptions of acute hepatitis C are rare. The apparent overlap with chronic hepatitis C makes difficult the characterization of any unique attributes. Proliferation of bile ductules can occur on the background of submassive necrosis, but lymphocyte aggregates are not prominent.

5.5.3.4 Hepatitis E

Histopathological features seem to resemble hepatitis A but classical descriptions are lacking.

5.5.3.5 Herpesviruses and exotic viruses

Hemorrhagic necrosis with viral inclusions may be prominent, especially in herpes hepatitis. Acidophilic degeneration of hepatocytes (Councilman bodies) are a signature feature of yellow fever but can be prominent in other arboviral infections.

5.5.3.6 Hepatitis of indeterminate etiology

Although etiologies probably vary, most explants show a 'map-like' pattern of random multiacinar collapse of the parenchyma alternating with random areas of regeneration (macronodules). Individual cases have multinucleated giant cells, especially in children. These are non-diagnostic, but have been reported in paramyxoviral (measles) infections. Massive hemorrhagic necrosis with swelling of the graft is prominent in those cases with recurrent liver failure following transplantation.

5.5.3.7 Electron microscopy

This technique is of limited value due to the widespread destruction of hepatocytes, delays in processing of tissue and limited numbers of viral particles seen with the main hepatotropic agents. A prospective study of utrastructural changes in liver in 23 patients with fatal ALF compared 14 with viral hepatitis (four hepatitis B, 10 of indeterminate etiology) with acetaminophen hepatotoxicity. Changes in fatal fulminant viral hepatitis were similar to those seen in acute uncomplicated viral hepatitis, including swollen mitochondria and dilated endoplasmic reticulum with disaggregation of polyribosomes from the cisternae. Perinuclear membranous vesicles, nuclear fibrillar rods and cytoplasmic tubular structures were found in the liver and more abundantly in the grafts of patients who developed recurrent liver failure within 2 weeks of transplantation for ALF indeterminate etiology. HBsAg accumulates in the endoplasmic reticulum in FCH that can have a fulminant clinical presentation and viral particles may be abundant in the immunosuppressed host with adenovirus infection. The ultrastructural characteristics of virus-like particles visualized in some cases of indeterminate etiology resembled the togaviridae (60–70 nm, enveloped with surface spikes). No virus particles were found for HBV and EBV.

5.5.4 Other tests

Rapid changes in synthetic function can be assessed by measuring regularly levels of Factor V, the PT and International Normalized Ratio (see section 1.6.2.4).

Leukopenia and thrombocytopenia are common in ALF regardless of etiology. A relative lymphocytosis may suggest infections such as EBV but can be found in HIV infection and non-viral causes, especially malignant infiltration. A polymorphonuclear leukocytosis can be found in liver failure complicated by microbial infection, including the mycoses. Hematological features of disseminated intravascular coagulation occur regardless of etiology. A hemolytic anemia can occur with hepatitis A but should prompt exclusion of Wilson's disease or accompanying hemoglobinopathies such as sickle cell anemia or glucose-6-phosphate dehydrogenase deficiency. Serum levels of liver enzymes alanine aminotransferase (ALT) and aspartate aminotransferase (AST) are high (>10 times upper normal limits) early in fulminant hepatitis A and B. Such levels typically are falling by the time most patients present with hepatic encephalopathy and do not reflect the degree of hepatocellular necrosis, especially when low. Trends in liver tests should be interpreted in conjunction with rising or falling PTs that measure liver synthetic function. Low aminotransferase levels typically are found in the terminal phases and may indicate lack of regeneration of hepatocytes and, consequently, a poor prognosis. Blood levels of ammonia are elevated invariably but are typically lower than those seen in decompensated cirrhosis. High levels of ammonia may reflect impending elevations of raised intracranial pressure but do not correlate well with the degree of coma, etiology or prognosis.

5.6 Management

ALF is a rare and complex medical emergency requiring early discussions and referral to a specialized center with facilities for managing multiorgan failure and liver transplantation. The principles of management are similar regardless of etiology.

There are no standardized protocols for managing ALF. Most experience comes from the European centers and is derived from care of patients with acetaminophen hepatotoxicity and indeterminate etiology (formerly termed non-A, non-B hepatitis). The reader should consult standard textbooks of hepatology for more details.

Early transfer is essential regardless of etiology. Grade III encephalopathy is associated with clinical onset of cerebral edema and exacerbated by travel. Any of the following features should prompt discussion and referral to a specialized center: hepatic encephalopathy, elevated PT (> 2 times normal control), renal impairment, metabolic acidosis (blood pH <7.3), hypotension, hyponatremia and thrombocytopenia. An alkalosis (respiratory) is more common than acidosis in viral causes of ALF (contrast metabolic acidosis in acetaminophen hepatotoxicity).

Early diagnosis is essential because prognosis without grafting depends on etiology (*Figure 5.1*). In clinical practice, results of serological assays often are not available in time to impact on decisions about treatment options, particularly transplantation. Frequently, critical information regarding potential risk factors, such as exposure to a hepatotoxic agent, is unavailable or unreliable in the presence of encephalopathy.

N-acetyl cysteine is being used increasingly in ALF due to viral hepatitis (not only for acetaminophen hepatotoxicity) to replenish hepatic glutathione stores that seem to be depleted in ALF, regardless of etiology. Aspirin is contraindicated in suspected herpesvirus infections because hemorrhagic complications are common. A Reye's-like syndrome has been associated with VZV (*Table 5.8*).

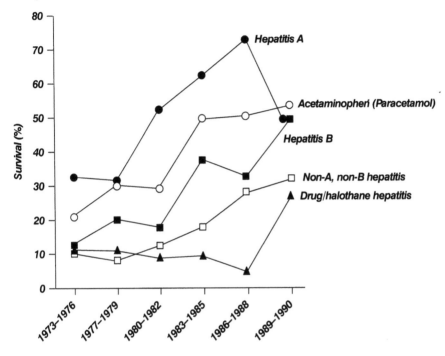

Figure 5.1. Survival, by etiology, of patients with ALF and grades III and IV encephalopathy admitted to the Liver Failure Unit, King's College Hospital, London, 1973–1990. (Reproduced from Hughes *et al.*, *Gut* 1991; **32** (Suppl.): S86–91, with permission from the BMJ Publishing Group).

Table 5.8. Poor prognostic indicators in acute liver failure (For guidance only; other factors must be considered (after O'Grady *et al.* 1988))

Viral and other causes	Acetaminophen (Paracetamol)
Any three of these: 0.96[a]	All three: 0.67[a]
● Etiology (e.g. non-ABC)	● Grade III encephalopathy
● Age <10 or > 40 years	● PT >100 s[b]
● PT >50 s[b]	● Creatinine > 300 μmol/l
● Serum bilirubin > 300 μmol/l	
● Jaundice > 7 days before encephalopathy	
or: 1.0[a]	and/or: 0.95[c]
● PT > 100s[b]	Arterial pH < 7.3

[a]Positive predictive value.
[b]Based on UK thromboplastins. These differ from USA (see section 1.6.2.4) and PTs are not comparable. A PT (UK) of 50 s compares approximately with 20–30 s (USA). A PT (UK) of > 100 s compares approximately to > 30–40 s (USA).
[c]With normovolemia.

5.6.1 Antiviral therapies

Data from controlled clinical trials are lacking. Historically, sparse data from small series and anecdotal reports would seem to indicate only a limited role for antiviral therapies. The consensus view, based mostly on studies in fulminant hepatitis B and B + D, is that virus replication is less than in uncomplicated acute infection and may have ceased before presentation. Also, the toxicity of the conventional antiviral agents such as interferons, gancyclovir, acyclovir among others precluded their use in such seriously ill patients except, justifiably, for herpesvirus infection.

Key Notes Complications of acute liver failure

- Cerebral edema
- Cerebral blood flow variable
- Epileptic seizures
- Systemic circulation abnormalities
- Respiratory distress syndrome
- Renal dysfunction
- Bacterial infection > 80%
- Fungal infection > 25%
- Viral infections (CMV etc.)
- Coagulation abnormalities/hemorrhage
- Metabolic disturbances: acidosis, alkalosis, hypoglycemias, hypo Mg^{2+}, hypo PO_4^{3-}
- Pancreatitis

Interest in using antiviral therapies has been rekindled, at least in fulminant hepatitis B, with reappraisal of the importance of viral factors in pathogenesis and the advent of safer drugs, such as the nucleoside analogs. Regeneration of the liver may require complete cessation of virus replication to terminate the immune attack by the host. In fulminant viral hepatitis, unlike acetaminophen hepatotoxicity, the likelihood of repeated or protracted hepatic injury may impair the ability of the liver to regenerate in time to avoid complications, particularly sepsis.

Individual reports support the safety of nucleoside analogs, such as lamivudine, in fulminant hepatitis B but do not allow objective assessment on their role in favoring liver regeneration and the time delay in achieving maximum effect.

Acyclovir and vidarabine have been used successfully when given early in herpes simplex hepatitis, including during pregnancy, although mortality probably exceeds 90%. Antiviral therapies against CMV, such as ganciclovir, should be considered as pre-emptive therapy following grafting, especially if the CMV status of the donor is unknown at the time of transplantation.

5.6.2 Immunosuppressive therapies

Theoretically, these should benefit cases of ALF attributed to an exaggerated immune attack, such as in fulminant hepatitis B. Clinical trials are lacking and concern remains over using any immunosuppression in viral infections. Corticosteroids have been shown to be of no proven benefit in clinical trials of assorted cases of ALF and may exacerbate complications such as infection. Also, FCH in the graft has been linked to use of high dose immunosuppression in patients transplanted for chronic hepatitis B and, very rarely, for fulminant hepatitis B.

5.6.3 Outcomes following liver regeneration

Spontaneous survivors of fulminant hepatitis A and hepatitis B can anticipate a full clinical and histological recovery and are immune for life. Clearance of HBV and histological resolution is anticipated in the majority of survivors of ALF due to hepatitis B. Whether the exceptions who pursue a chronic course with persistence of virus truly were cases of ALF, is undecided. Recovery from fulminant hepatitis E would be expected to parallel that of HAV, but follow-up data are limited. Data are insufficient for hepatitis C, those with more than one infection (B and C, B + D and C) and other viruses such as the herpesvirus group.

Symptomatic CMV infection, including hepatitis, is common (> 50%) in seronegative recipients who receive a liver transplant from a CMV positive donor, especially if grafted for ALF. Multiple episodes of infection may ensue. Untreated viremias in this group are associated with significant morbidity and mortality. These patients must be considered high-risk for significant infection and should receive prophylactic antiviral therapy from the time of transplantation.

5.6.4 Indications for transplantation

These are discussed in section 6.1.1. The challenge is to assess those patients most likely to recover without recourse to grafting. Indicators for a poor prognosis (*Table 5.8*) give a high positive predictive value (> 95%) in viral ALF, including indeterminate etiology (non-A–E).

5.6.5 Infection of the graft

Recurrent ALF is extremely rare despite common infection in the graft (see section 6.3.1).

5.6.5.1 Fibrosing cholestatic hepatitis

Individual case reports document severe hepatitis in the graft (FCH) following transplantation for ALF due to hepatitis B and despite hepatitis B immune globulin (HBIG) immunoprophylaxis (see sections 3.2.4.1 and 6.3.3).

5.6.6 Auxiliary orthotopic liver transplantation

This has been carried out successfully in ALF.

5.7 Hepatitis A and B immunization

All transplant recipients should be offered immunization against hepatitis A and B. Unfortunately, their efficacies are reduced due to the immunosuppression. The hepatitis B vaccine is not contraindicated in transplant recipients who survived fulminant hepatitis B. High anti-HBs levels may allow discontinuation of HBIG and/or antiviral therapies aimed at preventing graft infection.

5.8 Contact tracing

Screening of close contacts and family members is important to break the chain of infection but remains difficult in clinical practice as almost all cases of ALF due to viral hepatitis are sporadic and transmitted from a symptomless source. In the West, most cases of hepatitis A and E are sporadic; their sources are unknown. In many cases of hepatitis B the source is a symptomless sexual contact typically seropositive for HBsAg and anti-HBe. IVDU may account for some cases of hepatitis B and/or C.

Screening should include the conventional serological markers (*Table 1.1*). Seronegative individuals are susceptible, especially sexual contacts and children. These should be given HBIG (for hepatitis B) or IG (for hepatitis A) and commence a course of active immunization (vaccine) as soon as possible.

5.9 Healthcare personnel

Obvious transmission of hepatitis A, B, C, E and indeterminate etiology from cases of ALF does not seem to occur. Safety recommendations for all attendants include immunization against hepatitis A and B. Care should be taken handling feces and fomites from patients with hepatitis A and E and blood and body fluids in cases of B and C. Pregnant health care attendants should not be exposed to HEV and herpesviruses.

Further reading

Abzug, M.J., Levin, M.J. (1991) Neonatal adenovirus infection: four patients and review of the literature. *Pediatrics* 87: 890–893.

Acharya, S.K., Dasarathy, S., Kumer, T.L., *et al.* (1996) Fulminant hepatitis in a tropical population: clinical course, cause and early predictors of outcome. *Hepatology* 23: 1448–1455.
Prospective serological analysis of causes of ALF from one tertiary referral center in India of 430 consecutive patients (8% of hospital admissions), 1987–1993; one-third were hepatitis A, B,

B+D and two-thirds classified initially as non-A, non-B or drug reactions were HEV and HEV/HCV based on PCR testing.

Alexopoulou, A., Karayiannis, P., Hadziyannis, S.J., *et al.* (1996) Whole genome analysis of hepatitis B virus from four cases of fulminant hepatitis: genetic variability and its potential role in disease pathogenicity. *J.Viral Hepatol.* **3**: 173–181.

Ando, K., Moriyama, T., Guidotti, L.G., *et al.* (1993) Mechanisms of class I restricted immunopathology. A transgenic mouse model of fulminant hepatitis. *J. Exp. Med.* **178**: 1541–1554.
Implicates class I-restricted cytotoxic T lymphocytes and α-interferon in the massive necrosis in the transgenic model of fulminant hepatitis B.

Ascher, N.L., Lake, J.R., Emond, J.C., *et al.* (1993) Liver transplantation for fulminant hepatic failure. *Arch. Surg.* **128**: 677–682.
One year actuarial survival of > 90% of liver grafts in 35 patients (20 with non-A, non-B) transplanted for ALF in California.

Asano, Y., Yoshikawa, T., Suga, S., *et al.* (1990) Fatal fulminant hepatitis in an infant with human herpesvirus-6 infection. *Lancet* **335**: 862–863.

Bahn, A., Hilbert, K., Martine, U., *et al.* (1995) Selection of precore mutant after vertical transmission of different hepatitis B virus variants is correlated with fulminant hepatitis in infants. *J. Med. Virol.* **47**: 336–341.

Baumert, T.F., Rogers, S.A., Hasegawa, K., *et al.* (1996) Two core promotor mutations identified in a hepatitis B virus strain associated with fulminant hepatitis result in enhanced replication. *J. Clin. Invest.* **98**: 2268–2276.

Benador, N., Mannhardt, W., Schrantz, D., *et al.* (1990) Three cases of neonatal herpes simplex virus infection presenting as fulminant hepatitis. *Eur. J. Pediatr.* **149**: 555–559.

Bernuau, J., Rueff, B., Benhamou, J.P. (1986) Fulminant and subfulminant liver failure: definitions and causes. *Sem. Liver. Dis.* **6**: 97–106.
Definitions used predominantly in Europe based on clinical features and abnormal coagulation, overlap with other definitions and not necessarily excluding underlying chronic disease.

Brown, K.E., Tisdale, J., Barrett, J., *et al.* (1997) Hepatitis-associated aplastic anemia. *N. Engl. J. Med.* **336**: 1059–1064.
Retrospective analysis of 10 cases associated with non-A, non-B, non-C hepatitis between 1990 and 1996; seven recovered within 1 year with intense immunosuppression using cyclosporin, corticosteroids and anti-thymocyte globulin.

Buckwold, V.E., Zu, Z., Yen, T.S.B., *et al.* (1997) Effects of a frequent double-nucleotide basal core promoter mutation and its putative single-nucleotide precursor mutations on hepatitis B virus gene expression and replication. *J. Gen. Virol.* **78**: 2055–2065.

Carman W.F., Fagan, E.A., Hadziyannis, S., *et al.* (1991) Association of a pre-core genomic variant of hepatitis B virus with fulminant hepatitis. *Hepatology* **14**: 219–222.
The precore variant is not associated invariably with acute liver failure due to hepatitis B virus.

Carmichael, G.P., Zahradnik, J.M., Moyer, G.H. *et al.* (1979) Adenovirus hepatitis in an immunosuppressed adult patient. *Am. J. Clin. Pathol.* **71**: 352–355.

Chenaerd-Neu, M.P., Boudjema, K., Bernuau, J., *et al.* (1996) Auxiliary liver transplantation: regeneration of the native liver and outcome in 30 patients with fulminant hepatic failure—a multicenter European study. *Hepatology* **23**: 1119–1127.
Complete regeneration of the native liver occurred in seven out of 15 patients with viral hepatitis (all three HAV, four of six HBV and three of four indeterminate). Patients who were young (< 40 years), and showed a hyperacute course had the most favorable outcome.

Dits, H., Frans, E., Wilmer, A. *et al.* (1998) Varicella zoster virus infection associated with acute liver failure. *Clin. Infect. Dis.* **27**: 209–210.
Diagnosis in liver tissue by electron microscopy with fluorescent staining, immunohistochemistry and in situ hybridization.

Donati, M.C., Fagan, E.A., and Harrison, T.J. (1997) Sequence analysis of full length HEV clones derived directly from human liver in fulminant hepatitis E. In: *Viral Hepatitis and Liver Disease* (ed. M. Rizetto, R.H. Purcell, J.L. Gerin, and G. Verne). Edizioni Minerva Medica, Torino, pp. 313–316.
Lack of obvious virulence markers in a full length HEV sequence from fulminant hepatitis E.

Ehata, T., Omata, M., Chuang, W.L., *et al.* (1993) Mutations in core nucleotide sequence of hepatitis B virus correlate with fulminant and severe hepatitis. *J. Clin. Invest.* **91**: 1206–1213.

Fagan, E.A. (1994) Acute liver failure of unknown pathogenesis: the hidden agenda. (Editorial). *Hepatology* **19**: 1307–1312.
Editorial review of unsuspected viral causes of acute liver failure due to non-A, non-B, non-C non-D, non-E.

Fagan, E.A., Ellis, D., Tovey, G., *et al.* (1992) Toga virus-like particles in fulminant sporadic non-A, non-B hepatitis and after transplantation. *J. Med. Virol.* **38**: 71–77.
A prospective study of 99 native livers and 20 grafts with all varieties of acute, fulminant and chronic liver disease. Virus like particles resembling the togaviridae were found in seven of 17 patients with fulminant hepatitis of indeterminate etiology and in all cases with recurrent liver failure following grafting; the hepatitis F syndrome.

Fagan, E.A., Hadzic, D., Saxena, R., *et al.* (1999) Vertical transmission from early pregnancy of a naturally occurring hepatitis A virus variant: evidence for persistent infection in man. *Pediatr. Infect. Dis. J.* **18**: 389–391.
Unusual variant with deletions in 5' UTR detected in a mother and baby who developed severe hepatitis A, 4 weeks after birth.

Fagan, E.A., Harrison, T.J. (1994) Exclusion in liver by polymerase chain reaction of hepatitis B and C viruses in acute liver failure attributed to sporadic non-A, non-B hepatitis. *J. Hepatol.* **21**: 587–591.

Fagan, E., Menon, T., Valliammai, T., *et al.* (1994) Equivocal serological diagnosis of sporadic fulminant hepatitis E in pregnant Indians. *Lancet* **334**: 342–343.
Equivocal IgM anti-HEV test results and low HEV RNA levels in fulminant hepatitis E from India.

Fagan, E.A., Smith, P.M., Davison, F., *et al.* (1986) Fulminant hepatitis B in successive female sexual partners of two anti-HBe-positive males. *Lancet* **2**: 538–540.

Fagan, E.A., Williams, R. (1990) Fulminant viral hepatitis. *Br. Med. Bull.* **46**: 462–480.

Fagan, E.A., Yousef, G., Brahm, J., *et al.* (1990) Persistence of hepatitis A virus in fulminant hepatitis before and after liver transplantation. *J. Med. Virol.* **30**: 131–134.
Persistence of HAV in two patients transplanted for fulminant hepatitis A with infection of the graft without significant sequelae. HAV antigen was detected in the stool during clinical relapse as well as in liver.

Farci, P., Alter, H.J., Govindarajan, S., *et al.* (1996) Hepatitis C virus-associated fulminant hepatic failure. *N. Engl. J. Med.* **335**: 631–634.
Case report of hepatitis C (genotype 1b) in association with post-transfusion hepatitis complicated by coagulopathy, encephalopathy, and massive necrosis of the liver. Temporal rise and fall in HCV RNA levels and agonal seroconversion to anti-HCV antibodies. High HCV RNA levels and low genetic diversity contrast with findings in fulminant hepatitis B.

Feranchak, A.P., Tyson, R.W., Narkewicz, M.R., *et al.* (1998) Fulminant Epstein–Barr viral hepatitis: orthotopic liver transplantation and review of the literature. *Liver Transplant. Surg.* **4**: 469–476.
Review of 16 published cases as well as case report of 22-month-old girl assumed to be immunocompetent.

Feray, C., Gigou, M., Samuel, D., *et al.* (1993) Hepatitis C virus RNA and hepatitis B virus DNA in serum and liver of patients with fulminant hepatitis. *Gastroenterology* **104**: 556–562.
Low frequency in ALF.

Friedt, M., Gerner, P., Lausch, E., *et al.* (1999) Mutations in the basic core promotor and the precore region of hepatitis B virus and their selection in children with fulminant and chronic hepatitis B. *Hepatology* **29**: 1252–1258.
First report of analysis of BCP mutations in children who acquired HBV by vertical transmission.

Gimson, A.E.S., O'Grady, J.G., Ede R.J., *et al.* (1986) Late-onset hepatic failure: clinical, serological and histological features. *Hepatology* **6**: 288–294.
Sub-group with especially poor prognosis.

Gunther, S., Sommer, G., Von Breunig, F., *et al.* (1998) Amplification of full-length hepatitis B virus genomes from samples from patients with low levels of viremia: frequency and functional consequences of PCR-induced mutations. *J. Clin. Microbiol.* **36**: 531–538.

Hasegawa, K., Huang, J., Rogers, S.A., *et al.* (1994) Enhanced replication of a hepatitis B virus mutant associated with an epidemic of fulminant hepatitis. *J. Virol.* **68**: 1651–1659.

Hibbs, J.R., Frickhofen, N., Rosenfeld, S.J., *et al.* (1992) Aplastic anaemia and viral hepatitis: non-A, non-B, non-C? *J. Am. Med. Assoc.* **267**: 2051–2054.

Hughes, R.D., Wendon, J., Gimson, A.E.S. (1991) Acute liver failure. *Gut* **32** (Suppl.): S86–91.

Hussaini, S.H., Skidmore, S.J., Richardson, P., *et al.* (1997) Severe hepatitis E infection during pregnancy. *J. Viral Hepatitis* **4**: 51–54.
Report from the UK of two cases of severe hepatitis E in travelers returning from the Indian subcontinent. Both mothers and babies survived suggesting mortality can be reduced with expert care.

Johnson, J.R., Egaas, S., Gleaves, C.A., *et al.* (1992) Hepatitis due to herpes simplex virus in marrow-transplant recipients. *Clin. Infect. Dis.* **14**: 38–45.

Karayiannis, P., Alexopoulou, A., Hadziyannis, S., *et al.* (1995) Fulminant hepatitis associated with hepatitis B virus e antigen-negative infection: importance of host factors. *Hepatology* **22**: 1628–1634.

Klein, N.A., Mabie, W.C., Shaver, D.C., *et al.* (1991) Herpes simplex virus hepatitis in pregnancy. Two patients successfully treated with acyclovir. *Gastroenterology* **100**: 239–244.

Lee, W.M., Williams, R. (eds) (1998) *Acute Liver Failure*. Cambridge University Press, Cambridge.
First book devoted to the subject.

Liang, T.J., Hasegawa, K., Munoz, S.J., *et al.* (1994) Hepatitis B virus precore mutation and fulminant hepatitis in the United States. A polymerase chain reaction-based assay for the detection of specific mutation. *J. Clin. Invest.* **93**: 550–555.

Liang, T.J., Hasegawa, K., Rimon, N., *et al.* (1991) A hepatitis B mutant associated with an epidemic of fulminant hepatitis. *N. Engl. J. Med.* **324**: 1705–1709.
The variant associated with fulminant hepatitis had two stop mutations (G1898A and G1901A) in the precore region despite seropositivity for HBeAg.

Marsman, W.A., Wiesner, R., Batts, K.P., *et al.* (1997) Fulminant hepatitis B virus: recurrence after liver transplantation in two patients also infected with hepatitis delta virus. *Hepatology* **25**: 434–438.
Cautionary report of two patients transplanted for chronic HBV/HDV infection with fatal recurrent coinfection in their grafted livers despite HBIG immunoprophylaxis (contrast good results by Samuel et al. 1993).

McCaul, T.F., Fagan, E.A., Tovey, G., *et al.* (1986) Fulminant hepatitis: an ultrastructural study. *J. Hepatol.* **2**: 276–290.
Prospective study of ultrastructural changes in liver in 23 patients with acute liver failure including four from hepatitis B and 10 of indeterminate etiology. Comparisons with acetaminophen showed that for viral hepatitis there was more preservation of cell architecture but that polyribosomes were reduced and detached from the cisternae. No virus particles were found for HBV and EBV.

Missale, G., Brems, J.J., Takiff, H. *et al.* (1993) Human leucocyte antigen class I-independent pathways may contribute to hepatitis B virus-induced liver disease after liver transplantation. *Hepatol.* **18**: 491–496.
Case report of closely monitored episode of recurrent HBV infection with progression to FCH in the graft following transplantation for fulminant hepatitis B across ABO blood groups. Detailed immunological analysis of peripheral blood mononuclear cell responses.

Ogata, N., Miller, R.H., Ishak, K.G., Purcell, R.H. (1993) The complete nucleotide sequence of a pre-core mutant of hepatitis B virus implicated in fulminant hepatitis and its biological characterization in chimpanzees. *Virology* **194**: 263–276.

O'Grady, J.G., Alexander, G.J.M., Hayllar, K.M., *et al.* (1989) Early indicators of prognosis in fulminant hepatic failure. *Gastroenterology* **97**: 439–445.
Landmark paper. Clinical and laboratory indicators of adverse prognosis without transplantation in acute liver failure from acetaminophen and other etiologies.

O'Grady, J.G., Schalm, S.W., Williams, R. (1993) Acute liver failure: redefining the syndromes. *Lancet* **342**: 273–275.
Old and new definitions based on prognosis, not excluding underlying liver disease.

Omata, M., Ehata, T., Yokosuka, O., *et al.* (1991) Mutations in the precore region of hepatitis B virus DNA in patients with fulminant and severe hepatitis. *N. Engl. J. Med.* **324**: 1699–1704.

Oren, I., Hershow, R.C., Ben-Porath, E., *et al.* (1989) A common-source outbreak of fulminant hepatitis B in a hospital. *Ann. Intern. Med.* **110**: 691–698.
A nosocomial outbreak of fulminant hepatitis B and renal failure in six patients from Israel in 1986 attributed to contamination of a multidose heparin bottle used to flush heparin lock lines in a hemodialysis unit. Samples were anti-HCV seronegative on retrospective testing (see paper by Liang et al., 1991).

Pollicino, T., Zanetti, A.R., Cacciola, I., *et al.* (1997) Pre-S2 defective hepatitis B virus infection in patients with fulminant hepatitis. *Hepatology* **26**: 495–499.
Retrospective analysis of serum from a mother and son with fulminant hepatitis and comparison with acute uncomplicated hepatitis B and chronic infection showing pre-S2 defective genomes were common in fulminant hepatitis B.

Raimondo, G., Tanzi, E., Brancatelli, S., *et al.* (1993) Is the course of perinatal hepatitis B virus infection influenced by genetic heterogeneity of the virus? *J. Med. Virol.* **40**: 87–90.
Study of HBV heterogeneity in five babies born of carrier mothers; newborns infected with mixed infections cleared their infection whereas those with wild-type HBV became chronic carriers.

Rosh, J.R., Schwersenz, A.H., Groisman, G., *et al.* (1994) Fatal fulminant hepatitis B in an infant despite appropriate prophylaxis. *Arch. Pediatr. Adolesc. Med.* **148**: 1349–1351.

Samuel, D., Muller, R., Alexander, G., *et al.* (1993) Liver transplantation in European patients with the hepatitis B surface antigen. *N. Engl. J. Med.* **329**: 1842–1847.
Classic paper describing good outcomes for fulminant hepatitis B and D co-infection following liver transplantation and long-term (>6 months) immunoprophylaxis with HBIG (contrast paper by Marsman et al. 1997).

Santantonio, T., Mazzola, M., Pastore, G. (1999) Lamivudine is safe and effective in fulminant hepatitis B. *J. Hepatol.* **30**: 551.

Schiodt, F.V., Atillasoy, E., Shakil, O., *et al.* (1999) Etiology and outcome for 295 patients with acute liver failure in the United States. *Liver Transplant. Surg.* **5**: 29–34.
Retrospective analysis from 1994–1996 and 13 centers (12 with liver transplantation); indeterminate etiology (15%), hepatitis B (10%), hepatitis A (7%).

Schultz, D.E., Honda, M., Whetter, L.E., *et al.* (1996) Mutations within the 5′ nontranslated RNA of cell culture-adapted hepatitis A virus which enhance cap-independent translation in cultured african green monkey kidney cells. *J. Virol.* **70**: 1041–1049.

Sterneck, M., Gunther, S., Santantonio, T., *et al.* (1996) Hepatitis B virus genomes of patients with fulminant hepatitis do not share a specific mutation. *Hepatology* **24**: 300–306.
Complete genome analysis of eight patients with fulminant hepatitis B and one FCH after transplantation showed clustering of mutations in BCP/Enh II.

Sterneck, M., Gunther, S., Gerlach, J., *et al.* (1997) Hepatitis B virus sequence changes evolving in liver transplant recipients with fulminant hepatitis. *J. Hepatol.* **26**: 754–764.
Sequence analysis of complete viral genomes on three patients transplanted for chronic HBV/D coinfection who developed FCH showed no unique clustering of mutations; two of three had pre-S2 defective and HBeAg defective HBVs.

Sterneck, M., Kalinina, T., Gunther, S. *et al.* (1998) Functional analysis of HBV genomes from patients with fulminant hepatitis. *Hepatology* **28**: 1390–1397.
A study of functional phenotypes in fulminant hepatitis B showing comparable replication competence and a defect in expression of HBeAg in transfection studies of viral genomes in seven patients with fulminant hepatitis B. Only HBV genomes from the patient with FCH showed high replication competence in vitro that was associated with mutations in BCP (C1766T and T1768A). The presence of the precore stop mutation and T1762/A1764 had no effect on replication competence.

Sterneck, M., Kalinina, T., Otto, T., *et al.* (1998) Neonatal fulminant hepatitis B: structural and functional analysis of complete hepatitis B virus genomes from mother and infant. *J. Infect. Dis.* **177**: 1378–1381.

Tanaka, S., Yoshiba, M., Iino, S., *et al.* (1995) A common-source outbreak of fulminant hepatitis B in hemodialysis patients induced by precore mutant. *Kidney Int.* **48**: 1972–1978.

Trey, C., Davidson, L.S. (1970) The management of fulminant hepatic failure. In: *Progress in Liver Disease.* (eds H. Popper, F. Schaffner). Grune and Stratton, New York pp. 282–298.
Classical clinical definition of fulminant hepatic failure that excludes underlying liver disease.

Vento, S., Cainelli, F., Mirandola, F. *et al.* (1996) Fulminant hepatitis on withdrawal of chemotherapy in carriers of hepatitis C virus. *Lancet* **347**: 92–93.
Two patients with a fulminant exacerbation of chronic hepatitis C following withdrawal of chemotherapy for malignant lymphoma, characterized by low serum levels of HCV RNA, high alanine aminotransferase levels (6030, 3870 IU/l) and massive liver necrosis.

Verboon-Maciolek, M.A., Swanink, C., Krediet, T.G., *et al.* (1997) Severe neonatal echovirus 20 infection characterized by hepatic failure. *Pediatr. Infect. Dis. J.* **16**: 524–527.
Echovirus 20 isolated from stool, nasopharynx and blood within the first 2 weeks of life and rising titers of antibodies in four neonates who presented with a sepsis-like picture with fever, poor feeding and lethargy. Liver failure developed within 1 week of onset of symptoms. Outcomes were variable including full recovery, chronic cholestasis and death from multisystem failure.

von Weizsacker, F., Pult, I., Geiss, K., *et al.* (1995) Selective transmission of variant genomes from mother to infant in neonatal fulminant hepatitis B. *Hepatology* **21**: 8–13.
Three cases of neonatal fulminant hepatitis B with transmission of a subfraction of the maternal pool of HBV.

Williams R. (1996) Fulminant hepatic failure. *Sem. Liver. Dis.* **16**: 341–444.
Volume devoted to acute liver failure, predominantly of the experience at King's College Hospital, London.

Willner, I.R., Uhl, M.D., Howard, S.C., *et al.* (1998) Serious hepatitis A: an analysis of patients hospitalized during an urban epidemic in the United States. *Ann. Intern. Med.* **128**: 111–114.
Five cases of ALF (2%) among 256 symptomatic cases of hepatitis A during an epidemic in Tennessee during 1994–1995.

Yasmin, M., Bollyky, P., Holmes, E., *et al.* (1997) Genotype D variants with high replication efficiency predict good prognosis in fulminant hepatitis B. *Hepatology* **26**: 319A.
Prospective study from the UK of 16 fulminant hepatitis B patients showed genotype D variants with high replication efficiency were associated with A1896, early clearance and spontaneous recovery despite rapid progression of liver disease. Conversely, genotype A variants with low replication efficiency and wild-type (C1858/G1896) precore sequencing predicted death from progressive ALF.

Yotsumoto, S., Kojima, M., Shoji, I., *et al.* (1992) Fulminant hepatitis related to transmission of hepatitis B variants with precore mutations between spouses. *Hepatology* **16**: 31–35.

Chapter 6

Transplantation of Liver, Other Organs and Bone Marrow

6.0 Introduction

Liver transplantation is a rational therapeutic option for irreversible liver damage due to acute liver failure (ALF) and chronic liver disease. Of the approximately 4000 liver transplants carried out annually in the USA, chronic hepatitis C, often with alcoholic liver disease, accounts for up to 30% whereas hepatitis B accounts for around 5%.

Demand for transplantation is likely to increase with improvements in antiviral strategies against recurrence of infection in the graft. Before 1996, 5-year survival rates of patients and grafts following transplantation for chronic hepatitis B (around 50%) or chronic hepatitis C (around 65%) were reduced compared with non-viral chronic liver diseases such as autoimmune hepatitis (around 80% survival at 5 years). Recent data continue to show increasing rates in survival of graft and recipient (typically above 80% at 1 year), especially for hepatitis B. Overall, such improvements can be attributed to the more discerning selection of candidates, advances in surgical techniques, and discriminative use of immunosuppressive agents. In addition, the improved outcomes for recipients with hepatitis B reflect continuing optimization of antiviral therapies and immunoprophylaxis to prevent infection of the graft.

Clinical outcomes and prognosis following grafting for viral hepatitis reflect the impact of the initial infecting agent [hepatitis B virus (HBV) and hepatitis C virus (HCV)] and additional viruses, such as cytomegalovirus (CMV), that may be acquired through the donor organ and multiple blood transfusions. Also, transplant candidates may have coexistent diseases, especially alcoholic liver disease, that influence management and outcome.

The rising demand for organs continues to outstrip the supply of donors. Consequently, increasing numbers of transplant candidates die before a suitable organ becomes available. The shortage of organ donors has necessitated the use, in certain cases, of organs from patients seropositive for anti-HCV antibody and/or previous infection with hepatitis B virus (anti-HBc, anti-HBs). This use has been restricted primarily to patients unlikely to survive before a suitable donor becomes available and/or who are not considered candidates for transplantation, such as for large hepatocellular carcinoma (HCC). Grafting of kidneys and livers from HCV seropositive donors is being considered for recipients already seropositive for hepatitis C.

Key Notes Transplantation and viral hepatitis

Chronic hepatitis C is the etiology for around 30% of transplants annually in the USA

Viral hepatitis may coexist with other liver diseases, especially alcoholic liver diseases

Survival of grafts and recipients with chronic viral hepatitis are comparable to that of non-viral etiologies

6.1 Indications for liver transplantation

In general, candidacy and priority for transplantation are dictated by the severity of liver dysfunction and complications of liver disease rather than etiology, excepting malignancy (HCC). Conversely, virus etiology impacts on survival of graft and recipient. (See also section 5.6.4.)

6.1.1 Acute liver failure

Viral etiology affects the chances of survival without grafting. Spontaneous recovery can occur in around 30–50% of cases of ALF due to hepatitis A or B compared with less than a 10% chance for hepatitis non-A–E. The challenge is to predict early those most likely to recover without grafting. Nevertheless, most patients with ALF in coma (beyond Grade II; *Table 5.2*), regardless of etiology, are listed immediately for liver transplantation unless there are overriding contraindications (*Table 5.8*). Urgency to list is driven by the unpredictable availability of organs and limited window of opportunity to recover from major surgery.

Selection criteria based on prediction of poor prognosis include etiology (see section 5.6.4) and separate viral from non-viral [mostly paracetamol (acetaminophen)] causes of hepatotoxicity. Clinical and laboratory indicators for a poor prognosis give a high positive predictive value (> 95%) of death in viral causes of ALF, including unknown pathogenesis (non-A–E).

Overall, liver histology, obtained via transjugular biopsy, has had little impact on the decision to proceed to grafting, especially for viral hepatitis. Histopathological features of massive necrosis are non-specific and do not discriminate between the viral etiologies or non-viral causes, except in the rare patient with inclusion bodies due to herpesviruses (see Chapter 5).

6.1.2 Chronic liver disease

As with ALF, indications for transplantation (*Table 6.1*) are regardless of virus etiology, levels of viremia, virus genotype and subtype. Instead, indications are based on clinical criteria of end-stage, irreversible chronic liver disease and with cirrhosis of sufficient severity (Child's–Pugh classification, *Tables 1.22*, and *1.23*) to show decompensation with complications of portal hypertension, such as encephalopathy, bleeding from esophageal varices and ascites with risk of spontaneous bacterial peritonitis.

In the USA, around 5% of transplants in adults are due to end-stage cirrhosis attributable to chronic hepatitis B. Patients typically are seropositive for anti-HBe (rather than HBeAg) as well as HBsAg. A significant number (>10%) will be viremic (detectable HBV DNA), by branched chain (b)DNA assay or dot-blot hybridization, and more by sensitive polymerase chain reaction (PCR). Detectable HBV DNA is no longer a contraindication to liver transplantation, but, instead forewarns of an especially high risk of reinfection of the graft.

Patients with end-stage chronic hepatitis C have cirrhosis and anti-HCV antibodies. Viremia (HCV RNA detectable on reverse transcription (RT)–PCR) is almost invariable. Up to 10% of patients who present with 'cryptogenic' (unknown cause) cirrhosis, based on undetectable anti-HCV antibodies and no markers of other viral infections, have HCV RNA detectable in serum and/or liver.

6.1.3 Liver transplantation for hepatocellular carcinoma

This is reserved for the few patients unlikely to have metastases at the time of transplantation. Possible candidates have a small tumor burden (< 5cm total mass) limited to the liver and assessed typically by ultrasound and computed tomography (CT) imaging. In addition, the few are self-selected for being well enough also to endure the long waiting lists—typically more than 1 year in the USA. Meanwhile, some centers

Table 6.1. Indications for transplantation in chronic viral hepatitis

- Selection is regardless of virus etiology
- Indications are for severe end-stage liver disease:
    - Serum albumin <2.5 g/dl
    - Prothrombin time >5 s prolonged
    - Serum bilirubin >5 × normal
    - Recurrent spontaneous bacterial peritonitis
    - Hepatorenal syndrome
    - Uncontrolled bleeding varices (portal hypertension)
    - Major extrahepatic complications (e.g. bone disease)

undertake to reduce temporarily the tumor burden with chemotherapy, cryotechniques and embolization.

Prior to 1996, most patients transplanted for HCC had chronic hepatitis B; graft loss from severe hepatitis B preceded recurrence of tumor. Ultimate outcome and prognosis following transplantation for HCC associated with hepatitis C is determined by recurrence of tumor and spread of malignancy rather than recurrent HCV infection.

In patients with hepatitis B or C in whom HCC is discovered coincidentally at transplantation, outcome is determined by the impact of virus infection on the graft rather than malignancy. Such tumors tend to be small (< 3 cm diameter) and extrahepatic spread less likely than for declared malignancy.

6.2 Contraindications to liver transplantation

These are dictated by the severity of illness and likelihood of not surviving the operation rather than viral etiology. In general, patients with sepsis and multiorgan failure fare poorly with liver transplantation (*Table 6.2*). Most centers will not consider regrafting a patient who has developed severe fibrosing cholestatic hepatitis (FCH; see section 6.3.3) as this is likely to recur in successive grafts.

> **Key Notes** Liver tranplantation and HBV
>
> Virus replication is no longer a contraindication to transplantation for hepatitis B

Table 6.2. General contraindications to liver transplantation

---
- Advanced intercurrent disease (cardiac, pulmonary)
- Sepsis
- Human Immunodeficiency Virus seropositivity
- Severe metabolic derangements:
  e.g. acidosis (pH < 7.3 with normovolemia)
- Extrahepatic malignancy
- Liver tumor (HCC) >5 cm diameter
- Cholangiocarcinoma
- Major psychiatric illness (major psychosis)
- Active intravenous drug use
- Active alcohol abuse

6.3 Rejection and infection of the liver graft—immunosuppressive therapies and outcomes

Episodes of cellular rejection in liver allografts may occur less frequently following transplantation for HBV-related liver disease than for other etiologies. This observation has led to the judicious reduction of immunosuppressive therapy to avert the development of FCH. Presumably, the defects in immune-mediated mechanisms that prevent clearance of HBV also are limited in the rejection process. Conversely, cellular rejection in renal grafts may be enhanced following use of interferon (IFN) for concurrent viral disease.

Prevention of infection of the graft remains a major challenge as long-term survival and quality of life for the patient transplanted because of chronic viral hepatitis become comparable to recipients transplanted for non-viral etiologies.

The selection and doses of immunosuppressive therapies are important, but complex, determinants of graft injury and consequently survival, especially with viral hepatitis. The known adverse effects on virus replication, especially of corticosteroids, must be judged against the protection afforded against acute cellular rejection and other immune-mediated mechanisms of liver injury. Studies on the influence of immunosuppressive agents on virus replication are hampered by the lack of *in vitro* systems for studying HBV and HCV. A corticosteroid responsive element has been documented in the genome of HBV. However, the impact of cyclosporin, FK506 and azathioprine, and, especially, their potential for synergy when used in combinations with corticosteroids, require further study.

6.3.1 Acute liver failure

Data are available for hepatitis A virus (HAV), HBV and hepatitis D virus (HDV). Infection of the graft is mild and viral etiology does not dictate survival of recipient or graft despite the inevitable immunosuppressive therapy that favors enhancement of virus replication (*Table 6.3*).

Table 6.3. Infection of the graft following transplantation for ALF

Hepatitis	Infection of graft	Recurrent ALF in graft	Outcome
A	Common	No	Mild hepatitis and clearance
B	Common[a]	No	Mild hepatitis, some chronic liver disease
B+D	Uncommon	No	Mild hepatitis, some chronic liver disease
C	NK	No	NK
E	NK	NK	NK
F	Common	Yes	Recurrent Alf; hemorrhagic features
Herpesvirus	NK	NK	NK
Adenoviruses	NK	Yes	NK: ALF can occur following insidious hepatitis before grafting

[a]Without HBIG; NK; Not known.

6.3.1.1. Hepatitis A

Graft infection may occur following transplantation for fulminant hepatitis A. ALF does not recur. The clinical course, including timing of clearance of virus from stool and seroconversion from IgM to IgG anti-HAV antibodies, can be similar to uncomplicated hepatitis A, despite immunosuppressive therapy and detectable HAV in liver tissue. HAV RNA and virus antigen have been detected in grafts up to 9 months after liver transplantation for fulminant hepatitis A without clinical symptoms.

Virus infection following transplantation for ALF:

- Related to levels of viremia at transplant
- Can occur with HAV, HBV and HBV+HDV
- Typically mild disease
- Uncommon for HBV with antiviral therapy and hepatitis B immune globulin
- Recurrent ALF does not occur
- Data are insufficient for HCV and HEV

6.3.1.2 Hepatitis B and D

The graft becomes infected in around 20% of cases (*Figure 6.1*) but this is rare if hepatitis B immune globulin (HBIG) and/or lamivudine are used as antiviral intervention.

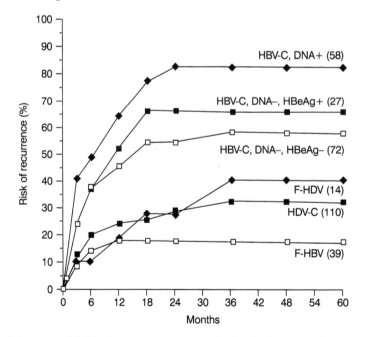

Figure 6.1.   Actuarial risk of recurrence of HBV infection, indicated by the reappearance of HBsAg according to initial liver disease and virus replication before transplantation. HBV-C, HBV-related cirrhosis; F-HDV, fulminant HDV; HDV-D, HDV-related cirrhosis; F-HBV, fulminant HBV and DNA+ and DNA−, denotes the presence and absence of HBV DNA. Values in parentheses represent the number of patients. Reproduced by permission from Samuel, *et al.*, 1993 *New Engl. J. Med.* **329**: 1842–1848. © 1993 Massachusetts Medical Society. All rights reserved.

Liver failure does not recur with graft infection despite immunosuppression that favors virus replication. Outcomes at 5–10 years following transplantation for ALF are good, including with graft infection, as most cases have mild chronic hepatitis on histology. Rapid progression to cirrhosis or development of FCH is rare in the infected graft.

The discordance between the initial severe illness (ALF due to HBV) and relatively mild consequences in the grafts that become infected, is unclear (see section 5.6.5.1) but is related to variations in virus virulence as well as the immunological response of the host. Immune clearance in fulminant hepatitis B may be directed at select quasispecies within a heterogeneous population of HBV transmitted by the symptomless contact, leaving seemingly less virulent species to infect the graft.

6.3.1.3 Hepatitis C, E and herpes viruses

Numbers are insufficient to assess rates of infection and outcomes following grafting for ALF.

6.3.1.4 Unknown pathogenesis (including hepatitis F syndrome)

In the West, the majority of patients categorized as non-A, non-B hepatitis also are non-C, non-D, non-E. In the Indian subcontinent, Sudan, Ethiopia and Egypt (Chapter 2) a substantial proportion of sporadic cases of severe hepatitis probably are hepatitis E.

These indeterminate cases probably reflect various non-viral and viral etiologies. In the West, a subset (up to 10%) of cases of presumed viral, non-A–E are characterized (hepatitis F syndrome) by the presence of virus-like particles (VLPs: 60–70 nm diameter) resembling the *Bunyaviridae* and *Togaviridae* and in native liver and graft with recurrence of ALF typically within 2 weeks of grafting. The clinical illness after grafting resembles a hemorrhagic fever. The abundance of VLPs in the graft in all cases of recurrent liver failure may reflect the adverse effects of immunosuppression on virus replication.

Hepatitis G virus (HGV)/GBV-C has been detected in some cases of ALF of unknown etiology but seems an unlikely pathogen. Most cases have been found, on retrospective analysis, to have received blood products prior to testing.

6.3.2 Chronic viral hepatitis

Despite the potential for almost universal infection of the graft with hepatitis B and C viruses, survival is determined by etiology of the disease simply because effective antiviral strategies are available to prevent HBV, but not HCV, infection of the graft.

> **Key Notes** Graft infection by HBV and HCV
> Chronic hepatitis B: graft infection rates rise with level of viremia (HBV DNA)
> Chronic hepatitis C: graft infection is universal regardless of level of HCV RNA

The role of other viral factors, such as predominance of genotype, distribution of quasispecies and possible variations in virulence under the influence of immuno-suppression, remains unclear (see section 6.3.2.2). As donors and recipients are not tissue matched for HLA haplotypes, the extent of HLA match/mismatch will affect the ability of MHC-restricted cellular immune responses to eliminate liver cells infected by viruses such as HBV and HCV.

6.3.2.1 Hepatitis B and B + D infection

Until around 1996, most liver transplant centers were reluctant to transplant patients with chronic hepatitis B because survival was poor; 55% died within 60 days. There-after, severe infection with loss of the graft was common, especially with detectable HBV DNA prior to grafting (and regardless of HBeAg versus anti-HBe). In fact, out-comes were worse than for hepatitis C.

The instigation of effective antiviral strategies against graft infection with HBV has improved outcomes to levels (above 80% 1-year survival) comparable to those seen with other (non-viral) etiologies. Nevertheless, clinical and histological outcomes remain unpredictable when the graft becomes infected. Progression from mild non-specific hepatitis (historically called chronic persistent hepatitis: CPH) to cirrhosis can be complete within months in the immunosuppressed transplant recipient and con-trasts with the more indolent course prior to grafting.

In 1993, a pivotal study by Samuel *et al.* from 17 European centers monitored impact of HBIG immunoprophylaxis (less, or more, than 6 months) for up to 3 years follow-ing transplantation in 334 recipients transplanted for ALF-HBV and chronic hepatitis B. Results showed that HBV DNA level (by dot-blot hybridization) and/or HBeAg were predictive factors for recurrence of hepatitis B following transplantation. In patients with undetectable HBV DNA at grafting, long-term (>6 months) HBIG following transplantation reduced recurrent infection to around 60%, compared with 70–80% with detectable HBV DNA at the time of grafting.

Recurrent infection with HBV overall was less (around 70%) following transplantation for HDV- with HBV-related cirrhosis in the absence of HBIG therapy (see *Figure 6.1*). Typically HBV DNA levels are low or undetectable in transplant candidates with dual infections. Also, HDV exerts an inhibitory effect on HBV replication. The fate of the graft (and recipient) infected with HDV and HBV is improved compared to that with HBV alone: FCH is less common and significant numbers show eventual clearance of viruses (undetectable HBV DNA and HDV RNA) with seroconversion from HBsAg to anti-HBs.

6.3.2.2 Chronic hepatitis C

Infection of the graft is universal with viremia regardless of level of HCV RNA and genotype, but remains clinically and histologically mild in around 50% of cases (*Table 6.4*). HCV infection of the graft is evident within 2–4 weeks of transplantation and may

Table 6.4. Histological outcomes following HCV infection in the graft ($n = 130$) (Gane et al., 1996)

Severity of hepatitis in graft	n	%	Months of follow-up mean (range)
None	15	12	20 (6–103)
Mild	70	54	35 (6–130)
Moderate	35	27	35 (6–127)
Cirrhosis	10[a]	8	51 (24–138)
Graft loss	5 of 10[a]	50	32–106

[a] Liver failure developed in five out of 10 patients with HCV-induced cirrhosis following liver transplantation between 32 and 106 months following grafting; three listed for a second liver transplant (two grafted), and two died from decompensated cirrhosis at 61 and 84 months.

be confused with cellular rejection. Early acute infection is characterized by a mild lymphocytic infiltrate with interface hepatitis (piecemeal necrosis) and acidophilic bodies that may indicate apoptosis. Hepatocytes undergoing early necrosis typically show minimal ballooning. Damage to bile ducts is mild, in contrast to that seen with acute cellular rejection. The hepatitis becomes more obvious over 2–3 months with the typical inflammatory infiltrates in portal tracts and lobules. Thereafter, the necro-inflammatory activity tends to improve, but persists in most cases. Around 10–20% of infected grafts develop progressive liver disease (fibrosis, even cirrhosis) within two years of transplantation (*Table 6.5*).

Although overall cumulative patient survival rates are similar (around 70% at 5 years) for recipients transplanted for hepatitis C versus other diseases, follow up beyond 5 years may be necessary to show discordance.

Indices, such as age, ethnic origin, choice of immunosuppression (cyclosporin versus FK506; data are insufficient to assess OKT3), including adjunct corticosteroids to treat acute cellular rejection and number of rejection episodes, do not predict development of severe hepatitis C post-grafting. Similarly, laboratory measures such as levels of serum alanine aminotransferase (ALT) and HCV RNA, and level of HLA mismatch for class I antigens between donor and recipient, also are unhelpful. Serum ALT levels remain within the normal range in around 50% of graft recipients with severe

Table 6.5. Comparison of histological severities in grafts with scheduled biopsies at 1 and 5 years after transplantation (after Gane et al., 1996)

Severity of hepatitis in graft	HCV positive at 1 year		HCV positive at 5 years		HCV negative at 5 years	
	n	%	n	%	n	%
None	7	8.5	2	5.1	70	76.9
Mild	51	62.2	20	51.3	11	12.1
Moderate	24	29.3	9	23.1	5[a]	5.5
Cirrhosis	0	0.0	8	20.5	5[a]	5.5
Total	82	100.0	39	100.0	91	100.0

[a] Of these 50% were related to HBV.

recurrent hepatitis C, including cirrhosis. Transplant candidates with high levels of serum ALT and bilirubin may be at some risk but other factors confound correlation.

HCV RNA levels typically are very high [sometimes exceeding 10^{11} genome equivalents (geq)/ml] following grafting and may continue to rise presumably due to the adverse effect of immunosuppressive therapy on virus replication. High levels of HCV RNA (> 10^7 geq/ml) prior to grafting are associated with especially high levels afterwards. The clinical significance of this relationship is unknown because level of viremia *per se* does not predict outcome. Every spectrum of liver disease and histological activity (HAI) has been recorded with low, as well as high, levels of serum HCV RNA. The HAI tends to fall in the months following transplantation in recipients without obvious acute cellular rejection, whereas HCV RNA levels tend to rise, sometimes to 5–10 times their level pre-transplantation. Genotype 1b has been implicated as causing especially severe liver disease with and without transplantation. The spectrum of diversity within a quasispecies as well as levels of virus replication seem to vary according to virus genotype. Consequently, the quality and quantity of the immune response probably varies and may account, at least in part, for such adverse outcomes. Other possibilities include the relation between genotype and level and control of virus replication and the bias in reporting of predominant genotypes in a population.

The impact of concurrent viral infections, such as HBV, CMV and GBV-C/HGV, requires evaluation. Preliminary data suggest that HCV RNA levels and outcomes are comparable at 2 years after grafting with and without HGV (GBV-C) infection. Concurrent (detectable HBV DNA), or previous exposure (anti-HBc seropositive) to HBV is common in patients transplanted for hepatitis C and should be included in analyses of outcomes.

6.3.3 Fibrosing cholestatic hepatitis

A few patients grafted for chronic hepatitis B or C present with marked cholestatic jaundice and rapid clinical deterioration progressing to ALF within weeks of transplantation. FCH can occur following grafting for chronic hepatitis B in recipients considered 'low-risk' for infection (anti-HBe seropositive without detectable serum HBV DNA by dot-blot hybridization) as well as high-risk recipients (HBeAg/HBV DNA seropositive at grafting). The risk of developing FCH is reported to be greater with HBeAg-negative variants of HBV. However, these variants are common among transplant candidates, the majority of whom are anti-HBe positive.

Clinical features of FCH:

- Can follow grafting for chronic hepatitis B or C
- Uncommon with HBV and HDV co-infection
- Independent of HBV DNA level in transplant candidate
- Rapid deterioration (fulminant presentation):
  - Markedly prolonged prothrombin time
  - Encephalopathy, liver failure
- Graft loss in typically <3 months
- Almost invariably fatal

Pathogenetic mechanisms remain unclear. Some of the histopathological features of FCH due to HBV resemble those seen in a transgenic mouse model of chronic HBV infection. In contrast with immunohistochemical features of ALF due to hepatitis B, the combination of extensive expression of viral antigens (HBsAg, HBcAg) and disproportionately mild cellular infiltration would suggest that damage to hepatocytes occurs predominantly by cytolytic and non-cellular immune mechanisms, that are independent of HLA class matching.

Laboratory findings in FCH:

In serum

- Markedly elevated bilirubin levels
- Disproportionately low elevation AST/ALT
- High levels HBV DNA or HCV RNA

Histological features (extensive)

- Cytoplasmic staining for HBsAg with HBV
- Nuclear staining for HBcAg and HBeAg with HBV
- Periportal fibrosis
- Extension of bile duct epithelium into acinus
- Cellular and canalicular cholestasis
- Ballooning degeneration of hepatocytes
- Cytolytic hepatocellular necrosis with lobular collapse
- Ground glass transformation of hepatocytes with HBV
- Mild–moderate mixed inflammatory cell infiltrate
- Rapid development of fibrosis, cirrhotic nodules

HBIG and hepatitis B vaccine do not prevent development of FCH (see immunoprophylaxis, below). The presence of hepatitis delta antigen (HDAg) in the graft is associated with less severe liver disease. FCH due to HBV has become rare with the more judicious use of immunosuppressive therapies and especially the early withdrawal of corticosteroid therapy.

A picture resembling FCH, but sometimes without fibrosis, has been described in up to 1% of graft recipients with hepatitis C but data remain limited. This accelerated course contrasts with chronic HCV infection in the native liver that often follows an indolent course, taking years to present with cirrhosis and decompensated liver disease. In contrast to chronic rejection, proliferation of bile ducts is common and widespread fibrous septa develop as the liver disease progresses to chronic hepatitis. The inflammatory activity remains mild, suggesting a direct cytolytic effect of virus.

HCV RNA levels rise almost invariably following transplantation, regardless of outcome. Why some patients develop FCH due to HCV, remains obscure. Others may show minimal damage of the graft despite HCV RNA levels exceeding 10^{11} geq/ml. The density of inflammatory infiltrate typically is disproportionately low for the degree of hepatocellular necrosis and may indicate a cytopathic role for the viruses.

Graft loss from FCH due to HBV and HCV is almost invariable including following re-transplantation. Attempts at reducing the high levels of virus replication with IFNs have been unsuccessful. Individual publications document successful control of virus replication with use of nucleoside analogs for FCH related to hepatitis B.

6.4 Preventing infection of the liver graft

Historically, approaches focused on hepatitis B and involved passive immunoprophylaxis with polyclonal immunoglobulins (HBIG) given from the anhepatic phase of transplantation surgery.

Until the mid 1990s, IFN-α-2b, the main antiviral therapy, was reserved for treating hepatitis B and C infections that recurred following transplantation. Overall, eradication remains unlikely once HBV and HCV are established in the graft. Also, side-effects are considerable and concern remains over the tendency to precipitate cellular rejection.

Unfortunately, there is no consensus as to when and how to initiate and optimize antiviral strategies before and after grafting. Subsequently, protocols for monitoring and definitions of recurrence differ between centers. HBsAg is a late marker of recurrent hepatitis following transplantation and usually indicates established HBV infection. Detection of viral nucleic acids in serum is preferable but the prognostic value depends on sensitivity of assays and frequency of monitoring.

6.4.1 Passive immunization

Parenteral administration of HBIG, containing high titers of polyclonal anti-HBs and anti-HBc antibodies, has been the mainstay immunoprophylaxis against graft infection with HBV and the major intervention responsible for improved outcomes for patients transplanted for hepatitis B. No immunoprophylaxis is available for hepatitis C.

HBIG doses typically are administered via intramuscular or intravenous routes. The aim is to maintain levels of anti-HBs above an arbitrary titer of 100 IU/l (100 mIU/ml), and preferably above 500 IU/l.

There is no consensus as to the optimal strategy for administering HBIG (*Table 6.6*). Some centers choose to administer HBIG in regular dose and intervals rather than titrate according to serum anti-HBs levels. Neither approach is ideal. Anti-HBs levels may fluctuate significantly within, and between, patients for many reasons including pharmacokinetic principles and levels of viremia. Close titration of dose to recent serum level of anti-HBs is difficult to achieve in clinical practice. Experience is less with administration of monoclonal anti-HBs and results must be interpreted with caution. Infection of the graft may be more common than with polyclonal anti-HBs (HBIG) because fewer epitopes are selected for neutralization. HBV variants, which can escape the neutralization by antibodies, may be selected by the monoclonal or polyclonal anti-HBs, and have been documented as a cause of recurrence in transplant patients treated with HBIG (see section 3.2.5.4).

Table 6.6. HBIG following transplantation for hepatitis B

- Administer minimum of 6 months from anhepatic phase of grafting
- Duration of therapy unknown, probably indefinite
- Delays, rather than prevents, infection of the graft with HBV
- Is ineffective in clearing HBV once the graft becomes infected

Ideally, titrate doses against anti-HBs levels in each patient

- Maintain anti-HBs levels above minimum of 100 mIU/ml (ideally > 500 mIU/ml)
- Regular, close monitoring is essential with individual protocols
- Anti-HBs levels may fluctuate greatly
- Increasing HBIG dose requirements may herald increasing virus replication and graft infection, despite undetectable HBsAg and HBV DNA (by hybridization) in serum
- Also monitor HBsAg and HBV DNA following transplantation
- HBsAg and HBV DNA may coexist with anti-HBs especially with the emergence of 'antibody escape' variants

HBIG is expensive, in limited supply and can double the cost of liver transplantation. HBIG is not licensed for routine use in liver transplant recipients in the USA. Side-effects with the intravenous preparation include allergy (anaphylaxis, urticaria), myalgia and back pain but long-term sequelae have not been documented. Accumulation of mercury has occurred as part of the manufacturing process but this is excluded from preparations of HBIG after 1999. Until the late 1990s, HBIG was predominantly of polyclonal origin and produced by harvesting plasma from patients who had recovered from natural infection and from vaccinees following booster doses of plasma-derived hepatitis B vaccine. The increasing use of recombinant vaccine to boost anti-HBs levels results in anti-HBs which reacts with a restricted set of epitopes (the *a* determinant) and this may impact on the frequency of emergence of 'escape' mutants (see section 3.2.5.4).

6.4.2 Prophylactic antiviral therapies

The shift towards using antiviral therapies prior to grafting was fuelled by the observation that, in chronic hepatitis B, low or undetectable HBV DNA levels (on dot-blot hybridization) prior to grafting correlated with good survival of the graft and recipient. Early attempts were made to reduce HBV DNA with IFN-α-2b in doses of 1.5–10 MU three times a week for a minimum of 8 weeks prior to grafting. Results showed a reduced rate of infection of the graft when serum HBV DNA was undetectable by dot-blot hybridization (25%) at transplantation compared with patients who remained HBV DNA seropositive (80%). At that time, these results were comparable to outcomes in patients given HBIG (without IFN) for a minimum of 6 months after grafting.

The optimal duration of antiviral therapy is difficult to predict given the uncertainties of waiting for a donor. In end-stage liver disease, IFNs are tolerated poorly and side-effects are substantial (see section 1.13). Patients with cirrhosis can decompensate with development of ascites, encephalopathy and sepsis. Deterioration of liver function may ensue with 'successful' treatment, HBV DNA becoming undetectable and sero-conversion from HBeAg to anti-HBe. Thrombocytopenia and leukopenia can increase with ensuing risks of bleeding and infection.

The use of IFNs to prevent graft infection with HBV has been surpassed by HBIG (see above) and the recent introduction of the nucleoside analogs (see section 6.6). The subsequent improvements in survival have led to renewed enthusiasm for transplanting patients with hepatitis B in centers previously opposed to such a policy. In contrast, different antiviral strategies must be used for HCV as infection of the graft is universal with viremia. No true prophylaxis is available. IFN-α-2b and ribavirin, at best only can 'control', rather than prevent and eradicate, infection in the immunosuppressed recipient.

Ganciclovir and famcyclovir have antiviral activity against HBV as well as the herpes viruses. Their limitations are severe following transplantation due to renal toxicity and the need for intravenous administration. Lamivudine may be used to reduce (HBV) viral load prior to transplantation and in an attempt to prevent infection of the graft. Resistant variants arise in some patients (section 3.8.2.3) and combination therapy with HBIG also is available.

6.4.3 Vaccines

The hepatitis A and B vaccines would seem ideal following transplantation for ALF-HAV and ALF-HBV respectively; a secondary immune response should occur, with rapid rise in IgG antibodies. This should be of especial benefit in the setting of ALF-HBV to prevent early infection of the graft and to obviate the cost and inconvenience of repeated injections of HBIG.

Unfortunately, more than 50% of transplant candidates make insufficient levels of anti-HBs to be considered protected from future infection. Further, immunosuppression reduces the efficacy of immunization following transplantation, including in some recipients deemed good responders prior to grafting.

Vaccines containing HBsAg, including with the pre-S2 region, have been used in an attempt to boost the immune response of the host in patients transplanted for chronic hepatitis B. Overall, results are disappointing and difficult to interpret because most protocols have included HBIG.

Key Notes All transplant candidates should be:
- Screened for HBV and HCV
- Immunized against hepatitis A and B

6.5 Management of acute liver failure

Normal immunoglobulin (NIG) containing anti-HAV antibodies has been administered anecdotally in some cases of ALF-HAV. This need is debatable because the outcome of infection in the graft is mild with clearance of infection. HBIG seems to protect against most reinfections of the graft following transplantation for ALF-HBV.

The relative rarity of ALF has precluded controlled studies on efficacies of antiviral therapies, immunosuppressive drugs and other strategies. Early anecdotal reports used IFNs, acyclovir or adenine arabinoside monophosphate (ARA-AMP) (see section 3.8.2.4) prior to grafting, mostly for fulminant hepatitis B and non-A, non-B hepatitis. These showed no obvious benefit on outcome but results were confounded by selection of a heterogeneous population with unlikely spontaneous liver regeneration and deaths from non-viral complications such as uncontrolled cerebral edema.

Individual case reports suggest that lamivudine is safe in fulminant hepatitis B although some concern remains over the incidence of pancreatitis. Use of HBIG and a nucleoside analog seems sensible as residual virus replication, albeit at low level, may be one factor in preventing spontaneous regeneration of the liver in fulminant hepatitis B.

One management strategy that requires formal study could be to use HBIG and a nucleoside analog from admission to transplantation or spontaneous recovery. Following transplantation for ALF-HBV, long-term antiviral therapy may not be required as most survivors clear HBV. Accordingly, HBIG could be terminated within 6 months of grafting with the patient maintained on the nucleoside analog until immunization with hepatitis B vaccine results in protective levels of anti-HBs (> 100 MIU/ml).

Anecdotes based on survival report efficacy of acyclovir or ganciclovir administered early in the rare instances of infection of the graft with herpes simplex virus (HSV) and varicella zoster virus (VZV). Ganciclovir can reduce the severity of the primary infection due to CMV in the setting of a mismatched graft (seronegative recipient receiving a graft from a seropositive donor).

Data are lacking for efficacies for ALF due to hepatitis A, C or E.

6.6 Chronic hepatitis B and B+D

HBIG has been used successfully to reduce the rate of graft infection when given during the anhepatic phase of the hepatectomy and maintained after grafting for a minimum of 6 months (*Figure 6.2*). Early studies showed administration of polyclonal HBIG for a minimum of 6 months following transplantation for chronic hepatitis B reduced the actuarial risk of recurrence in the graft at 3–5 years from > 70% to around 35–50%. The presence of HDV seems to reduce further the actuarial risk of recurrence (to around 10–23%) and improve the chances of clearance of both viruses if HBIG is given for a minimum of 6 months following grafting.

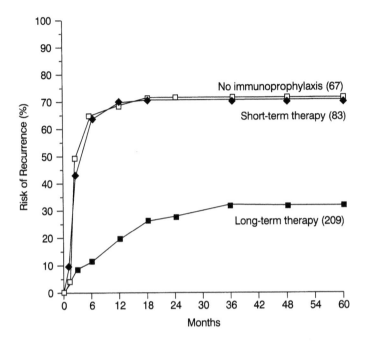

Figure 6.2. Actuarial risk of recurrence of HBV according to the duration of passive immuno-prophylaxis with HBIG. The risk of recurrence in the group given long-term therapy was significantly lower than in either of the other groups. Values in parentheses represent the number of patients. Reproduced by permission from Samuel *et al.*, 1993 *New Engl. J. Med.* **329:** 1842–1848. © 1993 Massachusetts Medical Society. All rights reserved.

A popular protocol used at many transplant centers involves intravenous administration of HBIG (10 000 IU) during the anhepatic phase of transplantation and daily (intravenously or intramuscularly according to the center) for 7 days and thereafter every month according to anti-HBs levels. Most centers advocate maintenance of minimal levels of anti-HBs of above 100 IU/l indefinitely, whereas some suggest 250 IU/l or 500 IU/l as the lowest level that prompts a further dose.

Lamivudine with and without HBIG is under clinical trial in transplant candidates and in recipients who become HBV DNA positive after grafting. Published reports, typically on less than 20 patients, show encouraging results with significant reductions in HBV DNA levels and minimal side-effects. Individual patients have shown sufficient improvement in liver synthetic function to warrant removal from the immediate transplant list.

An open label study from ten USA and Canadian centers involved 38 patients who proceeded to transplantation with a minimum of 6 months follow-up. Lamivudine monotherapy (100 mg orally per day, beginning 0–177 days prior to transplantation) was well tolerated and effectively suppressed HBV DNA to undetectable levels in around two-thirds of patients. Follow-up at 1 year in 11 patients showed that nine (82%) had undetectable HBV DNA. YMDD variants were detected in five patients (three prior to grafting) but without clinical deterioration.

Most authorities agree that HBIG with lamivudine holds the most promise for preventing graft infection following transplantation for hepatitis B. The high cost of HBIG and limited supplies have driven attempts to optimize antiviral strategies. Accordingly, in 1998, the National Institutes of Health (NIH) set the goal to reduce the recurrence of HBV following grafting to below 5%. Multicenter studies extending beyond 5 years are underway across the USA to assess the efficacy of HBIG and lamivudine in reaching this goal.

6.6.1 Escape variants following transplantation

Long-term use beyond 6 months of (polyclonal and monoclonal) HBIG has been associated with emergence of HBV variants with amino acid substitutions in the *a* determinant of the surface gene that allow escape by reducing the affinity of HBIG to bind to HBsAg. Often HBV DNA levels are low prior to grafting and remain below pretreatment values; also, most of the patients remain symptomless. However, the clinical course can be unpredictable, especially in the immunosuppressed transplant recipient. Individual case reports document severe exacerbations in hepatitis with marked elevations (>10 times baseline level) in liver enzyme tests and prothrombin times, as well as HBV DNA.

Mutations affecting the *a* determinant of HBsAg, particularly the aspartic acid at residue 144 (usually substituted by alanine) and the glycine at residue 145 (usually substituted by arginine), have been associated with antibody escape in transplant recipients. Maintenance of high antibody titers may help to prevent selection of escape mutants. HBIG made using plasma from naturally immune individuals is preferable to that from vaccine recipients, because the former should contain a broader spectrum of antibody specificities, including against pre-S epitopes.

The nucleoside analogs famciclovir and lamivudine have been found to select HBV variants with mutations causing amino acid substitutions in the polymerase gene, including in transplant recipients. Lamivudine resistance is associated with changes in the YMDD motif (see section 3.8.2.3), as well as mutations in the 'B domain', further upstream. Resistance to famciclovir also is associated with substitutions in the B domain. The surface open reading frame is overlapped entirely by that encoding the polymerase (*Figure 3.1*) and sequences encoding the *a* determinant of HBsAg and the B domain of the polymerase overlap. Despite this proximity, thus far there are no examples of a situation (as for human immunodeficiency virus (HIV)) where resistance to one antiviral agent increases sensitivity to another. Nonetheless, experience of combination therapy for HIV infection provides lessons for HBV. Sequential monotherapy may select for multiply resistant variants.

There is no consensus on how best to 'rescue' patients who become symptomatic with HBV variants resistant to nucleoside analogs such as lamivudine and famciclovir. In some cases, discontinuation of the nucleoside analog or administration of a nucleotide analog (such as adefovir dipivoxil) has resulted in reversal of abnormal tests and clinical remission on reversion to wild-type HBV. Unfortunately, some patients have pursued a relentless course and died, especially with sequential use of famciclovir after lamivudine that targets similar regions of the DNA polymerase. Rescue with drugs such as adefovir dipivoxil, that target other regions of the polymerase gene of HBV seems sensible although large scale data on their efficacy are not available.

Management following grafting for hepatitis B/ and D or C:

- Serial levels of HBV DNA (dot-blot), HCV RNA (RT–PCR)
- Conventional markers such as HBeAg and anti-HCV may be unreliable
- Multiple infections may co-exist
- Reduce to a minimum immunosuppressive therapies, especially corticosteroids
- Vigilant surveillance for bacterial and fungal sepsis
- Serial urine cultures for CMV
- Herpesvirus infections can mimic rejection and sepsis
- Use acyclovir in high-risk patients for HSV and VZV; ganciclovir for CMV

6.7 Chronic hepatitis C

The decision to treat infected graft recipients remains controversial. Infection of the graft is almost invariable but long-term outcome is variable and unpredictable. Up to 50% of transplant recipients will have only mild inflammatory activity on liver histology 3–5 years later. Data on the outcome of long-term follow-up (> 5–10 years) studies are too few for comment. Elimination of HCV RNA is difficult to achieve following grafting because levels of viremia tend to be high and continue to rise. Trials are underway to assess the efficacy of IFN-α with and without ribavirin.

Protocols and doses (3 MU of IFN-α three times a week for 6 months and ribavirin up to 1500 mg daily) are similar to those without the transplant setting. Overall, results are disappointing on achieving clearance of HCV RNA. Levels of HCV RNA fall with IFN-α but return to baseline almost invariably following cessation of therapy. A few patients achieved longer durations of normalization of ALT levels following cessation of therapy with a combination of both drugs.

Monitoring HCV infection after transplantation:

- Infection of the graft occurs in >90%
- Detection of HCV RNA levels remains the gold-standard
- HCV RNA levels tend to rise
- Anti-HCV antibodies may become undetectable (EIA, RIBA)
- Seroconversion to anti-HCV antibodies may not occur, or be delayed many months if the virus is acquired from an infected donor or blood transfusion

Most studies can be criticized for having too few patients and using normalization of serum ALT levels rather than clearance of HCV RNA by sensitive (RT–PCR) techniques. This is especially relevant to monitoring effects of ribavirin. Also, doses do not take into account body weight or surface area, which correlate with distribution of drug and the especially high, often rising, levels of HCV RNA following transplantation.

Controversies with antiviral strategies following transplantation for chronic hepatitis B and C; there is no consensus on:

- Definition of recurrence of HBV and HCV
- Optimal timing for initiating antiviral therapy
- Optimal doses and duration of therapies
- Who to treat:
  infection of the graft with HCV is > 90% but half have mild hepatitis at 5 years
  infection of the graft with HBV is 60–80% with unpredictable outcome
- Toxicity risks with long-term therapies
- Increased risk of rejection episodes

6.8 The infected graft

HBIG and hepatitis B vaccine are ineffective in clearing HBV from the graft once HBV DNA becomes detectable in serum and/or liver. Detection of HBsAg in hepatocytes by immunohistochemical techniques is diagnostic of HBV infection. Following grafting for hepatitis D, HDAg may become detectable in the graft before HBsAg.

Serum ALT levels may remain within the normal range in around 50% of recipients with grafts infected with HCV. Histopathological features of HCV infection in the graft and cellular rejection overlap sufficiently often to make impossible discrimination between the two diseases.

Eradication of virus is more difficult to achieve in the post-transplant setting of immunosuppression and higher than baseline viremias. The risks of treatment and side-effects of antiviral therapy must be balanced against likely success in achieving eradication of the virus.

6.8.1 Regrafting for viral hepatitis

Graft loss exceeds 50% for second and subsequent transplants for presumed viral, non-A–E and for FCH-HBV. Outcomes are more variable following retransplantation for chronic hepatitis B and chronic hepatitis C. Around 5–10% of grafted patients require regrafting for severe chronic HCV infection that can develop within 1–2 years following transplantation. Histological features show an accelerated rate of progression to advanced fibrosis and cirrhosis. Graft loss from cirrhosis has been reported as early as 1–2 years after transplantation. While graft loss due to hepatitis C is not invariable following successive transplantations for severe hepatitis C, survival rates are low typically due to additional complications such as renal failure, sepsis and acute respiratory distress syndrome.

6.9 Other infections following transplantation

Infectious complications are a leading cause of death. Infections with CMV and bacteria are the most common complications following liver transplantation.

6.9.1   Cytomegalovirus

This is the most important viral pathogen infecting the graft following organ transplantation for non-viral disease. Symptomatic CMV infection, including hepatitis, occurs in over 25% of transplant recipients especially (> 50%) in seronegative recipients who receive a liver transplant from a CMV-positive donor. Multiple episodes of infection may ensue. The majority of symptomatic infections involve the liver and lung. Untreated viremias in this group are associated with significant morbidity and mortality. These 'mismatched' patients must be considered high-risk for significant infection and should receive prophylactic antiviral therapy from the time of transplantation. Also, there is an increased risk of allograft rejection and superinfection with other opportunistic agents, such as *Pneumocystis carnii* and *Cryptococcus neoformans*.

Infection is uncommon in seronegative recipients receiving an organ from a donor seronegative for CMV. Blood and blood products including white blood cells should be screened for CMV. Significant infection can follow administration of seropositive donations into a seropositive (as well as seronegative) recipient. Reactivation of latent infection or with new subtypes is more common than primary infection because many patients have pre-existing antibody. The hepatitis usually occurs in the setting of multisystem failure and sepsis. An acalculous cholecystitis has been described.

Use of IFN therapy to reduce shedding of CMV has led to conflicting results and concern remains of the likelihood of precipitating acute cellular rejection of the graft.

Risk factors for developing symptomatic CMV infection following liver transplantation; Mayo clinic 1988–1990 (after Paya, *et al.*, 1993):

- Receipt by a seronegative patient of an organ from a CMV seropositive donor
- Severe liver disease (ALF) requiring grafting—prolonged prothrombin time
- OKT3 immunosuppression
- Hepatic artery thrombosis

6.10 HBV- and HCV-positive liver donors

HBV and HCV infections are more likely to be acquired via a positive organ donor than by blood transfusion. The risk of acquiring HCV from a blood transfusion or blood products has fallen considerably since the introduction of routine screening of blood donors in the West in 1990. Current estimates of risk are around 1 : 62 000 per unit of blood despite donor screening for anti-HCV antibodies by second generation enzyme immunoassay (EIA).

Organ donors should be screened for HCV as well as HBV and CMV, among other viruses. Assays are being developed to detect GBV-C/HGV but the significance of a positive test remains to be defined. Comprehensive screening is difficult to achieve in clinical practice given the need to test for panels of antibodies and high false reactivity rates of 'emergency' testing. A negative anti-HCV antibody test does not exclude HCV infection. Detection of HCV RNA by RT-PCR is impractical in the context of emergency transplantation. HCV RNA and HBV DNA have been detected in some cases in the donor serum and liver despite undetectable serological markers. Reactivation in the graft can occur under these conditions but, in general, the ensuing hepatitis B is mild. Diagnosis depends on detection of nucleic acids because conventional serological markers may remain undetectable.

The continuing shortage of organ donors has lead to a reappraisal of the use of organs from donors positive for HBV, HCV and CMV. Most organ procurement centers restrict the use of organs from positive donors to selected recipients. Importantly, the requirement of a donor liver to be free of HBV and HCV is desirable but not essential for a long-term, successful outcome.

6.10.1 Anti-HBc positive donors

Up to 1% of organ donors in the USA and UK have serological markers for past exposure to HBV (anti-HBc and anti-HBs) or HCV [seropositive using sensitive second-generation EIAs and recombinant immunoblot assay (RIBA)]. This prevalence is five to ten times higher than for blood donors in the UK. Hepatitis B infection (*de novo*) has been documented in 5–10% of liver transplant recipients grafted for non-hepatitis B-related conditions. Transmission is assumed to have occurred at, or around, the time of transplantation from transfusions (blood and blood products) and/or the donor liver.

Donors with anti-HBc only (seronegative for HBsAg and anti-HBs):

- This denotes previous exposure to HBV
- Low levels of virus replication (HBV DNA) can be detected in the liver/ extrahepatic sites in <5 % of cases
- Some tests are non-specific, false positives under emergency testing conditions

Before the early 1990s, donors with normal liver tests who tested positive for anti-HBc alone (seronegative for HBsAg, anti-HBs and IgM anti-HBc) occasionally were used for emergent liver transplantation. Without HBIG, around 70% of recipients became viremic, compared with a 20% infection rate in those recipients already immune to the virus (anti-HBs seropositive from prior immunization or natural immunity). Importantly, seroreversion from anti-HBs to HBsAg (and even with detectable HBV DNA and HBeAg) can occur following transplantation of a liver graft from an anti-HBc seropositive donor. Nevertheless, although outcomes from large-scale studies are not available, individual transplant centers report favorable survival rates for patients and grafts over 1–3 years, despite recrudescence of HBV DNA in the serum and reversion to HBsAg seropositivity in some cases. Interestingly, obvious declaration of HBV, with presence of ground glass cells and positive staining of HBsAg and HBcAg in liver tissue may be delayed for several years.

Anti-HBc and anti-HBs seropositive donors with normal liver tests in the USA:

- Have had previous exposure to HBV
- Have very low potential risks for transmission
- Rarely may harbor HBV DNA in the liver
- Amount to more than 100 donors a year
- Could account for as many as 500 transplants annually if safe

6.10.2 Anti-HCV seropositive donors

Reports of 0–35% acquisition of HCV, based on the non-specific first-generation EIAs and non-routine screening of potential blood donors probably overestimated the risk. Although blood donors are believed to represent the general population, the sero-prevalence rate is around 5–10% for organ donors by second-generation RIBA, and 3.4% by RT–PCR, compared with less than 0.5% in volunteer blood donors in the USA.

Outcomes following grafting of a liver between seropositive donor and recipient depend on the dominance of virus quasispecies. Although data are limited, dominance by the donated virus (excepting genotype 1b) is associated with favorable outcome, whereas dominance from recipient strains may herald progressive liver disease. These observations seem to mirror the behavior of HCV in the donor and recipient.

6.11 Issues concerning transplantation of other organs

HBV and HCV have been reported following transplantation of kidney, heart and bone marrow. Studies are in progress to assess the impact of the many variables, other than level of viremia, such as type of organ, ischemia times during procurement, preservation fluids and immunosuppressive therapy regimes. Estimates in the USA suggest that the risk of seroconversion from anti-HCV seronegative to anti-HCV seropositive status is between 5 and 15% following kidney and bone marrow transplantation.

HCV infection is predicted to be common among recipients of heart and lung transplants because exclusion of donors based on screening has been limited. Also, the impact of HCV infection in these recipients is unknown due to the relatively small numbers of long-term survivors. Estimates of rates of infection based on detection of HCV RNA, rather than anti-HCV, are likely to be higher and more accurate in the immunosuppressed than immunocompetent host. Outcome is variable but loss of the graft has been reported for hepatitis B and C.

6.11.1 Heart transplantation

Up to 1999, only around 60 recipients of hearts from HCV-infected donors were reported in the literature. Data are too few to assess the risks related to lung transplantation.

In a poll in the USA published in 1997, 71% of heart transplant centers stated that they would use hearts from HCV seropositive donors in certain clinical situations. Transmission rates were around 80% based on detection of HCV RNA in the recipient. Before HCV RNA was sought, HCV infection was under-reported due to the falsely negative anti-HCV tests in the immunosuppressed transplant recipient. False-negative EIA tests were especially common following heart transplantation due to the high doses of immunosuppression (compared with kidney and liver transplantation) that delayed any seroconversion.

The largest study by Lake *et al.* reported clinical outcomes in 96 heart transplant recipients. This group showed evidence of liver dysfunction in over half of the recipients of hearts from HCV seropositive donors. Further, liver disease was a major cause of death despite similar actuarial survival rates compared with recipients of hearts from HCV seronegative donors.

6.11.2 Renal transplantation

The seroprevalence for HBsAg remains high in patients with renal failure, especially those receiving hemodialysis (see section 3.8.5). Many infected patients have significant liver disease and this is prone to exacerbation especially following transplantation. In fact, the HBV-related liver disease can show accelerated progression to fibrosis and cirrhosis following renal transplantation of the HBsAg seropositive chronic carrier.

Anti-HCV seropositivity has been recorded in up to 30% of patients with end-stage renal failure requiring hemodialysis but numbers with viremia may be higher due to their immunosuppressed state. *De novo* infection with HCV also was relatively common with renal transplantation, prior to the advent of screening donors. The incidence of hepatitis C among hemodialysis patients has been reduced significantly with exclusion of HCV seropositive blood donors, reduction in transfusion requirements with introduction of recombinant erythropoietin to correct the chronic anemia and judicious use of immunosuppression before and after transplantation.

Some renal transplant centers justify the use of renal grafts from anti-HCV seropositive donors into previously uninfected recipients based on comparable short-term outcomes. Clinically significant hepatitis with accelerated progression to cirrhosis and decompensation has been reported in some HBsAg seropositive patients receiving renal allografts. Overall, interpretation of the data is difficult for the usual limitations of too few patients and short (< 5 year) follow-up, lack of histology and reliance on liver tests, rather than detecting viral nucleic acids, to reflect severity of the underlying liver disease.

6.11.3 Bone marrow transplantation and viral hepatitis

Exacerbation of pre-existing chronic liver disease may occur in association with withdrawal of immunosuppressive therapy or chemotherapy in patients treated for malignancies, including in bone marrow recipients. Severe hepatitis C and B have been recorded in individual cases. These can progress to liver failure and a picture resembling FCH. Levels of viremia tend to be high and virus is detectable in most hepatocytes by immunochemical or molecular techniques, suggesting an immunological and/or cytopathic role for HCV or HBV, respectively. This response is fatal and recurrent disease is common in the few patients who have proceeded to liver transplantation.

HCV, and HBV, infections have been recorded in around 5% of patients transplanted for leukemia. This incidence is much higher (up to 50%) in adults with hemoglobinopathies due to their requirement for repeated blood transfusions predating universal screening of blood donors. Numbers are insufficient to determine the impact long-term on severity of any liver disease.

> **Key Notes** Donation by anti-HCV seropositive individuals
>
> The Centers for Disease Control (CDC) recommend that anti-HCV seropositive individuals do not donate blood, organs and other tissues, or semen

6.12 Future strategies

Several strategies focus on supporting liver function in ALF until there is spontaneous regeneration or a graft becomes available. Auxiliary liver transplantation should be considered in ALF due to viral hepatitis.

6.12.1 New antiviral agents

Antiviral agents are required that do not depend on an intact immune response of the host to effect eradication of the virus from the graft. Antiviral therapies should be considered before transplantation in chronic hepatitis B or C to reduce the viral loads presented to the graft. Antiviral therapies should be implemented as soon as possible after grafting to prevent infection of the graft. Combinations, including nucleoside analogs and IFNs, may be more efficacious than solo therapies in reducing levels of viremia (HCV RNA, HBV DNA). The duration of therapy is likely to be long.

6.12.2 Adoptive transfer

Individuals with chronic hepatitis B have cleared their infection and become immune (anti-HBs seropositive) following bone marrow transplantation from a donor immune to hepatitis B either naturally acquired from infection or following immunization (see also section 6.11.3). Clearance was achieved despite immunosuppression. These data add strong support to the concept that hepatitis B virus infection can be cleared from a chronically infected liver following encounter with a competent immune system.

This model is relevant to an infected liver (e.g. from an anti-HBc donor) introduced into a host immune to hepatitis B (see section 6.10.1). Whether the disparate MHC expression between donor and recipient will impair viral clearance, is not known.

6.12.3 Xenobiotic (pig, baboon) livers

This approach takes advantage of the species specificity of human hepatitis viruses. Disadvantages include the potential for introducing animal viruses into man as well as the inevitable problems of graft rejection across species. Recipients of porcine xenografts could be susceptible to zoonoses such as leptospirosis and microbial agents implicated in erysipelas in pigs as well as porcine endogenous retroviruses.

Transgenic species are being developed with modifications in genes to limit activation of the complement cascade and other systems involved in the pathogenesis of hyperacute rejection.

Further reading

Bradley, A.D., Patel, R., Portela, D., *et al.* (1996) Prognostic significance and risk factors of untreated cytomegalovirus viremia in liver transplant recipients. *J. Infect. Dis.* **173**: 446–449.
CMV viremia is a poor predictor of CMV organ involvement and should not be used to select patients for prophylactic antiviral therapy following transplantation. CMV viremia is common in CMV-negative recipients who receive a liver graft from a CMV-positive donor.

Cacciola, I., Pollicino, T., Squadrito, G., *et al.* (1999) Occult hepatitis B virus infection in patients with chronic hepatitis C liver disease. *N. Engl. J. Med.* **341**: 22–26.
Study of 200 Italian patients with HCV and liver disease; 33% had HBV DNA detectable in liver compared with 14% controls without HCV. HBV DNA was detected twice as frequently among those who had failed IFN-α-2b monotherapy and was more often associated with cirrhosis than milder liver disease. No specific gene rearrangements were noted on sequencing but the low level viremias and undetectable antigens suggest possible suppression of gene expression by HCV.

Chazoulleres, O., Mamish, D., Kim, M., *et al.* (1994) 'Occult' hepatitis B virus as source of infection in liver transplant recipients. *Lancet* **343**: 142–146.
Mild outcomes despite infection.

Davies, S.E., Portmann, B.C., O'Grady, J.G., *et al.* (1991) Hepatic histological findings after transplantation for chronic hepatitis B virus infection, including a unique pattern of fibrosing cholestatic hepatitis. *Hepatology* **13**: 151–157.
First description of FCH.

De Man, R.A., Bartholomeusz, A.I., Niesters, H.G.M., *et al.* (1997) The sequential occurrence of viral mutations in a liver transplant recipient re-infected with hepatitis B: hepatitis B immune globulin escape, famciclovir non-response, followed by lamivudine resistance resulting in graft loss. *J. Hepatol.* **29**: 669–675.

Dodson, S.F., Issa, S., Araya, V., *et al.* (1997) Infectivity of hepatic allografts with antibodies to hepatitis B virus. *Transplantation* **64**: 1582–1584.
Mild outcomes despite infection.

Douglas, D.D., Rekela, J., Wright, T.L., *et al.* (1997) The clinical course of transplantation – assocated with *de novo* hepatitis B infection in the liver transplant recipient. *Liver Transplant. Surg.* **3**: 105–111.

Fagan, E.A. (1994) Acute liver failure of unknown pathogenesis: the hidden agenda. (Editorial). *Hepatology* **19**: 1307–1312.
In ALF in the West, non-A, non-B also is non-C, non-D, non-E. Describes the hepatitis F syndrome.

Fagan, E.A., Yousef, G., Brahm, J. *et al.* (1990) Persistence of hepatitis A virus in fulminant hepatitis before and after liver transplantation. *J. Med. Virol.* **30**: 131–134.

Gane, E.J., Portmann, B.C., Naoumov, N.V,. *et al.* (1996) Long-term outcome of hepatitis C infection after transplantation. *N. Engl. J. Med.* **334**: 815–820.
Retrospective study from King's College Hospital of outcomes in 130 patients with persistent HCV RNA post-transplantation with follow up of between 6 months and 3 years. Only 12% had no hepatitis; 45% had moderate hepatitis/cirrhosis. Although cumulative 5-year survival rates for patients transplanted for HCV- (70%) and non-HCV-related (69%) liver disease were similar, longer follow-up times will be required to assess the adverse impact of HCV infection on the graft.

Ghany, M.G., Ayola, B., Villamil, F.G., *et al.* (1998) Hepatitis B virus S mutants in liver transplant recipients who were reinfected despite hepatitis B immune globulin prophylaxis. *Hepatology* **27**: 213–222.
A significant correlation occurred between emergence of mutations in the S gene, subtype (adw2) and duration of HBIG. Mutants were undetectable in the majority of pre-transplant sera and reversion to wild-type was recorded in all but one patient on withdrawal of HBIG.

Gish, R.G., Lau, J.Y.N., Brooks, L., *et al.* (1996) Ganciclovir treatment of HBV infection in liver transplant recipients. *Hepatology* **23**: 1–7.
Long-term (4–10 months) intravenous ganciclovir was safe and efficacious in the treatment of the infected graft but costs of parenteral therapy remain high.

Jain, A., Demetris, A.J., Manez, R., *et al.* (1998) Incidence and severity of acute allograft rejection in liver transplant recipients treated with alfa interferon. *Liver Transplant. Surg.* **4**: 197–203.
Retrospective study following transplantation for hepatitis B (32 patients) and C/NANB (73 patients) using IFN-α 5 MU three times a week. Although there was no significant increase in the number of rejection episodes compared with 132 controls, the overall high level of immunosuppression may have protected against rejection.

Lake, K.D., Smith, C.I., LaForest, S.K.M., *et al.* (1997) Policies regarding the transplantation of hepatitis C-positive candidates and donor organs. *J. Heart Lung Transplant.* **16**: 917–920.
A poll study in the US showed that 71% of cardiac transplant centers would use hearts from HCV seropositive donors under certain circumstances.

Lake, K.D., Smith, C.I., Milfred-Laforest, M.R., *et al.* (1997) Outcomes of hepatitis C (HCV +) heart transplant recipients. *Transplant. Proc.* **29**: 581–582.
Study of 96 recipients of hearts showing similar actuarial rates of survival whether related to HCV positive and negative donors but significant liver dysfunction in 51% of recipients. Liver disease was a primary cause of death.

Laskus, T., Wang, L.F., Rakela, J., *et al.* (1996) Dynamic behavior of hepatitis C virus in chronically infected patients receiving liver graft from infected donors. *Virology* **220**: 171–176.
Patients retaining their strain developed significantly more active liver disease in their graft than those infected by donor strain. Genotype 1b became the dominant strain in all recipient/ donor pairs when present.

Lin, H.-M., Kauffman, M., McBride, M.A. (1998) Center-specific graft and patient survival rates. *J. Am. Med. Assoc.* **280**: 1153–1160.

Lok, A.S., Liang, R.H., Chung, H.T. (1992) Recovery from chronic hepatitis B. *Ann. Intern. Med.* **116**: 957–958.

Markowitz, J.S., Martin, P., Conrad, A.J., *et al.* (1998) Prophylaxis against hepatitis B recurrence following liver transplantation using combination lamivudine and hepatitis B immune globulin. *Hepatology* **28**: 585–589.
Encouraging results in 14 patients with end-stage chronic hepatitis B treated with lamivudine 150 mg/day before, or within 11 days of transplantation; 93% patient and graft survival at 1.1 year with no serological evidence of recurrence including five patients with detectable HBV DNA prior to grafting.

Melegari, M., Scaglioni, P.P., Wands, J.R. (1998) Hepatitis B virus mutants associated with 3TC and famciclovir administration are replication defective. *Hepatology* **27**: 628–633.

O'Grady, J.G., Alexander, G.J.M., Hayllar, K.M., Williams, R. (1989) Early indicators of prognosis in fulminant hepatic failure. *Gastroenterology* **97**: 439–445.
Prognostic indicators in viral and non-viral causes of liver failure; based on etiology, prothrombin time, age, acidosis. More reliable in viral than non-viral causes.

Oldhafer, K.J., Gubernatis, G., Schlitt, H.J., *et al.* (1994) Auxiliary partial orthotopic liver transplantation for acute liver failure: the Hannover experience. In: *Clinical Transplants* (eds P.I. Terasaki, J.M. Cecka), UCLA Tissue typing Laboratory, Los Angeles, CA, pp. 181–187.
Regeneration of native liver in three out of four patients, including within 2 weeks in a child with ALF of indeterminate etiology.

Paya, C.V., Weisner, R.H., Hermans, P.E., *et al.* (1993) Risk factors for cytomegalovirus and severe bacterial infections following liver transplantation: a prospective multivariate time-dependent analysis. *J. Hepatol.* **18**: 185–195.
CMV and bacterial infections are a leading cause of morbidity and mortality following liver transplantation. Mayo Clinic survey 1988–1990: the risk factor identified for bacterial infection by multivariate analysis was transfusion of large quantities of fresh frozen plasma. The risk factor for CMV was transplanting a seronegative recipient with a liver graft from a seropositive donor.

Protzer-Knolle, U., Naumann, U., Berg, T., *et al.* (1998) Hepatitis B virus with antigenically altered HBsAg is selected by high-dose hepatitis B immune globulin after liver transplantation. *Hepatology* **27**: 254–263.
Nine of 30 patients receiving long-term high dose HBIG to maintain anti-HBs above 100 U/l developed specific 'escape' variants: emergence of aa144 (exclusive to genotype A) variant or

aa145 (exclusive to genotype D) variant. Graft failure (44%) was higher than for other patients (23%). Importantly, the increase in incidence of escape variants (from 5% to 52%) after 1992 was attributed by the authors to use of the different immunogen (recombinant antigen replaced plasma-derived, in the vaccine) used to boost plasmapheresis donors.

Pruett, T.L. (1997) Primary transplant treatment of hepatitis B: hepatitis B immunoglobulin (passive immunization). *Liver Transplant. Surg.* **3**: S13–S15.

Radomski, J.S., Moritiz, M.J., Armenti, V.T., Munoz, S.J. (1996) Hepatitis B transmission from a liver donor who tested negative for hepatitis B surface antigen and positive for hepatitis B core antibody. *Liver Transplant. Surg.* **2**: 130–131.

Samuel, D., Muller, R., Alexander, G., *et al.* (1993) Liver transplantation in European patients with the hepatitis B surface antigen. *N. Engl. J. Med.* **329**: 1842–1847.
Seminal paper of European experience of outcomes following grafting for acute and chronic HBV and HBV/HDV infection.

Sawyer, R.G., McGory, R.W., Gaffey, M.J., *et al.* (1998) Improved clinical outcomes with liver transplantation for hepatitis B-induced chronic liver failure using passive immunization. *Ann. Surg.* **227**: 841–850.

Shouval, D., Ilan, Y. (1995) Transplantation of hepatitis B immune lymphocytes as means for adoptive transfer of immunity to hepatitis B virus. *J. Hepatol.* **23**: 98–101.
Special review article on adoptive transfer using bone marrow transplantation and peripheral blood lymphocytes from HBV immune donors.

Terrault, N.A., Zhou, S., McCory, R.W., *et al.* (1998) Incidence and clinical consequences of surface and polymerase gene mutations in liver transplant recipients on hepatitis B immunoglobulin. *Hepatology* **28**: 555–561.
Recurrent HBV infection occurred with failure of fixed doses of intravenous polyclonal HBIG in 8.3% of 24 patients followed for around 28 months post-transplantation for chronic hepatitis B. Only half of the failures were associated with emergence of HBV escape variants with mutations in the a determinant and most of these occurred more than 6 months post-transplantation.

U.S. Scientific Registry of Transplant Recipients and the Organ Procurement and Transplantation Network. (1998) *Transplant Data 1988–1996.* Department of Health and Human Services. Health Resources and Services Administration, United Network for Organ Sharing, pp.1–55.

Wachs, M.E., Amend, W.J., Ascher, N.L., *et al.* (1995) The risk of transmission of hepatitis B from HBsAg (−), HBcAb (+), HBIgM (−) organ donors. *Transplantation* **59**: 230–234.

Weight, T.L., Combs, C., Kim, M., *et al.* (1994) Interferon α therapy for hepatitis C virus infection after liver transplantation. *Hepatology* **20**: 773–779.
Typical disappointing results when IFN therapy is commenced for persistent HCV infection following transplantation. The number of episodes of acute rejection were not increased but this was an uncontrolled pilot study with a small number (18) of patients.

Wreghitt, T.G., Gray, J.J., Allain, J.P. *et al.* (1994) Transmission of hepatitis C virus by organ transplantation in the United Kingdom. *J. Hepatol.* **20**: 768–772.
Almost invariable transmission of HCV and significant liver disease in recipients by organs from anti-HCV seropositive donors.

Chapter 7

Viral Hepatitis in Pregnancy and Pediatric Practice

7.0 Maternal and pediatric issues

Viral hepatitis remains the most common cause of jaundice in pregnancy. Clinical presentation is most common in the third trimester. The spectrum of viruses and presentations are no different from the non-pregnant population. In regions of the world with high rates of immunity to hepatitis A following exposure in infancy, acute viral hepatitis in adults is most likely to be due to hepatitis B or E. In general, the anticipated outcome for the mother with viral hepatitis is the same as her non-pregnant counterpart except for hepatitis E and herpes hepatitis. Clinical and epidemiological features overlap and diagnosis relies on detecting specific serological markers of acute and chronic infection. The differential diagnosis includes various diseases peculiar to pregnancy that can mimic closely viral hepatitis but require different management strategies, including prompt delivery.

Children are susceptible to the same viruses as adults. Neonates also are exposed to maternal infection via vertical and perinatal transmission. Pathogenetic mechanisms must differ between neonates and children and adults to account for differences in clinical presentations and outcomes. The neonate and the young baby, like the immunosuppressed host, may develop severe hepatitis complicating a common childhood infection. Also, they show a propensity for persistent infection as exemplified by hepatitis B.

This chapter focuses only on those issues related directly to pregnancy and pediatric practice that are not addressed specifically in other chapters. We focus on the impact of viral hepatitis on the mother, vertical transmission, impact and outcomes following infection in the new-born and child and immunoprophylaxis against hepatitis A and B.

7.1 Prenatal issues

7.1.1 Fertility

This is not affected in acute, uncomplicated hepatitis. In fact, women can become pregnant subsequent to malabsorption of the contraceptive pill due to significant diarrhea, nausea and vomiting. Infertility and subfertility with amenorrhea and oligomenorrhea occur with advanced liver disease, especially cirrhosis, and this occurs regardless of etiology. Consequently, women with severe chronic viral hepatitis/cirrhosis are unlikely to become pregnant. Fertility returns in most young women following uneventful liver transplantation.

7.1.2 Sexual transmission of hepatotropic agents

This aspect is covered in the Chapters 2, 3 and 4 on the individual viruses.

7.1.3 In vitro *fertilization*

HBV DNA has been detected regularly in fractions of seminal fluid, including round cells, leukocytes and (rarely) spermatozoa. Seminal fluid should be considered a potential source of transmission and all donors should be tested for hepatitis B markers.

The identification of hepatitis C virus (HCV) in semen remains controversial. HCV RNA was not detectable using polymerase chain reaction (PCR) techniques in fractions of seminal fluid, round cells and spermatozoa from 56 viremic men attending a tertiary referral centers in Italy with especial expertise in processing seminal fluid from donors with human immunodeficiency virus (HIV). Thirty-four of the men had inhibitors to the PCR present in their fractions. Interestingly, hepatitis G virus (HGV)/GBV-C was found in 50% of the fractions without PCR inhibitors, but the significance of these discordant findings remains unknown.

7.2 Impact of viral hepatitis on the mother

7.2.1 Acute viral hepatitis during pregnancy

Clinical features, morbidity and maternal outcomes with acute hepatitis A, B, B+D and C do not differ from those of non-pregnant women. Most studies show some excess in intrauterine death, stillbirths and prematurity if viral hepatitis is contracted during pregnancy. However, overall the risk seems to be small. In contrast, data from developing countries show conflicting outcomes and may reflect the preponderance of hepatitis B and E, varying standards of medical care and bias towards reporting severe cases. Congenital abnormalities are not increased for hepatitis A–E.

Key Notes Acute and chronic viral hepatitis in pregnancy

- Most common cause of jaundice in pregnancy
- Diagnoses rely on specific serological markers

Maternal outcome
- Good for acute hepatitis A, B and C
- Poor for severe herpes simplex and hepatitis E

Fetal outcome
- Intrauterine deaths and premature births in severe cases
- Vertical transmission is maximal third trimester
  - 20–95% chronic HBV infection without immunoprophylaxis
  - 0–18% for HCV

Maternal features
- Hepatitis A, B+D, and C do not differ from non-pregnant state
- Hepatitis E and herpes simplex hepatitis may be severe
- May resemble liver disorders peculiar to pregnancy
  - Acute fatty liver, intrahepatic cholestasis
- May resemble hereditary and metabolic disorders
- Consider unusual infections (see *Table 7.3*)

Neonatal features
- Congenital infection may be severe with herpesviruses
- Virus shedding may be protracted despite clinical recovery
  - HAV, herpesviruses, congenital rubella

7.2.2 Acute liver failure

Acute liver failure (ALF) is covered extensively in Chapter 5. There is no predilection for ALF in pregnancy except for hepatitis E virus (HEV) (see section 7.2.3) in some countries and herpesviruses (see section 7.2.4). Management is the same as for the non-pregnant population. Fetal loss is high (> 50%); intrauterine death and stillbirth are common due to the severe metabolic derangements including intractable hypoglycemia and acidosis. Individual reports document successful outcomes for mother and child following liver transplantation during pregnancy for acute liver failure, including for viral hepatitis. Numbers of survivors are insufficient to draw conclusions on risks of vertical transmission and outcomes for the child.

7.2.3 Acute hepatitis E

HEV is the only major hepatitis virus with a predilection for severe hepatitis during pregnancy. The attack rate during epidemics is highest in young adults and maternal deaths of up to 20% have been reported for some developing countries, particularly the Indian subcontinent. Mechanisms remain unclear. (See also section 2.1.5).

7.2.4 Herpesviruses

Herpes simplex viruses (HSV-1 and -2), varicella zoster virus (VZV, causing chickenpox and shingles), cytomegalovirus (CMV) and Epstein–Barr virus (EBV) are causes of subclinical, anicteric, hepatitis. Individual cases of severe hepatitis, including ALF have been reported for HSV, human herpes virus 6 (HHV-6) and VZV (including an illness resembling Reye's syndrome in children, see section 7.8).

7.2.4.1 Herpes simplex hepatitis

HSV is the only virus in the West that seems to cause especially severe hepatitis in pregnancy. Mortality from ALF exceeds 90%. Pathogenetic mechanisms remain unclear. Presenting features of herpes infections are non-specific. Mucocutaneous stigmata and jaundice may be absent and diagnosis requires a high index of suspicion. Apart from pregnancy, most patients have an underlying risk factor, particularly immunosuppression, malignancy (lymphomas, leukemias and bone-marrow recipients with graft-versus-host disease), HIV and acquired immune deficiency syndrome (AIDS). (See also section 5.2.8).

7.2.4.2 Epstein–Barr virus

Acute hepatitis with minor elevation in liver enzymes [aspartate aminotransferase (AST) and alanine aminotransferase (ALT)] during infectious mononucleosis (glandular fever) is common but there is no predilection for severe hepatitis in pregnancy.

7.2.4.3 Cytomegalovirus (CMV)

CMV hepatitis in the pregnant and immunocompromised host (e.g. liver transplant recipient, or with HIV) can have serious consequences. CMV is the most important viral pathogen infecting the graft following liver transplantation for non-viral disease (see section 6.9.1) and following cardiac surgery. Women who have received a graft and later become pregnant are at especial risk of transmitting CMV infection to their offspring (see sections 7.5.4 and 7.8.1).

7.3 Chronic viral hepatitis

The management of women with severe liver disease during pregnancy and following liver transplantation is regardless of etiology and outside the scope of this text (see further reading). There is no evidence that pregnancy *per se* affects the outcome of chronic viral hepatitis in the absence of cirrhosis and portal hypertension. The outcome of the pregnancy and well-being of the child depend on the integrity of liver synthetic function and presence of portal hypertension in the mother, rather than virus etiology. Pregnant women with cirrhosis (regardless of etiology) and portal hypertension are at increased risk of bleeding from varices (esophageal and gastric) due to the increased circulating blood volume, and this risk rises progressively through to term. The risk of fetal loss remains high throughout pregnancy with advanced cirrhosis and is regardless of etiology. The management of mother and baby are regardless of viral or non-viral etiology.

7.3.1 Chronic hepatitis B and B+D

Viral markers of current or previous infection with HBV are found in up to 15% of pregnant women, depending on the background rate of endemicity. Diagnosis and general management are the same as for non-pregnant patients. Antiviral agents are contraindicated during pregnancy (see sections 1.13 and 4.7). The overriding issue is the need to protect the new-born from vertical and horizontal transmission.

7.3.2 Chronic hepatitis C

In a study of 1700 consecutive women seen for high-risk pregnancy in a university department of obstetrics in Italy, 29 (1.7%) were anti-HCV seropositive; eight had concurrent HIV infection. Only two showed any elevation in liver enzyme tests during pregnancy and within a 6-month follow-up and there were no adverse maternal and fetal outcomes. In Ireland, after 17 years follow-up of a cohort of 232 young women infected with HCV after receiving contaminated anti-D globulin, only 2.4% had developed cirrhosis; the vast majority have remained symptomless. In contrast, reports of non-pregnant patients from liver transplant centers tend to emphasize the severity of chronic liver disease with progression to cirrhosis and hepatoma especially in patients infected after 50 years of age.

7.4 Hepatocellular carcinoma

Development of hepatocellular carcinoma (HCC) has been linked with estrogens in oral contraceptive drugs, hepatitis B virus (HBV) and HCV infection, and smoking. Pregnancy is associated with especially rapid growth of the tumor (in response to estrogens) and an abysmal prognosis. Serum levels of carcinoembryonic antigen and beta-human choriogonadotropin may be raised in HCC, a finding that can cause confusion with gestational trophoblastic disease.

7.5 Vertical and perinatal transmissions and neonatal outcomes

All of the major hepatotropic viruses can be transmitted from mother to child (*Table 7.1*). HCV (and HIV) are transmitted via this route less frequently than HBV, despite sharing many epidemiological characteristics. Differences in transmissibility and

Table 7.1. The major hepatotropic agents and vertical/perinatal transmission

Virus	Mother	Features in neonate
HAV	IgM-anti-HAV	Possible parenteral exposure, outbreaks in intensive care units, severe protracted disease with shedding of virus in stool
HEV	Poorly defined	Poorly defined
HBV	HBeAg positive	Frequent mild, persistent infection
HBV	HBeAg negative	Can develop severe hepatitis associated with HBeAg-negative variants
HBV+HDV	HBsAg positive, anti-HDV positive	Perinatal transmission of HDV seems rare; may depend upon maternal HBeAg positivity
HCV	HCV RNA positive, anti-HIV positive	Transmission not invariable but seems to be more common from HIV-positive than HIV-negative mothers
HCV	HCV RNA negative, anti-HCV positive	Vertical and perinatal transmission unclear; infection uncommon in neonates
GBV-C/HGV	HGV RNA	Not defined
Non-A–E	Not defined	Not defined

outcomes for vertical and sexual routes may be explained, in part, by the compartmentalization of virus quasispecies. Virus variants may be selected to populate different sites such as semen (see section 7.1.3), blood, vaginal fluid and cells. Also, the diversity in clinical outcomes may be explained, in part, by selection of variants of HBV and HCV for clearance and persistence in the recipient.

7.5.1 HAV and HEV

Fecal contamination during delivery is limited by the short interval of viral excretion.

For HAV, vertical and perinatal transmission occur typically via contamination of maternal feces in a mother presenting with hepatitis A in the third trimester or soon after delivery. HAV can persist through pregnancy. The newborn remains susceptible to hepatitis A from the mother who continues to excrete virus in her stool for weeks. Neonatal immunoprophylaxis for hepatitis A rarely is necessary as most neonatal infections are mild and herald life-long immunity. In the West, outbreaks have been reported in neonatal intensive treatment units (ITUs) and, rarely, from parenteral transmission from blood products. Severe infection, including ALF has been documented in this setting accompanied by sepsis and multiorgan failure.

Data for HEV are scarce. Intrauterine death is common with severe hepatitis E in pregnancy and due probably to the profound hypoglycemia seen in ALF.

7.5.2 HBV and B+D

The risk of vertical transmission is maximal for acute viral hepatitis presenting in the third trimester (*Figure 7.1*). In California, the risk of transmission from mothers with

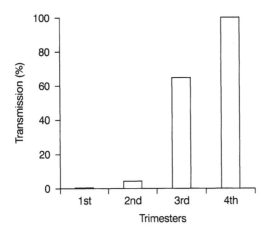

Figure 7.1. Risk of transmission of hepatitis B virus to the fetus in acute hepatitis B during pregnancy (from Tong *et al.*, 1981).

acute hepatitis B was found to be 0, 6, 67, 100% in the first, second, third and fourth trimesters, respectively. HBsAg-positive mothers who are viremic (detectable HBV DNA) around delivery almost invariably transmit the infection to their offspring. Overall, transmission is less frequent following acute hepatitis B presenting during pregnancy than chronic hepatitis B.

Worldwide, transmission from the chronic carrier mother to child around birth accounts for up to 40% of all chronic carriers and more than 1 000 000 deaths each year from HBV-related liver disease. For chronic hepatitis B, babies born of HBsAg positive mothers with virus replication have a 20–95% probability of becoming infected without immunoprophylaxis. Infectivity depends on the HBV DNA level and this relates to ethnic origin. In South-East Asia, around 30% of babies become infected within the first year of birth compared with around 5–10% in the African countries. Correspondingly, up to 50% of HBsAg seropositive Asian mothers also are HBeAg seropositive, compared with around 20% of African mothers. The high frequency of viremia among carrier mothers from the Far East and Southeast Asia explains the almost invariable (around 95%) infection of neonates. Vertical transmission occurs in up to 25% of HBeAg seronegative carrier mothers and 12% for anti-HBe seropositive mothers.

In highly endemic areas, especially the African countries, spread also occurs via horizontal transmission from close contact with a family member, household contact or peer who is infected with HBV. The probability of persistent infection declines with age to adult levels after around 7 years of age.

In the West, vertical and perinatal transmission of HBV is uncommon and restricted mostly to babies born of high-risk mothers, immigrants from countries of high endemicity, intravenous drug users and close contacts with high-risk behavior. Vertical transmission of hepatitis D virus (HDV) is rare and may depend upon HBeAg-positivity of the mother.

Table 7.2. Vertical transmission of hepatitis B: neonatal outcomes without immunoprophylaxis

Maternal serum markers	Maternal hepatitis B	Neonatal outcome
HBsAg, HBeAg IgM anti-HBc	Acute	Acute hepatitis, recovery and immunity
HBsAg, HBeAg	Chronic	Chronic carrier (20–95%)
HBsAg, anti-HBe	Chronic	Chronic carrier (0–20%)
HBsAg, neither HBeAg/ anti-HBe ('e' negative)	Chronic	Fulminant hepatitis (20%)

Neonatal outcome depends on whether the mother has acute or chronic hepatitis B during pregnancy and, in turn, on the likely heterogeneity of her virus population. Most neonatal infections are symptomless but persist to develop chronic infection in more than 90% of cases (*Table 7.2*). Factors that contribute to this high rate of persistence may include the immaturity of the immune system and a tolerizing effect of maternal HBeAg that may cross the placenta and circulate in the fetus. It has been suggested also that maternal anti-HBc may cross the placenta and modulate the fetal response to infection with HBV. Hepatitis B virions are not believed to cross the placenta except, perhaps with damage in the late stages of pregnancy. Reports of detection of the HBsAg or HBV in cord blood should be interpreted with caution because of the high risk of contamination with maternal fluids.

7.5.2.1 Fulminant hepatitis B in babies

Chronic hepatitis B is uncommon in babies born to HBsAg-positive mothers with anti-HBe (*Table 7.2*). Babies born to viremic mothers who are seronegative for HBeAg are at risk of severe acute hepatitis and liver failure following vertical transmission of HBV variants. The precore stop mutation is not sufficient to explain fulminant hepatitis B and other mutations, including in the core promoter (see sections 3.2.4.1 and 5.2.5.1), may be critical in determining the pathogenesis. Survivors clear virus and develop immunity (anti-HBs).

7.5.3 Hepatitis C virus

Vertical transmission is less frequent than for HBV. Early data, based primarily on first generation antibody tests and non-specific elevations in liver tests, probably overestimated the frequency of transmission. The consensus view is that vertical transmission is unlikely (0–18%) with chronic hepatitis C accompanied by low levels of viremia [defined arbitrarily as $<10^6$ genome equivalent (geq)/ml]. Transmission is more common with concurrent HIV infection (6–36%) but this may reflect the propensity for high viremia in HIV-infected women, especially at delivery. In a study from Italy, 30 mothers with anti-HCV antibodies (1.07%) were identified from 2980 consecutive pregnant women screened for anti-HCV antibodies. None was seropositive for anti-HIV antibodies. Ten mothers had HCV RNA detected in serum around delivery. All cord bloods had undetectable HCV RNA but three of these 10 high-risk babies subsequently had HCV RNA detected at around 3 months of age. Post-partum transmission is implicated also in a prospective study from Japan. Sequence analyses have confirmed vertical transmission through three generations. Data are insufficient to determine any variation of propensity for transmission with genotype and the effects of other factors such as HLA. As with vertical transmission of HBV, evidence is accumulating for selective transmission of quasispecies from the heterogeneous maternal population. Selection pressure on certain regions such as HVR1 may favor persistence of virus via antibody escape mechanisms. Transmission also may occur with acute hepatitis C arising during pregnancy but data are limited.

7.5.4 Herpesviruses

Maternal transmission of herpes viruses may be blood-borne via the placenta or from contamination with vaginal fluid and ulcers (herpetic) at delivery. Transmission from a symptomless mother may result in severe, disseminated infection in the neonate. Most patients with severe herpes hepatitis have an underlying risk factor, particularly immunosuppression, malignancy, anti-HIV positivity and pregnancy.

Characteristic features on liver histology include hemorrhagic necrosis and intranuclear inclusions. Individual cases of severe hepatitis, including ALF, have been reported in neonates and babies for herpes simplex (especially HSV-2), VZV (including an illness resembling Reye's syndrome), CMV and HHV-6.

7.6 Hepatotropic viruses in children

Children are susceptible to the same viruses as adults but outcomes may vary, such as asymptomless seroconversion in hepatitis A and a propensity for chronicity in hepatitis B.

7.6.1 Hepatitis A and E

In developing countries, over 80% of children above 3 years of age are seropositive for IgG anti-HAV, indicating previous exposure to hepatitis A. The vast majority of infections in childhood are symptomless or mild. Jaundice occurs in less than 5% of children aged below 5 years but in around 20% aged over 8 years. HAV RNA and HAAg are detectable in stool for months in some cases of neonatal infection. Symptoms in children are non-specific and include fever and indifference to feeding. Young children also may have fever, headache, loose pale stools and lose weight. Other findings include hepatomegaly, splenomegaly and lymphadenopathy. Recovery is rapid and complete, although relapses have been reported. Progression to ALF is very rare. Liver tests return to normal within 3 months but 5–10% will have elevations in AST and ALT for up to 9 months.

Sporadic cases of hepatitis E most likely occur in endemic regions. Studies in Egypt confirm antibody positivity and, by implication, previous exposure in children. The relative scarcity of reports of clinical hepatitis E in children suggests that these infections usually are mild or symptomless (resembling hepatitis A). The clinical features are non-specific with malaise and arthralgia.

7.6.2 Hepatitis B and D

Non-immunized HBsAg-negative children are at risk of horizontal transmission of HBV if a member of their family or peer group is a carrier. The risk may be as high as 40% in the first 5 years of life where a close household contact is viremic. The high probability of persistent infection of the neonate (> 90%) declines to adult levels (1–5%) over approximately 5 years.

Chronic hepatitis B in children tends to be mild, especially with immune tolerance to infections acquired very early in life. Up to 20% are likely to lose HBeAg during childhood (up to the age of 16 years). In one study of HBeAg-positive Asian-Americans, who were followed for 1–10 years, 15% of children (5–19 years) seroconverted spontaneously to anti-HBe, compared with 23% of adults aged 20–34 years and 17% of adults aged 35–50 years.

Although chronic hepatitis B is symptomless in the majority, children are at risk of severe disease. For example, the incidence of hepatocellular carcinoma (HCC) in Taiwanese children during 1981–1986 was approximately 0.7 per 100 000. Most tumors were associated with HBV infection. The incidence of HCC in this age group was roughly halved during 1990–1994, following the introduction of universal immunization against hepatitis B.

Studies of HDV infection in high prevalence areas show low rates in young children and an age-related increasing prevalence of markers of infection. Horizontal spread occurs among children in regions with especially high endemicity such as in Brazil. In the Amazon region, serological markers of HDV infection (IgM and IgG anti-HDV) can be detected in around 20–30% of children aged 10–14 years and reflect the very high carriage rate of HBsAg. HDV infection can account for a significant proportion of cases of severe hepatitis (liver failure), in children less than 15 years of age, in specific regions with high prevalence for HDV. Long-term chronic carriage of hepatitis D with B may increase the risk and rate of progression to cirrhosis compared to hepatitis B alone. Up to 20% of children presenting with chronic hepatitis D are diagnosed with cirrhosis.

7.6.2.1 Hepatitis B-related syndromes

Children with acute or chronic hepatitis B may develop a red (erythematous) papular rash—Gianotti-Crosti syndrome—over the face and limbs. The liver and lymph nodes usually are enlarged but jaundice is rare. The rash is not itchy (non-pruritic) and resolves within 2–4 weeks.

Renal dysfunction due to a membranous or mesangioproliferative glomerulonephritis has been reported in some children, especially in the East, in association with HBV infection. The pathogenetic mechanisms are unclear but the illnesses resemble immune complex disorders and deposits contain HBeAg.

Rarely, and as with adults, children may develop a vasculitis, aplastic anemia and serum-sickness like syndrome. Landry–Guillain–Barré syndrome has been reported.

7.6.3 Hepatitis C in pediatric practice

In the west, HCV infection is uncommon in pediatric practice except in the setting of the multiply transfused child with a hematological disorder, such as hemophilia, thalassemia or leukemia or the child with immunodeficiency. However, HCV infection may be under-diagnosed as most children are symptomless and not subject to screening programs. Horizontal transmission of HCV infection within the family is rare but can occur and relates to virus load in the mother. Early studies showed higher rates of transmission from HIV-positive mothers but not all studies have confirmed these findings.

A study of 458 patients who underwent cardiac surgery as children in Germany prior to 1991 (and the introduction of testing for HCV) revealed that 67 (14.6%) were positive for anti-HCV antibodies. Only 37 (55%) had detectable HCV RNA at a mean interval of 19.8 years after first surgery, and the remaining 30 were assumed to have cleared the infection. Only one patient had elevated ALT levels and only three out of 17 who underwent liver biopsy had histological signs of progressive liver damage.

In Asian and Oriental children, anti-HCV seropositivity by second generation enzyme immunoassay is detectable in around 10–35% of children with suspected viral etiology (non-A, non-B hepatitis) for their chronic liver disease.

As with adults, most presentations in childhood are with chronic rather than acute HCV infection. Symptoms usually are mild and non-specific. The outcomes in children are beginning to emerge and probably reflect reasons for, and routes of, transmission (see sections 7.7 and 7.7.1). One study from Spain compared histological outcomes for HCV infection acquired following transfusion between 15 neonatally infected children (mean age 12 ± 4 years) and 32 adults (mean age 40 ± 9 years) with persistently elevated liver enzymes and equivalent estimated duration of infection (around 11 years). Whereas all of the children had mild hepatitis, 56% of the adults had severe disease. Significant differences included lower levels of HCV RNA and ALT levels in children, despite the predominance of genotype 1b (80%) and male sex, in both groups. Interestingly, expression of several immunologically relevant molecules in the liver such as ICAM-1 and CD69 seemed to be down-regulated compared with findings in adults, suggesting a state of immune tolerance. Histopathological analysis of liver biopsies (1990–1996) from 80 Italian children (mean age 9 years) with chronic hepatitis C and elevated liver enzymes revealed severe hepatitis in 21.2% and moderate fibrosis in 16.2%. HCV was acquired via blood transfusion and vertical transmission and none had underlying systemic diseases. There was no correlation between severity of disease and genotype. Autoantibodies were common (11 children) including LKM-1 related to autoimmune hepatitis (one child) that responded to immunosuppression.

Apart from the overall mild hepatitis, prominence of lymphoid follicles and bile duct injury seem to be reported frequently in the liver biopsies of children with chronic hepatitis C.

In contrast, HCV infection may show significant progression in multiply transfused children with hemophilia, thalassemia and malignancies (see section 7.7.1).

7.7 Post-transfusion hepatitis

CMV is probably the most common cause of post-transfusion hepatitis in seronegative neonates and children in the West (compare hepatitis C in adults). More than 30% of the population acquire CMV during the first few months of life. After primary infection, the virus probably persists for life, and is, probably, the most common cause of post-transfusion hepatitis in seronegative neonates and children in the West (compare hepatitis C in adults pre-screening).

The matching of seronegative donors with recipients has reduced significantly the incidence of hepatitis and pneumonitis due to CMV in the multiply transfused, including recipients of leukocytes and other blood products.

Rarely, HAV has been associated with post-transfusion hepatitis (PTH), especially with fresh blood donations in neonatal ICUs. PTH from HCV is uncommon in the West where screening had occurred since the 1990s but is still a risk factor in multiply transfused patients.

7.7.1 Hemophiliacs, leukemics, hemoglobinopathies

The use of coagulation factor concentrates derived from pooled donations representing thousands of unscreened donors carried significant risks for the multitransfused patient. In a study of the deaths among hemophiliacs in the UK between 1977 and 1991 that predated routine screening for HCV (and some HIV) and introduction of heat treatment (in 1985), AIDS was the most common cause of death. However, liver disease was 16.7 times more frequent than in the general population and ranked third as a cause of death, especially from primary liver cancer.

Hepatitis C remains the most common cause of liver disease in hemophiliacs in the West. The majority of hemophiliacs infected via blood products have genotypes 1 and 3. The consensus of opinion is that liver disease is significant among hemophiliacs, although the inter-relation between the major hepatotropic viruses is difficult to resolve as long-term studies inevitably predated 1985 and screening in the 1990s. Although genotype 1b has been associated with especially severe hepatitis in non-hemophiliac as well as hemophiliacs, limited data show severe hepatitis in hemophiliacs infected with genotype 3 and also mixed genotypes. HCV RNA levels also tend to be higher in hemophiliacs compared with other patients but this correlates also with HIV co-infection.

In two studies totaling almost 400 hemophiliacs in England, the risk of developing decompensated liver disease/cirrhosis after approximately 20 years from first exposure was approximately 10–20%. Co-infection with HIV compounds the adverse prognosis. The cumulative risk of death over 25 years from liver disease for hemophiliacs was 6.5% if HIV positive compared with 1.4% if HIV negative. No excess association has been noted for extrahepatic manifestations of HCV infection among hemophiliacs except for antiphospholipid antibodies.

The prevalence of HBsAg seropositivity is below 3% in the West and represents mostly older patients who received blood products from unscreened donors.

Outbreaks of hepatitis A have been reported from several countries following introduction of concentrates subjected to only solvent/detergent (rather than heat) treatments. Solvent/detergent methods are virucidal for HIV and HCV but ineffective against HAV, parvovirus B19 and other viruses that lack a lipid envelope.

Patients with other hematological disorders and malignancies (thallassemia, hemoglobinopathies, leukemia) are subject to similar risks as hemophiliacs.

Evolution from chronic hepatitis C to cirrhosis also has been described in thalassemic children. A comparative study of 109 children (two-thirds had malignant disease) and 120 adults from Japan also reported mild hepatitis relative to adults. While all of the four children with the most severe liver disease (stage 3 fibrosis) had obvious confounding factors (hemophilia, fulminant hepatitis, lymphoma and sepsis), an additional 31 had stage 2 fibrosis at a mean of 3.7 years from exposure to biopsy. Such studies are difficult to compare because the majority of children studied have confounding factors that probably favor progression of the liver disease (such as repeated exposures and co-infections, high titer of HCV innoculum, HIV and other viruses, hemosiderosis) and mitigate against long-term survival and follow-up. In addition, treatment options are limited due to likely bone marrow suppression and anemia from antiviral drugs. All multiply transfused patients, especially hemophiliacs and thallassemics should be immunized against hepatitis A and B.

7.8 Common childhood infections and hepatitis

Pregnant women and children may resemble the immunosuppressed patient. Consequently, the hepatitis may be caused by untypical viruses such as HSV or HZV and adenoviruses, among other childhood infections.

Protracted infection with shedding of virus can occur following apparent clinical recovery from hepatitis A and rubella, among other viruses. These findings highlight the importance of host factors, especially the immune response, in determining the ability to clear, and consequences of clearing, any viral infection (*Table 7.3*).

7.8.1 Cytomegalovirus

The new-born may acquire CMV from an infected mother during delivery or via breast milk and other body secretions. Most infections are mild, anicteric illnesses.

Table 7.3. Childhood infections and hepatitis in pediatric practice

Virus	Epidemiology	Hepatitis
Rubella virus	Neonates, maternal infection	Congenital transmission, severe giant cell hepatitis
HSV	Neonates, maternal infection, immunosuppressed, HIV	Rare; severe giant cell hepatitis
VZV	Neonates and children	Common, mild in chickenpox; severe in 5% with disseminated infection
EBV	Children, adolescents	Common: 90% anicteric
Cytomegalovirus	Neonates, maternal infection	Congenital and perinatal transmission
	Immunosuppressed, HIV	Severe in immunosuppressed host e.g. transplant recipient
HHV-6	Neonates (sixth disease) Roseola infantum	Rare: giant cell hepatitis
Parvovirus B19	Children	Acute hepatitis and ? bile duct damage
Adenoviruses various subtypes	Children	Rare: wide spectrum—mild to severe— especially in immunosuppressed host e.g. transplant recipient
	Immunosuppressed, HIV	
Coxsackie A, B	Children	Rare
ECHO	Poorly defined	Rare
Measles	Poorly defined	Possible giant cell hepatitis
Mumps	Poorly defined	Possible hepatitis
Reovirus 3	Poorly defined, neonates	Neonatal hepatitis; ? biliary atresia

Severe hepatitis occurs most often in the pre-term infant with interstitial pneumonitis. CMV hepatitis should be considered in the diagnosis of any baby, especially if pre-term, showing failure to thrive, lymphadenopathy, anemia, hepatomegaly, abnormal liver enzyme levels and atypical lymphocytosis.

Up to 2% of neonates have signs of CMV infection. CMV infection is a major cause of congenital abnormality: 10% of new-borns will be severely handicapped following primary infection during pregnancy. Primary maternal infection in the first and second trimesters is associated with fetal brain damage. Various neonatal syndromes have been recorded following maternal reactivation with midbrain damage and cytomegalic inclusion disease.

Five percent of congenital infections with CMV include a neonatal hepatitis with hepatosplenomegaly, pneumonitis, chorioretinitis and a miriad of central nervous system disorders. The spectrum of illness is wide, from mild anicteric hepatitis to liver failure. Adverse outcomes are noteworthy following infections acquired early in life; vertical transmission (congenital CMV infection) and neonatal hepatitis. Also, there may be some association between the hepatosplenomegaly and later presentation

Key Notes Hepatitis attributable to common viral infections

- Anicteric hepatitis is common with CMV, EBV and VZV
- Overt hepatitis with herpes viruses and chickenpox
  - Suggests unusual pathogenesis

- Predilection for hepatic involvement in:
  - Congenital infection
  - Pregnancy
  - Immunocompromised hosts

- Maternal infection with HSV, CMV, rubella may be mild
  - Despite severe congenital infection

- Rising titers are delayed in many viral infections
- Laboratory isolation of virus may denote:
  - Reactivation of latent infection (e.g. herpes, adenoviruses)
  - Another cause for the hepatitis

- Protracted shedding of virus especially in:
  - Neonates
  - Immunosuppressed patients

with portal hypertension without cirrhosis. CMV can be detected in liver biopsies of up to 25% of patients with AIDS.

Severe hepatitis from cytomegalic inclusion disease, often accompanied by encephalitis, develops in around 5% of infected fetuses following maternal primary infection with CMV during pregnancy. Presenting features include hepatosplenomegaly with jaundice and petechiae in a pre-term infant with evidence of intrauterine growth retardation and microcephaly. The enlarged liver and spleen may show extramedullary hemopoiesis, intranuclear inclusion bodies and multinucleated giant cells. Changes in bile ducts, including proliferation, may occur along with fibrosis. Typical laboratory findings show elevated levels of AST and conjugated bilirubin, thrombocytopenia, atypical lymphocytosis and high levels of immunoglobulin (Ig)M.

Cholestasis and hepatomegaly may persist for months. Excretion of CMV into saliva and urine may persist for years. Prognosis is poor; survivors typically show signs of neurological damage and abnormal psychomotor function.

7.8.2 Epstein–Barr virus (human herpesvirus-4)

In the West, primary infection occurs during childhood in around 30% of children and usually is symptomless; around 50% of people acquire their primary infection as young adults. In developing countries more than three-quarters of babies become infected below 2 years of age.

Hepatic features tend to be mild especially in children and babies. Jaundice with fever and hepatosplenomegaly occurs in less than 20% of cases but elevated serum levels of liver enzymes occur in over 90%. Rarely, ALF may develop in association with erythrophagocytosis. Hepatitis can be severe in the immunosuppressed host, including with HIV infection and AIDS. Severe infections tend to arise in the setting of the immunosuppressed (transplant recipient), immunodeficient patient. Such patients are prone to post-transplant lymphoproliferative disease, especially those receiving chronic immunosuppressive therapies.

In infectious mononucleosis, the liver histology reveals focal necrosis of hepatocytes and infiltration of portal tracts and sinusoids by mononuclear cells. Proliferation of Kupffer cells and hepatocytes may be prominent. Cholestasis is not prominent but the hepatitis may persist for months; chronic liver disease does not occur.

IgG antibodies persist for life. Lymphocytes from patients who have recovered from EBV infection can be grown as cell lines.

Ampicillin and amoxycillin are contraindicated with EBV infections because they produce a characteristic maculopapular rash. Sometimes, inadvertently, this feature assists the diagnosis.

7.8.3 Varicella zoster

Hepatitis accompanying chickenpox is common and mild in most healthy children. Severe hepatitis presenting as ALF is rare (see section 5.2.8) but is recognized in the immunosuppressed, especially with malignancy (lymphomas, leukemias) and graft recipients. A Reye's-like syndrome with fever, hepatomegaly, acute encephalopathy and fatty degeneration of the liver has been described with chickenpox. The clinical features are similar to those of ALF with grossly elevated serum levels of liver enzymes and blood levels of ammonia and hypoglycemia.

7.8.4 Rubella (German measles)

Acute hepatitis with jaundice and hepatosplenomegaly can occur as part of the multi-system involvement of congenital rubella. Hepatosplenomegaly was frequent in the epidemic in North America in 1964. The liver dysfunction usually resolves. Rubella virus has been recovered from the liver in rare cases of fatal giant-cell hepatitis in neonates. Survivors can shed the virus for many months. Whether rubella involves the liver in acquired infection, is unclear. Development of extrahepatic biliary atresia has been linked circumstantially to previous visceral involvement by rubella virus.

7.8.5 Adenoviruses

Hepatitis due to adenoviruses seems to be rare although types 1, 2, 3, 5, 11, 16, among others have been recovered from feces and blood samples during sporadic and epidemic cases of acute hepatitis, including in North America. Adenoviruses have been detected by electron microscopy in the liver of isolated cases of ALF and severe hepatitis in children, including in the graft following liver transplantation, and in patients with HIV infection and AIDS. Proof of a causal effect remains difficult for any virus that can establish latency. Reactivation may occur especially on the background of immunosuppression.

7.8.6 Enteroviruses

Hepatitis has been reported rarely in neonates and children with Coxsackie A or B or various echovirus infections. Diagnosis mostly relies on seroconversion and recovery of virus from stool or, rarely, liver in isolated cases of disseminated infection in neonates.

7.8.7 Parvovirus B19

Individual cases report an association between infection with parvovirus B19 and the development of acute hepatitis, sometimes evolving to ALF, in association with aplastic anemia. No scientific evidence supports parvovirus as a hepatotropic virus.

7.9 Acute liver failure and Reye's-like syndromes

Individual cases of severe hepatitis, including ALF, have been reported for HSV, VZV, HHV-6 and parvovirus B19. A Reye's-like syndrome has been described in children below 15 years of age with chickenpox (varicella) and influenza A or B and from various viral vaccines. Clinical features include severe vomiting which may herald ALF with hypoglycemia, fever, hepatomegaly, acute encephalopathy and fatty degeneration of the liver. Young children may present with features of respiratory distress, seizures and profound hypoglycemia rather than vomiting. Gross elevations occur in serum levels of liver enzymes and blood levels of ammonia. Jaundice is rare because serum bilirubin levels remain within the normal range in most cases.

Pathogenetic mechanisms remain unclear. Many authorities attribute the illness to an underlying inborn error of metabolism, as yet undefined. The association with salicylates (aspirin) remains controversial. In the USA and UK numbers of convincing cases have decreased since the 1980s. Studies of Reye's syndrome and related disorders have been hampered also by the lack of consensus on definition, inadequate surveillance and incomplete clinical and laboratory data, especially histopathological findings.

Hepatitis of unknown etiology accounts for up to 40% of all categories of ALF in children in the West. HAV and HBV are uncommon causes of ALF in childhood because infection usually is symptomless especially under the age of 7 years. Individual cases of severe hepatitis A have been reported in newborns and children, including some due to unusual isolates. These have been associated with protracted and relapsing illness over several months, requiring liver transplantation. However, clearance of HAV can be achieved after transplantation despite immunosuppression. Severe hepatitis A in children is reported in some countries, such as Chile and Thailand, but the role of concurrent diseases is ill-defined.

Clinical features are not discriminative. Diagnosis relies on detection of serological markers of all the viral infections (see section 5.5) and exclusion of other causes such as Wilson's disease that may present with ALF. Management principles are the same as for adults (see section 5.6).

7.9.1 Viral hepatitis, acute liver failure and bone marrow aplasia

Viral hepatitis can be associated with depression of all elements of the bone marrow. Aplastic anemia has been documented with EBV, CMV, parvovirus B19 and non-A–E hepatitis. Tzakis *et al.* in 1988 reported development of aplastic anemia in nine out of 32 (28%) patients at 1–7 weeks following liver transplantation for fulminant non-A, non-B hepatitis. The causal agent(s) are unknown. Reliance on conventional immunological markers (antibodies) in such patients may yield false-negative results. In such cases, diagnosis relies on detection of viral antigens and nucleic acids (e.g. HCV RNA).

Recovery of bone marrow is favored by the use of additional immunosuppression (anti-thymocyte globulin) and cyclophosphamide as well as anti-rejection medications. This observation suggests suppression of a destructive element in the pathogenesis of aplastic anemia in these patients.

7.10 Diagnosis

Several clinical and laboratory findings overlap with other types of viral hepatitis, especially those due to CMV, EBV, adenovirus and rubella among many other infections. A non-specific lymphocytosis with atypical lymphocytes usually occurs 14–21 days after the onset of illness and is coincident with peak abnormalities in liver enzymes.

Diagnosis relies on testing for serological markers of the major hepatotropic agents (hepatitis A–E) and discrimination of acute from chronic infection (*Tables 1.3* and *1.8*). The spectrum of serological profiles is the same regardless of age and pregnancy. Patients with bone marrow suppression may not respond with rising titers of antibodies. In such cases diagnosis relies on detection of viral antigens and nucleic acids.

False positive and negative testing for vertically transmitted HCV:

- Maternal anti-HCV may persist (without HCV RNA) in babies for several months after birth
- Newborn babies infected with HCV may not seroconvert to anti-HCV antibody for up to 2 years
- HCV RNA testing is necessary to diagnose viremia in newborns

The monospot test can be falsely positive in other types of viral hepatitis. A positive Paul–Bunnell test is suggestive, but not conclusive, evidence for EBV infection. In EBV infection heterophil agglutination persists after absorption of the test serum by bovine erythrocytes.

Diagnosis of hepatitis due to HSV and VZV requires a high index of suspicion. Presenting features are non-specific and mucocutaneous stigmata and jaundice may be absent. Most infected women with HSV hepatitis have cervical and genital lesions. Tests include serological markers and the demonstration of inclusion bodies on histological examination of liver tissue. Herpes hepatitis in the immunosuppressed usually reflects a disseminated infection: virus can usually be recovered from cultures of blood and bone marrow. Hemorrhage in the liver can be marked in herpes hepatitis. Viral DNA detected by *in situ* hybridization in liver tissue can provide a rapid diagnosis if sampling is feasible.

Detection of virus in body fluids (urine and peripheral blood mononuclear cells) should be interpreted alongside results of serial serological tests, specifically, rising titers of antibody from paired sera.

Isolation of virus by culture of blood and tissues (e.g. liver) on human fibroblasts, along with a four-fold rise in titers of antibodies in paired samples, is sufficient to diagnose acute CMV infection in most cases. Analysis of single samples often is insufficient for making the diagnosis. Monoclonal antibodies and DNA probes can detect CMV in tissues, but as with isolated findings in saliva and urine, a positive result does not differentiate between acute from previous infection. CMV can be detected in around one-quarter of cervical swabs of healthy women and this does not correlate with risk and severity of congenital infection. Maternal CMV infection is symptomless despite transmission to the baby. Diagnosis relies on detection of virus in blood and rising titers of antibody.

CMV may persist in body fluids for months or years, especially following perinatal infection and CMV mononucleosis. Also, cultures may take weeks to show the characteristic cytopathic effects and rises in titers may be delayed. DEAFF tests (detection of early antigen fluorescent foci) are inferior to PCR for diagnosis of infection.

Congenital CMV infection:

- Found in 0.5–1.5% births (USA, 1978)
- Most common cause of hepatitis in babies
- A risk in immunosuppressed mothers (graft recipients, HIV positive)
- Antiviral agents contraindicated in pregnancy

Liver histology may show typical features of viral hepatitis (see section 1.7.3), nuclear inclusion bodies and giant cells. These features are non-specific and may be absent, particularly in adults with CMV mononucleosis.

Cultures of nasopharangeal secretions and urine from the new-born are more reliable for diagnosis than serological tests. CMV has been isolated from cultures of bile and body fluids. Virus can be detected in liver biopsy specimens by visualizing inclusions and using monoclonal antibodies and cDNA probes (*in situ* hybridization in liver tissue). Nuclear inclusions may be prominent on liver histology but these are not specific for HSV.

Diagnosis of other viruses such as rubella requires direct isolation of virus. IgM anti-rubella antibodies suggest recent infection but may persist for many months, especially following congenital infection. Conversely, antibodies may become undetectable after 2–3 years following neonatal infection making the retrospective diagnosis difficult.

7.10.1 Differential diagnosis

Specific considerations in maternal medicine are exclusion of disorders such as intra-hepatic cholestasis of pregnancy and acute fatty liver of pregnancy that require prompt delivery with fetal maturity to achieve resolution. The features of hemolysis, elevated liver enzymes, low platelet counts (HELLP) syndrome are common to many varieties of viral hepatitis. Underlying chronic liver diseases in pregnant women and children include Wilson's disease and autoimmune hepatitis that respond to specific treatments. Other conditions include α-1-anti-trypsin deficiency and a variety of metabolic disorders. Serological tests should include autoantibodies (anti-smooth muscle, anti-mitochondrial, antibodies) and screening for Wilson's disease. Plasma (separated from blood collected in lithium heparin) and urine samples should be collected on admission and stored frozen (-20°C, but preferably -70°C) for suspected metabolic disorders. A portion of any liver biopsy should collected fresh for copper studies (Wilson's disease) and for fat (oil red 'O') as for acute fatty liver. Some tissue (liver, skin) should be snap-frozen rapidly in liquid nitrogen and stored (-70°C) for testing for enzyme deficiencies. A tiny fragment fixed in gluteraldehyde can be processed for electron microscopy to exclude underlying storage disorders, Reye's syndrome and unusual viral infections.

The main differential diagnoses in the new-born include diseases that cause cholestasis, sepsis, malnutrition, biliary duct obstruction and several inborn errors of metabolism including α-1-anti-trypsin and galactosemia. Viral hepatitis is likely when liver histology shows a predominance of inflammatory cells in portal tracts and acini with necrosis of liver cells. Disarray of liver cell plates occurs without proliferation of bile ductules and in the absence of fatty infiltration. Paucity of bile ducts can occur with CMV infection as well as many other conditions such as biliary atresia, α-1-anti-trypsin deficiency and histiocytosis X.

Giant cell hepatitis remains a poorly defined collection of conditions. The outcome is very variable: recovery, liver failure, cirrhosis or chronic cholestasis. Some cases have been associated with paramyxoviruses such as measles as well as other viruses especially, rubella, herpes viruses and HIV-I (*Table 7.3*).

Although rare, children may present with microbial infections that mimic severe, including fulminant, hepatitis. Importantly, screening should include for syphilis, *Neisseria meningitidis*, *Shigella dysenteriae*, *Leptospira* spp., *Coxielli burnetti*, *Plasmodium* spp., *Entamoeba* spp., *Toxoplasma gondii* and other agents implicated in septic shock and hepatic necrosis.

Any patient, regardless of age, may present acutely as a 'flare' of chronic liver disease especially with chronic hepatitis B around the time of seroconversion from HBeAg to anti-HBe, superinfection with HDV or reinfection with virus variants. Also, acute viral hepatitis may disclose underlying chronic liver disease such as hepatitis A on alcoholic liver disease in the mother or α-1-antitrypsin deficiency in the child. Hepatitis A also may disclose autoimmune hepatitis and autoantibodies may be detectable concurrently with IgM anti-HAV antibodies. The approach in older children is similar to that seen in adults. Additional causes should be considered when the presentation is unusually severe.

7.11 Management during pregnancy

The general principles of managing viral hepatitis are the same regardless of age and pregnancy (section 1.13). Care is supportive. There is no benefit from early delivery of the baby (contrast acute fatty liver of pregnancy, HELLP syndrome and intrahepatic cholestasis of pregancy). Corticosteroids have been given to selected patients with protracted cholestatic hepatitis due to EBV (compare with protracted hepatitis A).

7.11.1 Amniocentesis

Amniocentesis, including with documented puncture of the placenta does not seem to add to the risk of vertical transmission of in HBsAg seropositive carrier mothers of Asian origin. Data are insufficient for hepatitis C.

7.11.2 Cesarean section

Cesarean section and early delivery of the baby are of no special benefit in preventing vertical transmission of acute and chronic viral hepatitis due to the major hepatotropic viruses, especially hepatitis B and/or D.

Transmission during labor and peripartum may result from mixing of maternal with fetal blood and contact with HBV DNA-positive cells and fluid such as cervicovaginal epithelial cells and by ingestion of amniotic fluid and breast milk. Transplacental transmission may be more frequent than previously recognized as HBV DNA was detected in 44% of livers of fetuses from HBsAg seropositive Chinese mothers. Transplacental transmission may explain failure to protect some neonates born of HBsAg-positive mothers despite immunoprophylaxis at birth.

Studies exceeding 800 mother–infant pairs are required to prove definitively that Cesarean section rather than vaginal delivery protects against vertical transmission of HCV. Around 6% (range 0–11%) of babies born of anti-HCV seropositive mothers test positive for anti-HCV or HCV RNA. Vertical transmission of HCV is around twice as likely if the mother is highly viremic (HCV RNA $>10^6$ geq/ml) as with immunosuppression or concurrent HIV infection. Serial testing for HCV RNA up to 2 years in the infant is necessary to exclude false negativity due to maternal transfer of anti-HCV antibodies.

Cesarean section should be considered for maternal primary infection with HSV with ulceration of the cervix and vaginal walls, to prevent contamination of the neonate during vaginal delivery.

7.11.3 General considerations

Isolation and barrier nursing are unnecessary for viral hepatitis with good nursing practice, especially washing hands between patients. Isolation of baby from HBsAg seropositive contacts is unnecessary for HBV and HDV, provided neonatal immunization is commenced at birth (see section 7.12.4).

Fecal contamination of food and water supplies can transmit hepatitis A and E. Handling of feces and bile containing HAV and HEV should be minimized by washing hands thoroughly and disposal of contaminated clothes and fomites by autoclaving and incineration. Pregnant personnel should avoid exposure to hepatitis E.

A high-risk mother who presents in labor with unknown HBV status should be managed as if infectious until hepatitis serologies are known. Neonatal combined immunoprophylaxis should be instigated without delay (see section 7.12.4). Babies born to mothers discovered to be HBsAg seropositive only at the postnatal visit should be screened for HBsAg, anti-HBc, and anti-HBs. Seronegative babies should complete an accelerated course (0, 1, 2 and 12 months) of hepatitis B vaccine.

Diagnosis of hepatitis B in the mother should prompt screening of all children and intimate household contacts. Seronegativity for HBsAg, anti-HBc, and anti-HBs will identify susceptible individuals who require immunization. Testing for HBsAg should be repeated in late pregnancy and postpartum in seronegative mothers considered to be at risk. Mothers with suspected acute infection should be tested for anti-HBc IgM.

Direct contact with blood-soaked dressings and pads should be avoided, especially with hepatitis B, C and B+D. All attendants should wear protective clothing, cover exposed cuts and abrasions and handle body fluids with care. Goggles and masks can prevent splashes into the eyes and mouth. Secondary spread to health care personnel is uncommon. The few reports of transmission of hepatitis B from attendants to patients have included obstetrical and gynecological surgery.

The risks of transmitting CMV infection to high-risk patients (immunosuppressed, infants) can be reduced by pre-screening blood, white blood cells and organ donors and matching seronegative donors and recipients. CMV hepatitis can occur in known seropositive recipients of seropositive donors who donated blood and organs. In children, the risks of spreading infection may be increased from protracted shedding of viruses (CMV) especially in congenital infections.

7.11.4 Breast-feeding

Breast-feeding is not contraindicated for viral hepatitis in pregnancy. Transmission of hepatitis B infection within the first 6 months is not prevented by avoiding breast-feeding and immediate separation from the infectious mother. Breast-feeding is not contraindicated with neonatal and maternal immunization. The hepatitis A and B vaccines are safe for breast-feeding mothers.

No precautions specific to hepatitis C have been recommended by the Centers for Disease Control and Prevention and American College of Pediatrics for infected pregnant women and their babies. Breast-feeding is not contraindicated. HCV RNA and anti-HCV antibodies have been detected in colostrum and breast milk but show no correlation with levels in maternal serum.

7.11.5 Antiviral therapies

No specific therapies are recommended for hepatitis A–E during pregnancy. Interferons, ribavirin and nucleoside analogs, such as lamivudine and famciclovir, are contraindicated in pregnancy. Ganciclovir and foscarnet (phosphonoformate) are useful in severe CMV infection but are contraindicated in pregnancy. Acyclovir can be successful if given early in severe HSV hepatitis.

Concerning ribavirin, category X: ribavirin is teratogenic; significant teratogenic and embryocidal effects have been documented:
- In all detailed animal studies: malformations of skull, eye, jaw, limbs and skeleton and gastrointestinal tract. Survival of fetuses and offspring was reduced
- At doses as low as 5% of recommended human doses
- Increased incidence and severity of defects at escalating doses
- Testicular degeneration in mice recovered in one to two spermatogenesis cycles

Recommendations for fertile men and women taking ribavirin are as follows:

- Double birth control is obligatory
- A negative pregnancy test should be documented prior to commencing ribavirin
- Monthly pregnancy tests are recommended throughout treatment
- Six months minimum 'wash-out' period prior to conception following cessation of ribavirin
- This recommendation assumes similar wash-out for females

Concerning interferons, category C: interferons are contraindicated in pregnancy and nursing mothers:

- Interferon-α-2b may impair human fertility
- Abnormal menstrual cycles noted in non-human primates
- Unknown kinetics in breast milk in humans

7.11.6 Maternal screening

7.11.6.1 Hepatitis A and hepatitis E

Pregnant women at especial risk of catching hepatitis A virus, such as those anticipating traveling to regions of high endemicity and those in contact with an index case, should be pre-screened for total anti-HAV antibodies to exclude unnecessary immunization. Pregnant women should be advised to avoid traveling to regions of high endemicity for HEV and to take extra precautions to minimize exposure to contaminated water (see section 2.10).

7.11.6.2 Hepatitis B and B+D

The American College of Obstetricians and Gynecologists and the Centers for Disease Control and Prevention continue to endorse universal prenatal screening in the USA. All pregnant women should be screened for HBsAg and IgM anti-HBc. This strategy is cost-saving and should identify 20 000 HBsAg seropositive women and prevent 3000 chronic infections per year in the USA. Following the identification of seropositive women, screening for HBsAg should be extended to the family and close (especially sexual) contacts and children in the USA. Susceptible contacts should be offered immunization with vaccine and hepatitis B immunoglobulin (HBIG; *Figure 7.2*).

Guidelines for selective screening of high-risk women extrapolated from the National Center of Health Statistics up to 1988 show less than 50% sensitivity. High-risk mothers include first- and second-generation immigrants and refugees from Indochina and Southeast Asia, the Philippines, the Arctic (Inuit), Haiti and the Dominican Republic, and the American Pacific Islands. Apart from Asian descent, no other risk factor was identified easily from histories.

7.11.7 Maternal immunization

There are no specific recommendations for passive immunoprophylaxis for hepatitis A during pregnancy. Immune globulin (IG), containing IgG anti-HAV antibodies, is safe

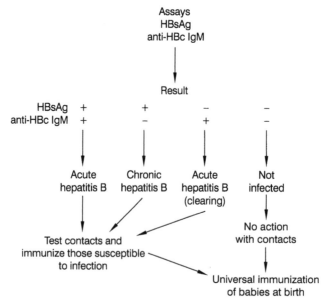

Figure 7.2. Screening for hepatitis B in pregnancy.

to administer to susceptible (IgM and IgG anti-HAV-negative) symptomless pregnant women prior to travel or when exposed to hepatitis A (in contact with an index case). Pre-screening for IgG anti-HAV antibodies avoids unnecessary immunization. IG may help prevent development of an icteric illness if given during the incubation phase of hepatitis A. IG is not effective as passive immunization against hepatitis E.

The hepatitis A and B vaccines are not licensed for use in pregnancy. In clinical practice, there is no specific contraindication or adverse consequence following administration of licensed killed vaccines such as those against HAV or the use of HBIG and a licensed hepatitis B subunit vaccine. Many women have been given hepatitis A and B vaccines (and HBIG) inadvertently (before pregnancy is declared) without excess adverse consequences on fetal and maternal outcomes. Relative risks must be assessed for each patient who must be closely monitored and counseled. No licensed immunoprophylaxis is available against hepatitis C or E.

7.11.7.1 Passive immunization against CMV

Specific, high-titer IG offers some protection against primary infection with CMV after accidental exposure and can reduce the severity of the infections in seronegative recipients of blood products and organs from seropositive donors. This is expensive and no information is available for efficacy against primary infection during pregnancy. No vaccines are available against herpes viruses. Recombinant vaccines are under development. Concern remains over establishing long-term latency for viruses with oncogenic potential.

7.12 Management of viral hepatitis in children

General principles of management do not differ from those for adults (see section 1.12).

7.12.1 Chronic hepatitis B and B+D

Overall, results with α-interferons (IFNs) have been disappointing probably due to the large number of children with immune tolerance to the virus when acquired via vertical transmission. IFNs generally are well tolerated in children and side-effects are comparable to those seen in adults.

Large-scale trials are in progress across the USA, Europe and Far East to assess the impact of lamivudine in hepatitis B in children, following safety profiles that have been completed for HIV-infected children as well as adults. The long-term goals are to reduce the frequency of progression to cirrhosis and liver cancer as well as reduction in virus replication. Concern remains over the lack of immunomodulatory effect of the nucleoside analogs especially as most of the subjects acquired their infection at/around birth. Also, long-term use is associated with emergence of resistant variants (see section 3.8.2.3).

Clearance rates of HDV with HBV infection are as disappointing in children as in adults with chronic hepatitis B and D.

7.12.2 Chronic hepatitis C

IFN-α-2b is used in similar protocols to those in adults (see section 4.7.8) except that dosages tend to be based on surface area (typically 3 MU m² three times a week). In Western children, rates of clearance of HCV are similar to those seen in adults and higher than those in Oriental children. Interpretation of response is limited by the small numbers of patients in each study and reliance on normalization of serum ALT levels rather than clearance of HCV RNA. Rates of graft infection and sequelae for children transplanted for viral hepatitis are similar to those seen in adults (see section 6.4).

Trials are in progress using ribavirin in combination with IFN-α-2b in children, including in hemophiliacs. Concern remains over the teratogenic effects of ribavirin, uncertainties of dose and limitations with hemolytic anemia. Ribavirin is contraindicated in hematological disorders associated with hemolytic anemia and thalassemia.

7.12.3 Cytomegalovirus

Ganciclovir (dihydroxypropoxymethylguanine: DHPG), a guanosine derivative, is effective in reducing the severity of several manifestations of CMV infection, including hepatitis, following organ and bone marrow transplantation. Typical dose regimens are 2.5–5.0 mg/kg daily. Some reports indicate improved benefit when ganciclovir is combined with specific IG. Clinical and virological relapses are common following discontinuation of therapy, especially in the immunocompromised host.

7.12.4 Pediatric immunization

The World Health Assembly has endorsed the universal immunization targets set for 1997 to integrate hepatitis B vaccine into the Expanded Program of Immunization (EPI). There is no contraindication to concurrent administration of vaccines against polio, diphtheria, tetanus and pertussis. Results with licensed recombinant and plasma-derived vaccines are similar. Universal immunization (*Figure 7.3*) replaces selective targeting of high-risk groups that failed to make any epidemiological impact on the incidence of HBV infection.

7.12.4.1 Expanded Program of Immunization

Other vaccines, such as against poliomyelitis, diphtheria, tetanus and pertussis, can be given at the same time. Separate sites, needles and syringes should be used. Multivalent vaccines (vaccine combinations) are under development.

The hepatitis A vaccine is not licensed for children less than 2 years of age. In neonates and children, emphasis is on protection against hepatitis B in view of the propensity for chronic infection. Also, naturally acquired HAV infection usually is symptomless in the very young and heralds life-long immunity.

7.12.4.2 Passive immunization against HAV, HBV

Normal immune globulin (NIG) contains anti-HAV antibodies (and anti-HBs antibodies) and should be considered in the pre-term neonate especially if exposed to HAV in

Data as of June 1998   ☐ National Program

                                              ☑ Planning

Figure 7.3. Countries implementing routine childhood hepatitis B immunization 1997 (data from the World Health Organization) at URL: www.who.int/gpv-surv/graphics/htmls/hepb.htm.

the intensive care unit. Doses are typically 0.06–0.12 ml/kg or 250–500 mg depending on age (above and below 10 years) and offer some protection for around 2–3 months. NIG may not be completely protective during outbreaks.

7.12.4.3 Active immunization (hepatitis A vaccine)

Use of HAV vaccine may be justified in countries where debilitating illness can occur from concomitant infections such as dengue and shistosomiasis, among others. The hepatitis A vaccines are licensed for infants above 2 years of age (*Table 2.5*). Whether universal immunization of very young children is warranted, requires assessment. The long-term duration of protective immunity (> 15 years) is unknown. The majority of infections are symptomless below the age of 7 years and this natural immunity is life-long.

7.12.4.4 Neonatal immunoprophylaxis against hepatitis B

The licensed HBV vaccines protect more than 90% of babies from HBV infection. Neonatal combined immunoprophylaxis (vaccine and HBIG) against hepatitis B is optimal if commenced within 12 h of birth. However, several authorities consider immunization with vaccine alone to afford sufficient protection from vertically trans-mitted HBV. Combination vaccines against hepatitis A and B cannot be used as part of neonatal immunoprophylaxis against hepatitis B as they are not licensed for use from birth.

The accelerated schedule of 0, 1, 2 and 12 months usually gives more rapid serocon-version and longer duration of detectable antibody (anti-HBs) than 0, 1 and 6 months. The accelerated schedule is preferred for babies born of HBsAg-positive mothers and should include HBIG given at birth with the first dose of vaccine (separate needles, syringes and intramuscular sites). The immunization schedule for babies born of HBsAg seronegative mothers is more flexible; the first dose of vaccine (without HBIG) should be given at birth (range 0–2 months), the second dose at 1–4 months after the first injection. The third dose should be given at age 12–15 months or interval after the second dose. HBV vaccine is safe and effective when administered along with other routinely recommended childhood vaccines.

For completion of the three dose basic course, the licensed single (non-combination) vaccines are interchangeable when administered in doses recommended by the man-ufacturer. The manufacturer's instructions should be read carefully as doses vary between the vaccines. [In 1999, the 2.5 μg (0.5 ml) pediatric dose of Recombivax was withdrawn to unify dosing schedules. The 5μg/0.5 ml dose of Recombivax is suitable for all vaccinees regardless of age (0–19 years) and regardless of HBsAg status of the mother.]

In July 1999, the American College of Pediatrics issued with the Public Health Service a joint statement that reinforced recommendations by the Food and Drugs Adminis-tration and other Agencies to eliminate use of vaccines containing mercury (thiomer-sal), especially among new-born babies, and in particular those born prematurely. If mercury-free vaccine is not available, the college recommends delaying the first dose of hepatitis B vaccine until 6 months, or until the small and premature baby has reached a size and developmental level corresponding to a full-term infant. The HBIG products available in the USA since 1999 do not contain mercury or thiomersal.

A combined Hib and hepatitis B vaccine (COMVAX) was licensed by the FDA in 1997 (*Table 7.4*).

Table 7.4. Hib and hepatitis B combination vaccine

COMVAX (MSD)
• 7.5 μg *Haemophilus influenzae* PRP[a]
• 125 μg *Neisseria meningitides* OMPC[a]
• 5 μg HBsAg (AlOH)
• Does not contain thiomersal (mercury-free)
Schedule for babies of HBsAg-negative mothers:
• 2, 4 and 12–15 months
• Contraindicated < 6 weeks (suppressed response)

[a]Hib component, PRP-OMP (polyribosylribitol phosphate polysaccharide) conjugated with a meningococcal outer membrane protein.

7.12.4.5 Increasing the coverage of hepatitis B vaccine

In the USA in October 1994, the Advisory Committee on Immunization Practices recommended universal immunization of all babies. They extended their recommendations to all adolescents (aged 11–12 years) and also young adults where intravenous drug use and sexually transmitted diseases are common. Their specific recommendations for immunization included:

- High-risk: immunization of all unvaccinated children aged < 11 years who are Pacific Islanders, Natives of Alaska and/or reside in households of first-generation immigrants from countries where HBV is of high (> 8% HBsAg-positive) or intermediate (2–8% HBsAg-positive) endemicity
  - Where high rates of HBV infection (>2% per annum)
  - There is high prevalence of chronic HBV infection (2–5%)
  - Since < 10% have received hepatitis B vaccine compared with 70% of US children
- All risks: immunization of all 11–12 year old children who have never received previously the hepatitis B vaccine (catch-up schedule)
  - 'Catch up' program combined with:
    - Two doses of measles–mumps–rubella (MMR) vaccine
    - Booster doses diphtheria and tetanus toxoids (DTP)
    - Assessment of immunity to varicella (chickenpox)
  - Recommendations
    - Are most effective if integrated into school and clinic practices
    - Should accelerate the decline in HBV infection in the US

7.12.4.6 Immunization failures

The licensed hepatitis B vaccines are highly immunogenic and efficacious, especially in neonates. Approximately 5% of neonates and children make a poor response (anti-HBs < 100 IU/l). Immunization failures occur with genetically predetermined poor response, infection *in utero*, immunosuppression from HIV, other diseases and drugs, and emergence of anti-HBs antibody escape variants of HBV containing mutations in the dominant epitope (*a* determinant) of the surface gene (see section 3.2.5.4).

Further reading

Advisory Committee on Immunization Practices (1996, 1997) Prevention of hepatitis A through active or passive immunization: Recommendations of the Advisory Committee on Immunization Practices (ACIP). *MMWR* (RR15) **45**: 1–38 and (RR45) **46**: 588.

Advisory Committee on Immunization Practices (1999) Recommended childhood immunisation schedule—United States, 1999. *MMWR* **48**(1); 12–13. *Regularly updated immunization schedule.*

Advisory Committee on Immunization Practices (1999) Combination vaccines for childhood immunization. *MMWR* (RR-5) **48**: 1–15.
Recommendations of the Advisory Committee on Immunization Practices (ACIP), the American Academy of Pediatrics (AAP), and the American Academy of Family Physicians (AAFP). Endorsement for the careful use of combination vaccines.

Alter, M.J., Mast, E.E. (1994) The epidemiology of viral hepatitis in the United States. *Gastroenterol. Clin. North Am.* **23**: 437–455.

Bacq, Y., Zarka, O., Brechot, J.-F., *et al.* (1996) Liver function tests in normal pregnancy: a prospective study of 103 pregnant women and 103 matched controls. *Hepatology* **23**: 1030–1034.
Normal ranges for AST and ALT levels are lower in young women than general populations.

Bahn, A., Hilbert, K., Martine, U., *et al.* (1995) Selection of precore mutant after vertical transmission of different hepatitis B virus variants is correlated with fulminant hepatitis in infants. *J. Med. Virol.* **47**: 33–41.

Caporaso, N., Ascione, A., Stroffolini, T. (1998) Spread of hepatitis C virus infection within families. Investigators of an Italian Multicenter group. *J. Viral Hepatitis* **5**: 67–72.
Evidence against significant risk of sexual transmission. Evaluation of 1379 household contacts of 585 HCV RNA-positive cases in Italy. Anti-HCV positivity was found in 15.6% of spouses compared with 3.2% of other relatives; 19.8% if married above 20 years compared with 8.0% for less time. Additional parenteral exposure accounted for the excess prevalence among spouses.

Centers for Disease Control. (1990) Protection against viral hepatitis. Recommendations of the Immunization Practices Advisory Committee (AICP). *MMWR* **39**: 5–22.

Centers for Disease Control and Prevention. (1996) Prevention of perinatal hepatitis B by enhanced case management. *MMWR* **45**: 584–587.

Chang, M.-H., Chen, C.-J., Lai, M.-S., *et al.* (1997) Universal hepatitis B vaccination in Taiwan and the incidence of hepatocellular carcinoma in children. *N. Engl. J. Med.* **336**: 1855–1859.
Reduction by >50% of HCC in children aged 6–14 years following instigation of immunization strategy.

Darby, S.C., Ewart, D.W., Giangrande, P.L.F., *et al.* (1997) Mortality from liver cancer and liver disease in haemophiliac men and boys given blood products contaminated with hepatitis C. *Lancet* **350**: 1425–1431.
UK Haemophilia Centre Directors' organization. Excess mortality from liver disease between 1977 and 1991.

Da Silva, O., Hammerberg, O., Chance, G.W. (1990) Fetal varicella syndrome. *Pediatr. Infect. Dis. J.* **9**: 854–855.

Delage, G., Montplaisir, S., Remy–Prince, S. *et al.* (1986) Prevalence of hepatitis B virus infection in pregnant women in the Montreal area. *Can. Med. Assoc. J.* **134**: 897–901.
Questionnaire to >30 000 pregnant women in nine hospitals in Canada. Demographic characteristics (risk factors) are unreliable: > 50% of carriers would be undetected.

Fagan, E.A. (1999) Diseases of the liver, biliary system and pancreas. In: *Fetal–Maternal Medicine* 4th Edn (eds R.K. Creasy, R. Resnik). WB Saunders Co., Philadelphia, pp. 1054–1081.
Comprehensive review of management of liver diseases, including liver transplantation, as well as acute and chronic viral hepatitis in pregnancy and immunoprophylaxis.

Fagan, E.A. (in press) Disorders of the liver. In: *Medical Disorders in Obstetric Practice*, 4th Edn (ed. M. de Swiet). Blackwell Scientific Press, Oxford.
Comprehensive review of acute and chronic viral hepatitis in pregnancy, including liver transplantation and immunoprophylaxis.

Fagan, E.A., Hadzic, D., Saxena, R., *et al.* (1999) Vertical transmission from early pregnancy of a naturally occurring hepatitis A virus variant: evidence for persistent infection in man. *Pediatr. Infect. Dis. J.* **18**: 389–391.
Severe viral hepatitis in a baby with identical virus variant in a mother who contracted hepatitis A during the first trimester.

Fagan, E.A., Meon, T., Valliammai, T., *et al.* (1994) Equivocal serological diagnosis of sporadic fulminant hepatitis E in pregnant Indians (Letter). *Lancet* **334**: 342–343.
Lack of specific diagnostic markers including IgM anti-HEV antibodies.

Fagan, E.A., Yousef, G., Brahm, J., *et al.* (1990) Persistence of hepatitis A virus in fulminant hepatitis before and after liver transplantation. *J. Med. Virol.* **30**: 131–134.
Severe protracted and relapsing hepatitis A in an adult and child that evolved into ALF. The child cleared HAV spontaneously despite infection of the graft.

Floreani, A., Paternoster, D., Zappala, F., *et al.* (1996) Hepatitis C infection in pregnancy. *Br. J. Obstet. Gynaecol.* **103**: 325–329.
A study of 1700 consecutive women seen for high -risk pregnancy in a university department of obstetrics in Italy; 29 (1.7 per cent) were anti-HCV seropositive; eight had concurrent HIV infection.

Friedt, M., Gerner, P., Lausch, E., *et al.* (1999) Mutations in the basal core promotor and the precore region of hepatitis B virus and their selection in children with fulminant and chronic hepatitis B. *Hepatology* **29**: 1252–1258.
Sequence analyses in nine babies with fulminant hepatitis B and their six mothers showing significant mutation rates in the basal core promoter (1762/1764) and precore stop mutation in anti-HBe positive fulminant babies compared with controls. Mothers showed more heterogeneity of sequences than their babies and all but one of the strains were genotype D.

Garcia-Monzon, C., Jara, P., Fernandez-Bermejo, M., *et al.* (1998) Chronic hepatitis C in children: a clinical and immunohistochemical study with adult patients. *Hepatology* **28**: 1696–1701.
A comparison of post-transfusion hepatitis C between 15 children and 32 adults followed for a mean of 11 years and later showing good outcomes in the children despite persistently elevated liver enzymes.

Giangrande, P.L.F. (1998) Hepatitis in haemophilia. *Br. J. Haematol.* **103**: 1–9.
Comprehensive review with 104 references, including liver transplantation and immunizations.

Guido, M., Rugge, M., Jara, P., *et al.* (1998) Chronic hepatitis C in children: the pathological and clinical spectrum. *Gastroenterology* **115**: 1525–1529.
Histopathological analysis of liver biopsies (1990–1996) from 80 Italian children with chronic hepatitis C and elevated liver enzymes. HCV was acquired via blood transfusion and vertical transmission. None had underlying systemic diseases. Chronic hepatitis was severe in 21.2% and 16.2% had moderate fibrosis. There was no correlation with genotype and 11 had circulating autoantibodies.

Halfon, P., Quentin, Y., Roquelaure, B., *et al.* (1999) Mother-to-infant transmission of hepatitis C virus: molecular evidence of superinfection by homologous virus in children. *J. Hepatol.* **30**: 970–978.
Perinatal and vertical transmission implicated in transmission from an HCV infected mother to her four sons.

Hans, H., Jacek, W., Olle, R., *et al.* (1997) Successful allogeneic bone marrow transplantation in a 2.5 year old boy with ongoing cytomegalovirus viremia and severe aplastic anemia after orthotopic liver transplantation for non-A, non-B, non-C hepatitis. *Transplantation* **64**: 1207–1208.
Three year follow-up after uneventful recovery with rapid engraftment of HLA identical bone marrow despite CMV viremia.

Hunt, C.M., Carson, K.L., Shahara, A.I. (1997) Hepatitis C in pregnancy. *Obstet. Gynecol.* **89**: 883–890.

Janjua, S., Hussey, M.J., Kaur, S., *et al.* (Submitted for publication) Management issues in survivors of severe liver disease presenting during pregnancy.
Includes liver transplantation for fulminant hepatitis B.

Kage, M., Fuyisawa, T., Shiraki, K., *et al.* (1997) Pathology of chronic hepatitis C in children. *Hepatology* **26**: 771–775.

Klein, N.A., Mabie, W.C., Shaver, D.C., *et al.* (1991) Herpes simplex hepatitis in pregnancy. Two patients successfully treated with acyclovir. *Gastroenterology* **100**: 239–244.

Ko, T.M., Tseng, L.H., Chang, M.H. (1994) Amniocentesis in mothers who are hepatitis B virus carriers does not expose the infant to increased risk of hepatitis B virus infection. *Arch. Gynecol. Obstet.* **225**: 25–30.

Lai, M.E., Virgilis, S., Argiolu, F., *et al.* (1993) Evaluation of antibodies to hepatitis C virus in a long-term prospective study of posttransfusion hepatitis among thalassemic children: comparison between first- and second-generation assay. *J. Pediatr. Gastroenterol. Nutr.* **16**: 458–464.
Progression to cirrhosis was significant in thalassemic children.

Laifer, S.A., Ehrlich, G.D., Huff, D.S., *et al.* (1995) Congenital cytomegalovirus infection in offspring of liver transplant recipients. *Clin. Infect. Dis.* **20**: 52–55.

Lau, W.Y., Leung, W.T., Ho, S., *et al.* (1995) Hepatocellular carcinoma during pregnancy and its comparison with other pregnancy-associated malignancies. *Cancer* **75**: 2669–2676.
Abysmal prognosis of HCC presenting during pregnancy.

Ngui, S.L., O'Connell, S., Eglin, R.P., *et al.* (1997) Low detection rate and maternal provenance of hepatitis B virus S gene mutants in cases of failed postnatal immuno-prophylaxis in England and Wales. *J. Infect. Dis.* **176**: 1360–1365.

Nordenfelt, E., Dahlquist, E. (1978) HBsAg positive adopted children as a cause of intrafamilial spread of hepatitis B. *Scand. J. Infect. Dis.* **10**: 161–163.

Ohto, H., Okamoto, H., Mishiro, S. (1994) Transmission of hepatitis C virus from mothers to infants. *N. Engl. J. Med.* **330**: 744–750.
Post-partum transmission is implicated also in a prospective study from Japan.

Pardi, D.S., Romero, Y., Mertz, L.E., *et al.* (1998) Hepatitis-associated aplastic anemia and acute parvovirus B19 infection: a report of two cases and a review of the literature. *Am. J. Gastroenterol.* **93**: 468–470.
Clinical association of aplastic anemia, acute hepatitis and acute liver failure with parvovirus B19 infection.

Peckham, C.S. (1991) Cytomegalovirus infection: congenital and neonatal disease. *Scand. J. Infect. Dis.* **80** (Suppl): 82–87.

Poles, M.A., Lew, E.A., Dieterich, D.T. (1997) Diagnosis and treatment of hepatic disease in patients with HIV. *Gastroenterol. Clin. North Am.* **26**: 291–321.

Rab, M.A., Rile, M.K., Mubarik, M.M., *et al.* (1997) Water-borne hepatitis E virus epidemic in Islamabad, Pakistan: a common source outbreak traced to the malfunction of a modern water treatment plant. *Am. J. Trop. Med. Hyg.* **57**: 151–157.

Raimondo, G., Tanzi, E., Brancatelli, S., *et al.* (1993) Is the course of perinatal hepatitis B virus infection influenced by genetic heterogeneity of the virus? *J. Med. Virol.* **40**: 87–90.

Sabatino, G., Ramenghi, L.A., di Marzio, M., *et al.* (1996) Vertical transmission of hepatitis C: an epidemiological study on 2,980 pregnant women in Italy. *Eur. J. Epidemiol.* **12**: 443–447.
Anti-HCV antibodies were identified in 30 (1.07%) of these consecutive pregnant women. None was seropositive for anti-HIV antibodies. Ten mothers had HCV RNA detected in serum around delivery. All cord bloods had undetectable HCV RNA but three of these high-risk babies subsequently had HCV RNA detected at around 3 months of age.

Semprini, A.E., Persico, T., Thiers, V,. *et al.* (1998) Absence of hepatitis C virus and detection of hepatitis G virus/GB virus C RNA sequences in the semen of infected men. *J. Infect. Dis.* **177**: 848–854.
Inability to detect HCV RNA in 56 samples of fractions of seminal fluid; 34 had inhibitors to the PCR. Interestingly, 50% of fractions had HGV RNA detectable if no PCR inhibitors were present. Supports safety of seminal fluid processed in specialized referral centers.

Sinatra, F.R., Shah, P., Weissman, J.Y., Thomas, D.W., Merritt, R.J., Tong, M.J. (1982) Perinatal transmitted acute icteric hepatitis B in infants born to hepatitis B surface antigen-positive and anti hepatitis Be-positive carrier mothers. *Pediatrics* **70**: 557–559.

Sterneck, M., Gunther, S., Santantonio, T., *et al.* (1996) Hepatitis B virus genomes of patients with fulminant hepatitis do not share a specific mutation. *Hepatology* **24**: 300–306.
Complete sequences of several HBV strains from fulminant hepatitis B showed no conserved mutations, but precore and core promoter mutations were common.

Tahara, T., Toyoda, S., Mukaide, M., *et al.* (1996) Vertical transmission of hepatitis C through three generations. *Lancet* **347**: 409.

Terazawa, S., Kojima, M., Yamanaka, T., *et al.* (1991) Hepatitis B virus mutants with precore-region defects in two babies with fulminant hepatitis and their mothers positive for antibody to hepatitis B e antigen. *Pediatr. Res.* **29**: 5–9.

Tong, M.J., Sinatra, F.R., Thomas, D.W., Nair, P.V., Merritt, R.J., Wang, D.W. (1984) Need for immunoprophylaxis in infants born to HBsAg-positive carrier mothers who are HBeAg negative. *J. Pediatr.* **105**: 945–947.
Severe viral hepatitis in babies born of HBeAg seronegative mothers with chronic HBV infection.

Tong, M.J., Thursby, M., Rakela, J., *et al.* (1981) Studies on the maternal–infant transmission of the viruses which cause acute hepatitis. *Gastroenterology* **80**: 999–1004.
Transmission of HBV is uncommon in acute hepatitis—may occur in the third trimester.

Torre, D., Tambrini, R. (1996) Interferon-alpha therapy in chronic viral hepatitis B in children: a meta-analysis. *Clin. Infect. Dis.* **23**: 131–137.
Six randomized trials involving 240 children with chronic hepatitis B resulted in clearance of HBV DNA in 36 of 126 followed up for a minimum of 6 months.

Tzakis, A.G., Arditi, M., Whitington, P.F., *et al.* (1988) Aplastic anemia complicating orthotopic liver transplantation for non-A, non-B hepatitis. *N. Engl. J. Med.* **319**: 393–396.

van Os, H.C., Drogendijk, H.C., Fetter, W.P., *et al.* (1991) The influence of contamination of culture medium with hepatitis B virus in the outcome of *in vitro* fertilization pregnancies. *Am. J. Obstet. Gynecol.* **165**: 152–159.

Viazov, S., Riffelmann, M., Sarr, S., *et al.* (1997) Transmission of GBV-C/HGV from drug-addicted mothers to their babies. *J. Hepatol.* **27**: 85–90.

Vogt, M., Lang, T., Frosner, G., *et al.* (1999) Prevalence and clinical outcome of hepatitis C infection in children who underwent cardiac surgery before the implementation of blood-donor screening. *N. Engl. J. Med.* **341**: 866–870.

von Weizsacker, F., Pult, I., Geiss, K., *et al.* (1995) Selective transmission of variant genomes from mother to infant in neonatal fulminant hepatitis B. *Hepatology* **21**: 8–13.

Weiner, A.J., Thaler, M.M., Crawford, K., *et al.* (1993) A unique, predominant hepatitis C virus variant found in the infant born to a mother with multiple variants. *J. Virol.* **67**: 4365–4368.

Wyatt, C.A., Andrus, L., Brotman, B., *et al.* (1998) Immunity in chimpanzees chronically infected with hepatitis C virus: role of minor quasispecies in reinfection. *J. Virol.* **72**: 1725–1730.

Yaziji, H., Hill, T., Pitman, T.C., *et al.* (1997) Gestational herpes simplex hepatitis. *Southern Med. J.* **90**: 347–351.

Young, F.J., Chafizadeh, F., Oliveira, V.L., *et al.* (1996) Disseminated herpesvirus infection during pregnancy. *Clin. Infect. Dis.* **22**: 51–58.

Zanetti, A.R., Tanzi, E., Paccagnini, S., *et al.* (1995) Mother-to-infant transmission of hepatitis C virus. *Lancet* **345**: 289–291.

Zhu, T., Wang, N., Carr, A., *et al.* (1996) Genetic characterization of human immunodeficiency virus type 1 in blood and genital secretions: evidence for viral compartmentalization and selection during sexual transmission. *J. Virol.* **70**: 3098–3107.

Chapter 8

HIV Infection and AIDS, Exotic Infections and Candidate Hepatitis Viruses

8.0 Issues

Viral hepatitis with the major hepatotropic agents (hepatitis A–D) and their serological markers are common in Human Immunodeficiency Virus (HIV) seropositive individuals and patients with the Acquired Immune Deficiency Syndrome (AIDS), reflecting shared risk factors such as sexual promiscuity, intravenous drug use (IVDU) and multiple transfusions. Management issues surrounding chronic hepatitis B and C, including treatment options and side-effects of anti-viral therapies, have become important subsequent to the improved survival of patients with HIV infection and AIDS.

Other infections of the liver occur as part of the systemic manifestations of many opportunistic organisms. Consequently, viruses, such as the herpes viruses, can be recovered from cultures of blood and bone marrow as well as liver in AIDS. Multiple microbial infections are common in HIV infection and AIDS.

Certain viruses, such as yellow fever virus (YFV), Rift Valley fever virus (RVFV), Ebola and Marburg viruses, commonly target the liver but infection is restricted by epidemiological factors, such as spread in tropical areas by insect vectors. Cases in the West arise mostly from imported infections. Exotic (non-indigenous) agents (see *Table 8.4*) should be considered in the differential diagnosis of many illnesses, including hepatitis, especially in the recent traveler.

Ebola virus, Marburg virus and RVF can target the liver with massive necrosis of hepatocytes resulting in severe acute liver failure (ALF). Lassa fever is highly contagious but seems to be endemic only in west Africa. Dengue is a leading cause of morbidity and mortality in children in Asia including China, Vietnam, Thailand and Indonesia and in the Carribean and Tahiti.

Novel viruses have been described recently, including GB virus C (GBV-C/HGV), a member of the *Flaviviridae* distantly related to HCV, and TT virus, which is related to the circoviruses. Although these viruses have been detected in individuals with hepatitis, their potential role as causative agents (and, indeed, whether they replicate in the liver) is uncertain.

8.1 HIV-I, viral hepatitis and the liver

Prior to the advent of highly active anti-retroviral therapy (HAART), data on HIV infected patients with hepatitis B and C suggested limited impact of chronic viral hepatitis on their natural history. These views are being revised in parallel with improvements in long-term prognosis and survival of this group of patients. Unfortunately most studies in this area do not allow for confounding factors that influence outcomes such as excess alcohol intake, nutritional status, increasing age, access to medical care, opportunistic infections, CD4 cell counts and stages of disease, and the impact of newer anti-retroviral medications.

HIV-1 does not cause a specific liver lesion and there is no evidence to indicate regular infection of hepatocytes. Instead, the virus infects CD4+ T cells, predisposing to other microbial infections. In AIDS, the p24 antigen of HIV is detectable in the liver, primarily in Kupffer cells, forming granulomata and giant cells and in endothelial cells lining the sinusoids. Consequently, the liver forms an important reservoir for HIV and has to be considered in any strategy for eradicating the infection.

Key Notes Viral hepatitis with HIV and AIDS

- Viral hepatitis is common in HIV-positive individuals
- Viral hepatitis is not stage-specific for HIV and AIDS
- Hepatitis B, hepatitis C, herpes- and adenoviruses show increased replication
- Multiple microbial infections coexist
- Differential diagnosis of viral hepatitis is wide:
  - Many other microbial infections of the liver
  - Veno-occlusive disease
  - Drug hepatotoxicity
  - Non-specific changes
  - Liver malignancies: lymphomas and Kaposi's sarcoma

8.1.1   Hepatitis A and E

Outbreaks of hepatitis A with progression to liver failure have been reported in the USA among IVDU and men who have sex with men. Clustering of infection reflects sharing of lifestyles and fecal–oral spread by anal–oral sexual contact and suboptimal sanitary conditions. There is no specific evidence for excess severity of hepatitis A virus

(HAV) infection in the absence of underlying liver disease from alcohol and chronic hepatitis B and C. No data are available for hepatitis E virus (HEV).

Key Notes Major hepatotropic viruses and HIV infection
- Hepatitis A–D are common; shared risk factors
- HIV influences HBV and HCV infections
  - Serological markers, reactivation and lower alanine aminotransferase levels
- Reduced CD4+ counts in HIV infection are associated with:
  - Higher HBV DNA levels in hepatitis B
  - Higher HCV RNA levels in hepatitis C
- Reduced survival in AIDS with HBV and HCV infection

8.1.2 Hepatitis B and B+D

Concurrent hepatitis B virus (HBV) and hepatitis C virus (HCV) infection are common in HIV seropositive patients and found in around 50% of patients with AIDS at postmortem. Rates of HBV infection vary according to risk groups for transmission of HIV and reflect the propensity for sexual transmission of HBV. Serological markers indicating current or past infection with HBV are evident in over 20% of men who have sex with men, 14% of heterosexuals and around 5–10% of IVDU. Also, following acute infection, HIV infected patients are less likely (< 50%) to clear HBsAg and HBV DNA than immunocompetent adults (> 90% chance of clearance). While actuarial survival rates for HBV infected patients are comparable between HIV seropositive and HIV seronegative groups, survival is reduced for AIDS patients with concurrent HBV infection.

Patients with dual HIV and chronic HBV infections tend to have higher serum levels of HBV DNA and lower levels of liver enzymes compared with those with HBV alone. In a study of 150 men (mean age 33 years) who had sex with men and who were infected with HBV, 82 (55%) were seropositive for HIV. Serum HBV DNA levels were around two times higher among HIV seropositive than negative men. Further, as for HCV infection (see section 8.1.3 below), falling CD4 cell counts were associated with an increase in viremia and falling alanine aminotransferase (ALT) levels. Whether the increase in hepatitis B viremia simply reflects an impaired immune response or interaction of the viruses, including through the production of transactivating proteins is unclear. Seroconversion from HBeAg to anti-HBe (over 337 months) also was more than twice as likely in HIV seronegative, compared with HIV seropositive, men.

Low levels of serum ALT and aspartate aminotransferase (AST) can be misleading because dual infection is associated with an increased risk of progression to cirrhosis. Alcohol excess and increasing age also are significant risk factors for developing cirrhosis. The risks of chronic liver disease remain high and therapy with α-interferons (IFNs) is less likely to eradicate HBV in HIV seropositive, than seronegative, patients.

Elevated serum liver enzymes (LFTs) (AST and ALT) may fall in parallel with CD4 cell counts, suggesting attenuation of liver damage. Conversely, restoration of CD4 cell counts with anti-retroviral therapy may be associated with a flare in hepatitis. Further, long-term treatment with nucleoside analogs such as lamivudine (3TC), as part of the HAART regimen, is associated with emergence of resistant variants of HBV (see section 3.8.2.3) in patients previously considered to have 'quiescent' HBV infection (anti-HBc and/or anti-HBs without detectable HBsAg).

Patients with HIV and AIDS who have markers of previous HBV infection (anti-HBs and/or anti-HBc) should be considered at risk of reactivation of infection. Seroreversions from anti-HBs to HBsAg and from anti-HBe to HBeAg with detectable HBV DNA have been recorded following discontinuation of HAART that included 3TC. Anti-HBs following natural infection with HBV, and through immunization, may become undetectable in individuals who later contract HIV infection and AIDS. A suboptimal response to the hepatitis B vaccine is common in anti-HIV seropositive vaccinees.

Hepatitis D virus (HDV) and HIV infection have been found in around two-thirds of hemophiliacs in Germany who received clotting factors prior to the advent of heat treatment. IgM and IgG anti-HDV antibodies may be detected less often than expected due to the immunosuppressive effects of HIV. As with HBV alone, liver disease may be more indolent in HBV/HDV patients with HIV and AIDS, presumably reflecting the immunosuppressive properties of HIV.

8.1.3 Hepatitis C

The prevalence of hepatitis C in HIV seropositive patients correlates with parenteral routes of infection (Table 8.1). Horizontal (sexual) and vertical transmission of HCV may occur more frequently from anti-HIV-positive, than negative, individuals, independently of co-transmission of HIV. Vertical transmission of HCV is favored with high levels of viremia that occur in HIV infection (see section 7.5.3). Development of severe chronic hepatitis with rapid decompensation of liver disease has been reported in HCV infection following seroconversion to anti-HIV antibody. As with HBV, levels of virus replication tend to be higher, and titers of anti-HCV antibodies lower, than without HIV infection.

Table 8.1. Serological markers (number of patients) of hepatitis B and C in 232 HIV infected German patients reflect sexual transmission for HBV and parenteral exposure for HCV (after Ockenga et al., 1997)

	Homosexual	Heterosexual	IVDU	Hemophiliacs
Anti-HCV	6	3	41	10
HCV RNA	5	2	31	9
HBsAg	15	2	3	1
Anti-HBc	73	14	43	7
HBV + HCV	2	–	1	1
Total number of patients	122	33	64	13

A retrospective study (1979–1993) of 183 HCV seropositive hemophiliacs (74 seropositive for HIV) with known date of exposure (median 15.1 years) to large donor pool concentrates showed that the relative hazard of developing liver failure after HIV infection was 21.4 compared with HIV seronegative patients with HCV infection. Also, hepatic decompensation increased after becoming positive for p24 antigen, with falling CD4 cell counts, greater use of factor concentrates and increasing age.

Alcohol in addition to HIV infection carries excessive risk of developing progressive liver disease in chronic hepatitis C. A retrospective histopathological study of liver biopsies was carried out on 210 consecutive patients with HCV infection (and without concurrent opportunistic infections) comparing outcomes for HIV status and excess alcohol intake. Patients in the HIV seropositive (60 subjects) and seronegative (150 subjects) groups had comparable mean age, gender ratio, durations of HCV infection, frequency of excess alcohol drinking and clinical presentations. Cirrhosis was higher in the dually infected group (HCV/HIV 30% versus 15.3% for HCV alone) for comparable estimates (mean around 12 years) of duration of infection. Cirrhosis was frequent with the combination of HIV + HCV with excessive alcohol (30.4%) compared with HIV + HCV without alcohol (29.7%). In the absence of HIV, cirrhosis was found in 24.5% HCV-positive excessive drinkers and 11.3% with HCV infection alone.

Data are conflicting for the impact of concurrent HCV and HIV-1 infections on liver inflammatory activity compared with HCV alone. In HCV infection, serum levels of AST, ALT and gammaglutamyl transpeptidase (GGT) are higher in HIV seropositive patients compared with HIV seronegative patients and regardless of alcohol intake. Histological evidence of liver damage may seem less in chronic hepatitis C with CD4 cell counts less than 400 cells/μl, suggesting that liver damage may be caused by immune mechanisms with HCV as well as HBV. However, the risk of progressive liver disease remains high.

One approach is to assess the impact of HIV according to estimated duration of infection. In a large multicenter study from Spain, comparable distributions of histological findings were documented in HIV seropositive and seronegative patient groups (*Table 8.2*) but with a shorter interval from time of exposure to biopsy in the HIV seropositive group. Most of the HIV seronegative patients with cirrhosis had been infected for more than 15 years. HCV RNA levels were twice as high in HIV infected patients with CD4 cell counts less than 500 cells/μl, compared with those with CD4 cell counts more than 499 cells/μl for comparable durations of HCV infection, genotype 1 status and ALT level.

Table 8.2. Relationship between cirrhosis and HIV status in 547 HCV infected patients (after Soto *et al.*, 1997)

Infection to biopsy (years)	HIV + HCV % (n=116)	HCV alone % (n=431)
1–5	11.9	0.0
1–10	14.9	2.6
1–15	12.4	4.7
>15	0.0	36.6

Long-term survival depends on HIV-related events and complications of AIDS rather than hepatitis C alone. However, as survival times continue to improve and CD4 cell counts normalize in significant numbers of HIV infected patients, the impact of HCV infection will have to be reviewed.

8.1.4 Herpesviruses

Hepatitis from the herpes viruses can be severe in patients with HIV infection and usually occurs in the setting of multisystem failure, sepsis and adrenal necrosis. Reactivation of latent viruses is favored with an impaired T-cell response. Cytomegalovirus (CMV) can be detected in around 25% of liver biopsies in AIDS. Disseminated infection can involve the retina, colon and esophagus as well as liver. CMV hepatitis also can be severe with HIV infection. Separate associations with a cholangiopathy and acalculous cholecystitis have been described (see also Chapters 6 and 7).

8.1.5 Other viruses

Adenoviruses have been implicated in causing hepatitis with significant liver cell necrosis in AIDS and other instances of immunosuppression such as following transplantation of bone marrow or liver.

8.1.6 Diagnosis in HIV infection and AIDS

Viral hepatitis must be considered along with many other microbial infections, alcoholic liver disease and other conditions that involve the liver. The diagnosis relies on assessment of results from panels of tests for viral antigens and antibodies (e.g. *Table 1.8*), assessing virus replication by serum levels of HBV DNA and HCV RNA, and recovery of virus (e.g. herpesviruses) from multiple sites including blood, bone marrow, and urine.

Clinical features of viral hepatitis are not diagnostic in the immunosuppressed individual as typified in HIV infection. Hepatomegaly with elevated liver enzymes (GGT and alkaline phosphatase, as well as AST and ALT) occur regularly in over 50% of patients with HIV infection and AIDS. Gross elevations in serum levels of alkaline phosphatase are suggestive of *Mycobacterium avium* in patients with AIDS without extrahepatic biliary obstruction, cholangiopathy from CMV, *Cryptosporidium* spp. or *Microsporidium* spp. and other fungal infections.

An elevated serum level of alkaline phosphatase is suggestive of:

- Fatty change: non-specific, alcohol, hepatitis C
- Granulomata: opportunistic infections, IVDU
- *M. avium*, fungi, *Cryptosporidium* spp., *Microsporidium* spp.
- Infiltrations: lymphomas, Kaposi's sarcoma of the liver
- Veno-occlusive disease
- Peliosis hepatis
- Bacillary angiomatosis
- CMV infection: acalculous cholecystitis and biliary strictures
- Biliary strictures, sclerosing cholangitis, ductopenia, papillary stenosis

Elevations in liver tests may be modest and jaundice uncommon despite significant hepatitis. Conventional serological markers, such as anti-HCV antibodies, may be falsely negative due to the immunosuppressant effects of HIV.

Hepatitis may be a presenting feature of non-Hodgkins lymphoma, Kaposi's sarcoma of the liver and hepatotoxicity due to drugs as well as to other microbial infections in the liver. General histopathological findings include giant cell hepatitis, especially in neonates and children, intranuclear inclusions (CMV and other herpesviruses) and granulomatous hepatitis including from tuberculosis or fungal infections. Cellular infiltrates (lymphocytes and plasma cells) may be misleadingly mild with low CD4/CD8 cell ratios and in Kaposi's sarcoma. Fatty change is common, non-diagnostic and often attributed to other causes including alcoholic liver disease and chronic hepatitis C. Fatty infiltration may cause elevations in serum liver tests, especially alkaline phosphatase. Veno-occlusive disease is common especially in HIV seropositive IVDU. Histological features of veno-occlusive disease, that may be superimposed on viral hepatitis, include congestion of sinusoids, degeneration and necrosis of perivenular hepatocytes and fibrosis. Discrimination between primary infection and reactivation of latent infection may be difficult. Iron overload in the liver may occur with chronic hepatitis C as well as in the multiply transfused individual.

CMV involving the liver may present with clinical features suggestive of biliary obstruction, acalculous cholecystitis, papillary stenosis or sclerosing cholangitis. Elevations in serum levels of alkaline phosphatase reflect damage to bile ducts and granulomata. Intrahepatic bile ducts may become sparse.

Hepatosplenomegaly, granulomata and formation of giant cells can be found with mycobacterial infections (*M. intracellulare, M. genavense*) as well as herpes infections. *M. intracellulare* involves the liver in up to 70% of patients with AIDS. Liver granulomata are common also in cryptosporidiosis, histoplasmosis and drug-related hepatitis. Listeriosis may cause an acute hepatitis with bacteremia and epithelioid granulomata of the liver. *Salmonella* hepatitis can be mistaken for viral hepatitis, especially in the cholestatic phase with falling ALT levels and rising alkaline phosphatase. In *Salmonella* hepatitis, the ratio of ALT/lactate dehydrogenase (LDH) typically is less than 4:1, in addition to the relative bradycardia and left shift (band forms) in white cell counts.

Ultrasound examination may show non-specific features such as altered echogenicity and hepatomegaly. Laparoscopy may reveal nodules in the liver, spleen and peritoneum in non-tuberculous mycobacterial infections. Liver biopsy may be informative in a small percentage of HIV infected patients with fever of uncertain origin but should be considered along with results of extensive testings using conventional microbiological approaches to diagnosis. A portion of any liver biopsy should be sent for culture of acid-fast bacilli among many other infections. Reduced synthesis of clotting factors from the liver in HIV-1 infection has been attributed to replication of HIV-1 in endothelial cells. The risk of hemorrhage may be increased with peliosis hepatis from rickettsial infection such as *Rochalimaea henselae* or *R. quintana*.

8.1.7   Management

Principles of management of viral hepatitis are the same as without concurrent HIV infection (see section 1.13).

8.1.7.1   Antiviral therapies

Overall, attempts to eradicate hepatitis B, C and D with IFNs have been disappointing in the immunosuppressed individual. Zidovudine does not inhibit replication of HBV DNA in man although studies *in vitro* of the triphosphate form show potent activity against the HBV DNA polymerase.

Many of the drugs used to treat infections in HIV and AIDS can cause elevations of liver enzymes (AST, ALT) to levels seen in viral hepatitis. Side-effects of nucleoside reverse transcriptase inhibitors such as abacavir, zidovudine, 2′3′-dideoxyinosine (didanosine), zalcitabine, stavudine and lamivudine include the rare potential for lactic acidosis and hepatic steatosis (as was seen with fialuridine (FIAU) hepatotoxicity). Elevated levels of liver enzymes also have been reported for the protease inhibitors such as ritonavir and saquinavir. As the protease inhibitors, especially ritinavir, are inhibitors of the family of cytochrome P450 isoforms (inhibitor of 3A4), the potential for interaction and hepatotoxicity arises with the many drugs that share these pathways for metabolism.

Hepatotoxicity, including hepatitis, can occur with drugs used to treat the opportunistic infections such as rifampicin and isoniasid for mycobacterial infections and trimethoprim–sulfamethoxazole for *Pneumocystis carinii* (see *Table 8.3*).

Clearance of HBV DNA and HCV DNA are very low with α-IFNs in HIV infected patients and those with multiple infections. Results with combination therapy (α-IFNs and ribavirin) and protease inhibitors are awaited in hemophiliacs but data from large-scale trials are lacking. Concern remains over the side-effects of these drugs, especially the propensity to cause anemia, leukopenia, thrombocytopenia and abnormal liver tests. Conversely, HAART can worsen liver disease in HCV infected/HIV infected patients with low (< 100 cells/μl) CD4 cell counts. HCV RNA levels increased temporarily and ALT levels increased in 19 patients treated with combinations of ritinovir, saquinavir, indinavir and reverse transcriptase inhibitors.

Table 8.3. Hepatotoxicity with drugs used in HIV infected patients

Delavirdine	Nevirapine
Efavirenz	Nucleoside reverse transcriptase inhibitors
Fluconazole	Protease inhibitors
Isoniazid	Rifabutin
Itraconazole	Rifampicin
Ketoconazole	

As protease inhibitors such as ritonavir, saquinavir, and indinavir show varying degrees of inhibition of the family of cytochrome P450 isoforms, the potential for interaction and hepatotoxicity arises with the many other drugs that share these pathways for metabolism.

8.2   Exotic infections—liver involvement

Yellow fever virus (YFV) and many other 'exotic' viruses, which are found usually only in restricted geographical locations, infect the liver (*Table 8.4*). Overt hepatitis usually arises in the setting of severe biphasic infection (e.g. YFV) with systemic illness including hemorrhage, fever and multiorgan failure such as in dengue hemorrhagic fever (DHF) and dengue shock syndrome (DSS), Congo Crimean hemorrhagic fever (C-CHF) and Rift Valley fever (RVF). In severe cases, there is a brief remission before the return of fever, vomiting and abdominal pain and the onset of hemorrhagic disease with hepatic and renal symptoms and evidence of shock, collapse and prostration and, occasionally, jaundice. ALF may develop with complications such as gastrointestinal hemorrhage and renal failure (see section 5.2.7). Surprisingly, jaundice is uncommon in many of these severe infections such as YFV and RVF, despite massive necrosis of the liver.

The spectrum of illness, especially of hepatic dysfunction, is wide. Significant variations in fatality rates occur during epidemics of YFV, including for the same virus variant. Differences in proportions of domestic and wild vector (mosquito) populations may lead to selection of variants with differing pathogenic potentials as well as variation in sizes of the inoculum among other variables for transmission.

Table 8.4. Exotic viruses that may target the liver

Virus	Liver	ALF	Jaundice
Yellow fever virus	Main target—fatty degeneration, mid-zonal necrosis and Councilman bodies	Yes	Yes
Lassa fever virus	Widespread parenchymal, focal necrosis, no zonal pattern	Rare	Uncommon
Marburg and Ebola viruses	Yes, AST > ALT Necrosis of single hepatocytes	Yes	Rare
Rift Valley fever	In less than 5%, resembles YF virus	Yes	Rare
Crimean-Congo hemorrhagic fever virus	Mid-zonal necrosis with Councilman bodies	Yes	Uncommon
Dengue virus (four serotypes)	Variable, necrosis of parenchymal cells in DHF	Uncommon	Uncommon

DHF, dengue hemorrhagic fever

8.2.1 Yellow fever virus

1992 saw a resurgence of yellow fever despite continuing attempts to control the vectors and immunize high-risk personnel. Most of reports were of young children in West Africa. The resurgence in South America, particularly in Peru and Bolivia, mostly was among adult forest workers. The principal threat to man occurs when forest mosquitoes move to adjacent areas of human habitation. Epidemics may then be sustained by transmission from man to mosquito to man.

YFV is spread by a number of species of mosquito, particularly *Aedes aegypti*, the dominant vector prior to the introduction of programs of control. In West Africa, urban outbreaks occur based on transmission by *A. aegypti*. In South America, vector control has eliminated the virus from most urban centers but YFV remains endemic in a sylvan cycle between monkeys and tree-breeding mosquitoes.

8.2.1.1 Biology of YFV

YFV is the prototype member of the flaviviruses, the prototype genus (with HCV and the pestiviruses, see sections 4.2.2 and 4.2.4) of the family *Flaviviridae*. Other flaviviruses include dengue viruses (four serotypes), Japanese encephalitis virus and tick-borne encephalitis viruses. Flaviviruses that infect humans are transmitted by biting mosquitoes and ticks.

The organization of the YFV genome is similar to that of HCV (*Figure 4.2*), the polyprotein encoded by a single open reading frame is processed to yield structural and non-structural viral proteins. The structural region encodes a nucleocapsid (core, C) protein and two surface glycoproteins M (matrix or membrane) and E (envelope). There is an additional non-structural protein, NS1, which has no equivalent in HCV. NS1 is unusual as a non-structural protein, being a glycoprotein expressed on the surface of infected cells and also secreted into the circulation. The function is obscure but antibodies to NS1 may be protective. The functions of the other non-structural proteins are believed to be similar to those of HCV.

Variants of YFV have been detected which differ in their immunological and biological properties, including virulence. Variants also have been detected in the vaccine (17D) that could account for variability in susceptibility to neutralization by antibodies. Unfortunately, no clear relation has been found between nucleic acid sequences, virulence *in vivo* and replication competence in mosquitoes.

Heterogeneity (around 4% for amino acids) has been noted within NS1 and envelope regions of YFV isolates from Africa and South America. Three geographic variants have been recognized within Africa; one from East and central Africa and two from West Africa (western and southern regions). Whether these variants have evolved following mass immunization remains unclear.

8.2.1.2 Liver involvement

YFV is a hepatotropic virus. The spectrum of illness is wide and can differ between children and adults. The liver becomes involved overtly typically during the second phase of the illness (4–5 days after insect bites). Post-mortem examinations reveal hemorrhagic necrosis and fatty degeneration with a disproportionately sparse inflammatory infiltrate. The hallmark feature of Councilman bodies (apoptotic cells) is not unique to yellow fever. Jaundice may be apparent without hemorrhagic sequelae in children with yellow fever. Jaundice is not invariable in children, including during epidemics. Albuminuria is a significant feature of yellow fever and helps to distinguish this from other causes of hepatitis.

8.2.2 Lassa fever virus, Marburg virus and Ebola virus

These viruses are endemic in west and central Africa, particularly Nigeria and Sierra Leone and are characterized by explosive outbreaks. Most cases are sporadic but epidemics have occurred in these regions. The infections are zoonotic. Lassa virus frequently has been isolated from the multimammate rat, *Mastomys natalensis*, in Sierra Leone and Nigeria. The African green monkey (*Cercopithecus aethiopo*) is the natural reservoir for Marburg virus. The source for Ebola virus is unknown. Transmission from person-to-person occurs primarily from blood and urine.

8.2.2.1 Biology

Lassa virus is an enveloped RNA virus which shows some serological cross-reactivity with other members of the arenavirus family. Pathogenetic mechanisms remain unclear but evidence points to a direct cytopathic effect of the virus rather than to host-mediated cellular and humoral immune mechanisms. Levels of viremia typically are high with little evidence for T-cell cytotoxicity and neutralizing antibody.

Ebola and Marburg viruses are RNA viruses of the family *Filoviridae*. Since its discovery in 1976, four subtypes of Ebola have been recognized; Ebola-Sudan and Ebola-Zaire can be fatal in humans whereas Ebola-Ivory Coast and Ebola-Reston usually are symptomless. The risk to man is from infection of imported animals from high-risk countries such as the Philippines. Although monkeys seem to be important in transmission of Marburg to humans, they are susceptible to fatal infections and may not be maintenance hosts of the virus.

8.2.2.2 Liver involvement

Elevations in enzymes (AST > ALT) are common in Lassa fever and correlate with adverse prognosis along with levels of viremia, degree of leukopenia and abnormal aggregation of platelets. Paradoxically, in Lassa fever, liver histopathology tends to show mild, non-specific features with sparse lymphocytic infiltrates. Hepatitis and liver failure are not the causes of death. Lassa fever tends to be mild in children, whereas adults can develop multiorgan failure with hemorrhage and ascites.

Ebola and Marburg viruses target the liver causing massive liver cell necrosis with Councilman bodies (see section 8.2.1.2). The ALF is associated with very high mortality; jaundice is rare.

8.2.3. Rift Valley fever

Rift Valley fever is a serious cause of hepatitis in domestic cattle and man in sub-Saharan Africa, Egypt and the Sudan. Spread is by mosquitoes. Man is the incidental host, usually when handling infected animals. Outbreaks seem to be confined to the African countries and are associated especially with flooding of habitats that favor breeding of mosquitoes. Repeated epidemics have been reported in Egypt, Sudan and West Africa. Sporadic cases have been reported from Senegal. The spectrum of clinical illness is wide. RVFV can replicate in mononuclear cells and may account for the lymphopenia.

8.2.3.1 Biology

RVFV is classified within the genus phlebovirus of the family *Bunyaviridae*. As with YFV, significant genetic variation occurs between isolates and may account for the diverse clinical outcomes in animal models inoculated with seemingly identical phenotypes.

8.2.3.2 Liver involvement

RVFV is a cause of severe hepatitis in cattle, sheep and other domestic animals particularly in sub-Saharan regions of East Africa. Liver injury is focal but in 1–2% of cases can be massive and occurs usually in the setting of encephalitis, damage to the eyes and hemorrhage in many organs. The liver probably is the major site of replication. Macrophages and IFN levels induced by the infection also play an important role in limiting the viremia and assisting recovery. In laboratory rodents, RVFV also produces a fatal hepatitis but the outcome can be diverse and paradoxical; animals that succumb to encephalitis seem to show limited liver cell necrosis. Conversely, hepatic necrosis and hemorrhagic necrosis can be widespread with numerous acidophilic bodies (see Yellow Fever) and inclusion bodies. Jaundice is rare.

8.2.4 Crimean-Congo hemorrhagic fever

C-CHF is widely distributed in eastern Europe, including Bulgaria, Greece, Hungary, the former Yugoslavia (and possibly Portugal, France and Turkey) as well as the Middle East, South Africa and African countries and the former Soviet Union. Spread to man predominantly is by tick bites and by contact with blood from infected domestic animals. C-CHF is an occupational hazard among farmers, workers in abattoirs and animal handlers such as butchers, veterinarians, farmers and hunters. This virus also has a propensity for nosocomial spread with fatalities among health care personnel involved in managing patients.

8.2.4.1 Biology

The C-CHF virus is a Bunyavirus belonging to the *Nairoviride* genus.

8.2.4.2 Liver involvement

Clinical features resemble those of yellow fever. Most cases have symptoms and mortality approaches 30% from hepatorenal failure. The duration of incubation is short (1–3 days) and onset of illness sudden. Deterioration into coma is rapid.

8.2.5 Dengue and DHF

Dengue is endemic in more than 100 countries in all continents except Europe and is spreading rapidly in many areas. The incidence of dengue hemorrhagic fever (DHF) and dengue systemic shock (DSS) have increased significantly in parallel with international travel, urbanization of tropical forests and the spread of mosquitoes. Regions of high endemicity include Asia, Africa and South and Central America and overlap somewhat with those of yellow fever. Dengue affects approximately 20 million people. In Southeast Asia, DHF and DSS are leading causes of hospitalization and death among children.

Primary dengue resembles a severe case of influenza with pronounced arthralgia (breakbone fever). DHF and DSS are complications which may arise following secondary infection in a person with circulating antibodies from previous exposure to a different serotype. These antibodies are non-neutralizing and enhance uptake by macrophages after binding to the virus. In infants, hemorrhagic fever and shock occurs following primary infection in the presence of maternal antibody. The characteristic clinical illness with hypotension and hemoconcentration reflects the underlying increase in vascular permeability, impaired hemostasis, thrombocytopenia and activation of the complement cascade with reduction in C3 and C5. Binding of heterotypic antibodies to macrophages is believed to enhance virus replication in these cells and lead to the release of vasoactive cytokines.

DHF occurs almost exclusively in the indigenous population, especially among children below the age of 6 years. Why DHF is uncommon in Westerners abroad, despite frequent occurrence of classical dengue, is unknown.

The clinical classification and grading of DHF is as follows:

I   Fever, non-specific constitutional symptoms, positive tourniquet (capillary fragility)
II Grade I and spontaneous bleeding: skin, gums, gastrointestinal tract

While that of DSS is as follows:

III Grade II of DHF and circulatory failure, agitation, pulse pressure < 21 mmHg
IV Grade II of DHF and profound shock (undetectable pulse and blood pressure)
  • Hemoconcentration (maximal hematocrit 20% above baseline) with thrombocytopenia ($< 10^5/\mu l$) are invariable
  • Plasma leakage (hemoconcentration, pleural effusion, ascites) invariable II–IV

8.2.5.1 Biology

The dengue viruses are classified within the genus flavivirus of the *Flaviviridae*. Four serotypes are described with significant cross-reactivity on laboratory testing. In contrast, in clinical practice, protective immunity is faithful to each serotype and accounts for the potential to be infected on more than one occasion. More recent analyses using molecular techniques have demonstrated significant variation in genetic sequences of nucleic acid and predicted amino acids but their clinical and serological significance is unclear.

8.2.5.2 Liver involvement

Elevations in serum levels of AST and ALT occur in around 10–30% of uncomplicated DHF but do not correlate with outcome (contrast Lassa fever). Jaundice is rare in DHF especially in the critically ill with rapid deterioration. Deepening jaundice, liver and renal failure may lead to coma and death in around 8 days. Mortality rates among those entering the second phase of illness may be as high as 50% but are difficult to estimate with the bias towards reporting the most severe cases. Survivors usually recover completely although a few pursue a chronic phase with prolonged jaundice and renal dysfunction that may last for several weeks.

8.3 Diagnosis of exotic infections

Unusual agents causing acute hepatitis should be considered in the differential diagnosis of many illnesses in the traveler returning from endemic areas, especially Africa, Central and South America and Asia.

Clinical and laboratory features overlap with many other infections which cause liver failure and renal dysfunction, in particular between malaria, typhoid fever,

Key Notes Diagnosis of exotic infections

Early communication with a Reference Center should be sought for advice on safety in handling specimens and optimal collection of appropriate clinical samples—such as freezing of tissue for detection of viruses leptospirosis, Kawasaki disease, rickettsial infections and West Nile fever, C-CHF, Lassa fever and Marburg and Ebola virus infections. Hepatitis has been reported in louse-borne and tick-borne relapsing fevers. Yellow fever should be included in the differential diagnosis of acute hepatitis in travelers returning from endemic areas, in particular Africa, Central and South America.

Leukopenia or relative lymphocytosis with reduction in numbers of polymorphonuclear cells are common. The prothrombin time, partial thromboplastin time and platelet counts may be near normal in dengue despite severe illness.

Disproportionately modest elevations in liver enzymes (AST and ALT) and bilirubin may occur despite severe hepatic damage. In Marburg and Ebola virus infections the serum level of AST is disproportionately elevated compared with that of ALT. Conversely, marked elevations in serum liver biochemistry may occur in Lassa fever despite mild histopathological changes in the liver.

8.3.1 Serological tests

IgM-specific enzyme immunoassays are the method of choice for diagnosis. Most jaundiced patients have high-titer antibodies. Prior to these tests, rising titers of antibodies were sought in paired sera with a four-fold increase required for definitive diagnosis. Diagnosis may require isolation of virus following inoculation of mice and tissue culture cells with clinical samples such as blood, peripheral blood white cells and bone marrow aspirates collected during the early phase of infection.

The diagnosis of recent infection with the individual flaviviruses may be difficult due to cross-reacting assays and secondary immune responses. In dengue, primary antibody responses are found in the non-immune and those without antibodies to other flaviviruses such as YFV, including the 17D vaccine. Secondary immune responses are more common than primary responses in indigenous populations living in regions endemic for dengue because antibodies to each of the four infecting serotypes persist for life following recovery. Accordingly, in primary infection, the IgM antibody response may be transient. A rapid rise in titers of immunoglobulin (Ig)G isotype reflects a secondary immune response following repeat infection in a person previously exposed to dengue or other flaviviruses, including in the yellow fever vaccine. Antibodies against Lassa fever virus can be detected by indirect immunofluoresence within 3 weeks of the onset of illness. However, neutralizing, and complement-fixing, antibodies are difficult to detect before 3–4 weeks. High seroreactivity using an immunofluorescent antibody test has been found for Ebola virus in Africa. Confirmation of this finding would indicate that non-lethal as well as lethal strains of Ebola exist in Africa and may pave the way for developing a vaccine.

8.3.2 Histopathological changes in the liver

These are not discriminative despite the diversity of viruses. Typical features seen with many arbovirus infections include midzonal to massive hepatic necrosis and cytoplasmic inclusions typified by acidophilic bodies seen in Kupffer cells and fatty degenera-tion of hepatocytes in yellow fever (Councilman bodies), Rift Valley fever and Argen-tine hemorrhagic fever among other infections. Activation of Kupffer cells may be prominent especially in Lassa fever and Argentine hemorrhagic fever. Inflammatory infiltrates may be sparse.

Other microbial infections, such as leptospirosis, malaria and bacterial and fungal sepsis, can mimic these illnesses. Furthermore, microbial infections are common in other liver diseases and add significantly to the morbidity and mortality. Transmission electron microscopy may detect virus particles in the liver, especially in macrophages for Ebola and Marburg virus infections.

> **Key Notes** Immunization against YFV
>
> Immunization against YFV is compulsory for all personnel involved in the man-agement of suspected cases of YF, including laboratory staff handling samples

8.4 Management of exotic infections

Patients should be attended in isolation with strict barrier nursing because person-to-person spread can occur via body fluids and aerosol. These restrictions should be maintained well into recovery, particularly for C-CHF virus that has a propensity for spread as nosocomial infections and for Lassa fever virus that can be detected in blood and urine over several months during convalescence in some survivors.

Management is supportive with strict attention to balance of fluids, electrolytes and osmolarities to prevent fluid overload secondary to increased permeability of capillar-ies. Respiratory distress syndrome is common and ventilatory support should be anticipated. Specific complications of ALF include severe hypoglycemia and sepsis (see Chapter 5). Care is optimal in specialized referral centers having expertise in intensive care, liver transplantation and infectious diseases. Lassa fever virus is the most contagious of the arenaviruses that infect man.

Uncontrolled studies claim some benefit with antiviral agents such as IFNs in dengue, and ribavirin in Lassa fever (*Table 8.5*). Survival is improved when therapy is com-menced within the first week of symptoms and for low levels of viremia. Cortico-steroids are of no benefit and may increase the viremia.

8.5 Prevention

This remains the top priority given the practical difficulties in managing severely ill patients and economic restraints of developing countries. The mandatory disease-

Table 8.5. Treatment of exotic infections

Infection	Antiviral therapies	Post-exposure prophylaxis
Yellow fever	NA	NA
Dengue fever	NA	NA
Ebola or Marburg	NA	NA
Lassa fever	Ribavirin	Ribavirin
Junin hemorrhagic fever	Immune plasma	(Ribavirin)[a]
C-CHF	(Ribavirin)[a]	(Ribavirin)[a]

[a]On clinical trial; data not available. NA, Not available; C-CHF, Crimean-Congo hemorrhagic fever.

control requirements updated by the Centers for Disease Control and Prevention, Atlanta have improved early diagnosis and seem effective in minimizing exposure to personnel in facilities handling non-human primates infected with Ebola viruses.

8.6 Vector control

Control of populations of mosquitoes remains the major method of preventing yellow fever and dengue in endemic areas. Spraying with insecticides reduces populations temporarily but has limited coverage over stagnant pools and water in rural dwellings.

Insect repellents, preferably containing N,N-diethyl-m-toluamide, should be applied to exposed skin between dusk and dawn and require repeated applications every 3–4 h. Electrical insecticide dispensers and burning coils which contain pyrethroids may be situated in the bedroom at night. Mosquito nets are effective if kept in good repair.

Control of populations of domestic rodents, the reservoir for Lassa fever and other arenaviruses, remains difficult especially with progressive urbanization of rural areas.

8.7 Immunization

Yellow fever is preventable above 6 months of age with a vaccine that provides protection for 10 years or more. An attenuated vaccine (strain 17D) was developed in the 1930s following passage of wild-type virus through mouse and chick embryo cultures and routinely is grown in embryonated hens' eggs. Immunization against yellow fever should be offered to persons (aged above 6 months) traveling to endemic regions. A single dose may give life-long protection, although international health regulations require booster doses every 10 years. The vaccine should not be given to babies of less than 4 months and, ideally, less than 9 months, the immunosuppressed and those with known allergy to eggs. Side-effects are uncommon in babies older than 4 months. The vaccine must be stored frozen and requires an effective cold chain for stability.

In endemic areas, such as the Gambia, YFV vaccine is being incorporated into the Expanded Program of Immunization. The vaccine may be given with measles vaccine to infants aged 9 months. Mass immunization also may be undertaken proactively in anticipation of an epidemic or in its early stages, although such a policy is not ideal.

Vaccines are being developed against all four serotypes of dengue virus (live-attenuated vaccines) and Lassa fever virus. Immunization against dengue viruses necessitates protection against all four serotypes to avoid immune enhancement of subsequent infections. Passive immunization against Lassa fever should be possible with locally harvested IG containing antibodies to Lassa fever virus. In clinical practice, results are unpredictable as batches vary in their content of neutralizing antibody and impact is disappointing with high viremias. An inactivated vaccine against RVFV is available for high-risk groups (e.g. laboratory staff and animal handlers living in endemic regions) and an attenuated live vaccine is available for livestock.

Also, concern remains over parenteral transmission of other viruses, especially HIV, HBV and HCV among many others.

8.8 Future issues

Progress with antiviral strategies for the exotic infections is hampered by late presentation of the patients and limited resources available in developing countries. IFNs seem to be of very limited efficacy, including in animal models. Compounds that stimulate macrophages and induce endogenous cytokines are under development but their value remains unproven in this setting. The exotic flaviviruses such as YFV and the dengue viruses and filoviruses (Ebola, Marburg) seem to be insensitive to IFNs. The reluctance to use IFNs in some exotic infections, such as those due to arenaviruses, is based on the finding of high levels of endogenous IFNs that correlate with severity of illness in some instances.

Immune plasma has been successful in some hemorrhagic fevers, such as Junin, but concern remains over quality control issues and especially safety of blood products including transmission of other hepatotropic agents and HIV.

8.9 GB virus C (HGV), TTV, SENV and HFV

The advent of sensitive molecular techniques that allow detection of novel virus genomes through comparison of potentially infected and control tissues has led to a revolution in discovery of viruses in patients with acute and chronic liver diseases. Whether these are hepatotropic or 'innocent bystanders' is difficult to resolve. Most patients with viral hepatitis share at least one risk factor for exposure to multiple transmissible agents via parenteral, sexual and fecal–oral routes. This dicotomy/paradox is illustrated well with the discovery of GB virus C (GBV-C), also known as 'hepatitis' G virus (HGV).

8.9.1 Biology of GBV-C/HGV

HGV/GBV-C and HCV are distinct members of the same family of RNA viruses (*Table 8.6*). They are related to the pestiviruses and flaviviruses and share several epidemiological features including transmission by parenteral routes. GBV-C and HGV, a closely related isolate, were cloned in 1995. They also cause acute and persistent infections but, in the absence of HCV, are not associated with liver disease. Whether the liver is a site for their replication remains unclear. The prevalence of GBV-C/HGV RNA rarely exceeds 30% in high-risk populations, whereas that for HCV RNA typically exceeds 75%. GBV-C/HGV may be cleared more frequently by the immune system than HCV.

8.9.1.1 Discovery of the GB viruses and cloning of the genomes

In 1963, a surgeon (GB) in the USA developed a moderately severe acute icteric hepatitis. Clinical features suggested a viral etiology and subsequent serological testing ruled out hepatitis A–E viruses. Transmissibility was demonstrated by elevation of serum ALTs in marmosets infected with acute phase serum collected on the third day of jaundice. In 1995, the cloning and sequencing of two agents, GBV-A and GBV-B, was reported, based on stored sera from the animal transmission studies.

Clearly, GBV-A is a simian virus and related viruses may be detected in many species of New World monkeys. Only the single isolate of GBV-B has been described. That this virus caused acute hepatitis in the surgeon cannot be ruled out.

8.9.1.2 GBV-C/HGV is a human virus

Novel sequences were detected using degenerate primers based on NS3 sequences of GBV-A, B and HCV initially in patients from West Africa with acute hepatitis. Independent investigators used polymerase chain reaction (PCR) techniques to amplify and sequence RNA from a patient with post-transfusion hepatitis and named this hepatitis G virus. Co-infection with HCV was recognized later.

Subsequent comparisons with published sequences between GBV-C and HGV revealed 85% homology at the nucleotide level and 95% homology at the amino acid

Table 8.6. Percentage amino acid identities among members of the *Flaviviridae*

	HCV	GBV-A	GBV-B	GBV-C	HGV	YFV	DEN-1
GBV-A	24.0						
GBV-B	28.7	23.4					
GBV-C	25.5	44.3	25.3				
HGV	24.9	44.5	25.3	94.7			
YFV	14.2	14.6	13.0	14.0	14.9		
DEN-1	13.9	13.3	13.8	14.9	14.5	45.1	
CSFV	17.1	15.4	16.9	16.0	16.1	15.3	15.8

YFV, Yellow fever virus; DEN-1, dengue virus serotype 1; CSFV, classical swine fever (hog cholera) virus.

level. The consensus is that GBV-C and HGV are similar, albeit distinct, isolates of the same member of the *Flaviviridae* (*Table 8.6*). The GB agents and HGV show limited homology with HCV.

8.9.1.3 Genotypes

Three or more major genotypes of GBV-C/HGV have been identified, based on more than 10–15% divergence of sequences. These genotypes cluster according to geographic regions but do not show the degree of sequence divergence seen for HCV.

8.9.1.4 Laboratory diagnosis of GBV-C/HGV

Progress in developing serological assays for viremia has been impeded by failure to identify any immunodominant epitopes in HGV and GBV-C, especially linear peptides. Diagnosis of infection depends on the detection of viral RNA by reverse transcription (RT)–PCR and does not discriminate between acute and chronic infection.

A test for antibodies to the envelope glycoprotein, E2, has been evaluated. Anti-E2 seems to be a marker of clearance of virus. Spontaneous clearance of persistent infection seems much more common than for HCV.

8.9.1.5 Prevalence of GBV-C/HGV

Worldwide, 1–4% of voluntary blood donors test positive for GBV-C/HGV RNA; others have cleared past infections (anti-E2 positive) (*Table 8.7*). The major route of transmission seems to be parenteral but viral RNA has been detected in saliva and semen. Perinatal transmission is more common than for HCV.

8.9.2 TT virus

A novel DNA virus (designated TT virus; TTV) was discovered, using representational difference analysis, in a patient with post-transfusion hepatitis of unknown etiology.

Table 8.7. Prevalence of GBV-C/HGV RNA (%)[a]

Volunteer blood donors (USA)	1.3
Commercial blood donors (USA)	12.9
Multiply transfused	20.8
IVDU	10.0
Chronic HCV	20.0
Non-A–E hepatitis (general)	6.9
Acute hepatitis	8.7
Cryptogenic cirrhosis	6.3
Fulminant hepatitis	9.1

[a] Data are based on small numbers of samples.

Key Notes GBV-C/HGV

- Novel flaviviruses: 25% homology (at the amino acid level) with HCV
- GBV-C and HGV are distinct isolates of the same virus
- Share high risk groups with HCV: less frequent viremia (< 20%) may reflect clearance
- 1.6% of volunteer blood donors in USA
- Viremia may persist for years
- Serum ALT may remain normal with viremia
- Anti-E2 indicates clearance of viremia
- Roles in ALF and chronic liver disease unclear

(Note that TT are the initials of the patient and not an acronym for 'transfusion transmitted'.) This discovery prompted much interest in potential disease associations, including viral hepatitis non-A–E, but has turned out to be another example of non-causal association of a virus with a disease.

Evidence of considerable sequence diversity among TT genomes came with the identification of more and more infected individuals. As PCR primers were refined to take account of this diversity, so increased the percentage of individuals who tested positive. At least 10% of volunteer blood donors in the USA test positive and studies of individuals in Africa (attending antenatal clinics) and South America (blood donors) report prevalences of over 50%. Thus, the virus seems to be ubiquitous in the human population and reports of associations with any pathology must be viewed with caution.

The entire nucleotide sequence of TTV has been determined. The genome is a 3852 nucleotide, single-stranded circular DNA molecule of negative polarity. There is considerable sequence divergence among isolates from humans and up to 13 'genotypes' have been described. Genetic organization and sequence homology measurements reveal that TTV is related to the *Circoviridae*, a family of viruses that infect mammals and birds (prototype: chicken anemia virus). In conclusion, TTV is the first circovirus known to infect humans, is widely distributed, but has not been shown to be the causative agent of any known disease.

8.9.3 SEN virus

At the time of writing the discovery of a novel DNA virus has been announced to the press but reports in the peer-reviewed scientific literature are lacking. The virus has been named after the initials of the patient from whom it was isolated. SEN virus is claimed to be detectable in around 20% of individuals with chronic liver disease but in only 1% of controls. SEN virus reportedly may be detected in 68% of patients with post-transfusion non-A–E hepatitis and is common in individuals infected with HBV and HCV. Confirmation of these data is required, particularly the absence of infection in controls. SENV may be a divergent genotype of TTV.

8.9.4 Hepatitis F virus and hepatitis F syndrome

Several claims have been made for hepatitis F virus (HFV), including in patients diagnosed with non-A–E hepatitis and who were later found to have occult hepatitis B (see section 3.6.3) Transmission of an agent associated with sporadic, enterically-transmitted non-A, non-B hepatitis [hepatitis F (French) virus] to Rhesus monkeys reportedly induced hepatitis. However, claims of the isolation, from the same source, of a virus with a 20 kb, double-stranded DNA genome have been discredited.

Another candidate, hepatitis F (fulminant) virus, has been implicated in acute liver failure of presumed viral (non-A–E) origin (see section 6.3.1.4). However, attempts to transmit this presumed virus to chimpanzees and other primates have been unsuccessful.

8.9.5 Future issues

The role of these viruses in disease remains ill-defined. Detection of viral nucleic acids by the RT–PCR does not imply pathogenesis and common risk factors may account for GBV-C/HGV viremia in individuals with acute, chronic and fulminant hepatitis. HGV RNA has been detected in around 1–2% of volunteer blood donors and can be more prevalent than HCV RNA. Assessment of the impact of these viruses in any disease will depend on serial studies that correlate changes in clinical outcomes and other laboratory tests. Claims that GBV-C/HGV replicates in the liver and can be detected in liver by *in situ* hybridization and immunostaining, require confirmation. Other studies point to GBV-C/HGV replicating in hemopoetic cells.

Further reading

Altfield, M., Rockstroh, J.K., Addo, M., *et al.* (1998) Reactivation of hepatitis B in a long-term anti-HBs positive patient with AIDS following lamivudine withdrawal. *J. Hepatol.* **29**: 306–309.
Case report of an anti-HBs seropositive male with recovery of HBV infection 18 years previously, who developed recurrent hepatitis B after triple anti-retroviral therapy was changed to exclude lamivudine. Serum ALT exceeded 1000 U/l, anti-HBe remained detectable without HBeAg but HBV DNA became detectable (around 10^9 copies/ml). Clinical and biochemical resolution with reversion to anti-HBs occurred with reinstatement of lamivudine. Sequencing of the a determinant region showed wild-type (not anti-HBs escape) HBV DNA.

Balfour, H.H. (1999) Drug therapy: antiviral drugs. *N. Engl. J. Med.* **340**: 1255–1268.
Comprehensive authoritative review (100 references) of 11 major antiviral therapies against herpes, respiratory syncitial virus and hepatitis but excluding antiretroviral agents.

Barritt, A.S. (1995) We blew it. *N. Engl. J. Med.* **332**: 945–949.
Case report of drug hepatotoxicity in AIDS.

Bodsworth, N., Donovan, B., Nightingale, B.N. (1989) The effect of concurrent human immunodeficiency virus infection on chronic hepatitis B: a study of 150 homosexual men. *J. Infect. Dis.* **160**: 577–582.

Buti, M., Jardi, R., Allende, H., *et al.* (1996) Chronic delta hepatitis: is the prognosis worse when associated with hepatitis C and human immunodeficiency virus infections? *J. Med. Virol.* **49**: 66–69.

CDC. (1990) Update: Ebola-related filovirus infection in non-human primates and interim guidelines for handling non-human primates during transit and quarantine. *MMWR* **39**: 22–24, 29–30.
Mandatory measures for controling imported diseases via transportation and quarantine of non-human primates.

Colin, J.-F., Cazals-Hatem, D., Loriot, M.A., *et al.* (1999) Influence of human immuno-deficiency virus infection on chronic hepatitis B in homosexual men. *Hepatology* **29**: 1306–1310.
A series of 132 French men showed lower levels of serum ALT and higher HBV DNA (despite comparable numbers with HBeAg and alcohol excess) in the 65 HIV seropositive compared with 67 HIV seronegative men. Distribution of severity of liver histology was comparable for HIV seropositive and seronegative groups, but cirrhosis was more common with HIV seropositivity. Alcohol excess and increasing age also contributed significantly to the development of cirrhosis in all patients.

Collier, A.C., Corey, L., Murphy, V.L., *et al.* (1988) Antibody to human immunodefi-ciency virus (HIV) and suboptimal response to hepatitis B vaccination. *Ann. Intern. Med.* **109**: 101–105.

Coppola, R.C., Manconi, P.E., Piro, R., *et al.* (1994) HCV, HIV, HBV, and HDV infections in intravenous drug addicts. *Eur. J. Epidemiol.* **18**: 964–968.

Eagling, V.A., Back, D.J., Barry, M.G. (1997) Differential inhibition of cytochrome P450 isoforms by the protease inhibitors ritonavir, saquinavir and indinavir. *Br. J. Clin. Pharmacol.* **44**: 190–194.

Fisher-Koch, S. (1995) Exotic viruses. In: *Antiviral Chemotherapy* (eds D.J. Jeffries, E. De Clercq). J. Wiley and Sons Ltd., Chichester, pp. 393–411.

Georges, A.J., Lesbordes, J.L., Georges-Courbort, M.C., Meunier, D.M.Y., Gonzalez, J.P. (1987) Fatal hepatitis from West Nile virus. *Ann. Institut Pasteur Virol.* **138**: 237–244.
First report of four patients from the Central African Republic with severe hepatitis with isola-tion of virus from blood and liver biopsies.

Howard, C.R., Ellis, D.S., Simpson, D.I.H. (1984) Exotic viruses and the liver. *Sem. Liver Dis.* **4**: 361–374.

International travel and health. World Health Organization, Geneva. 1996. 1–104.
Vaccination requirements and health advice, geographical distribution of health hazards.

Lacy, M.D., Smego, R.A. (1996) Viral hemorrhagic fevers. *Adv. Pediatr. Infect. Dis.* **12**: 21–53.
Review with 85 references.

Lefkowitch, J.H. (1994) Pathology of AIDS-related liver disease. *Digest. Dis.* **12**: 321–330.

Linnen, J., Wages, J., ZhangKeck, Z. Y., *et al.* (1996). Molecular cloning and disease association of hepatitis G virus: A transfusion-transmissible agent. *Science* **271**: 505–508.
First report of the discovery of HGV.

Monath, R.P. (1999) Ecology of Marburg and Ebola viruses: speculations and directions for future research. *J. Infect. Dis.* **179** (Suppl. 1): S127–138.
Hypothesis that filoviruses may be arthropod or plant viruses. High seroreactivity for Ebola virus in Africa, if confirmed specific would indicate non-lethal as well as lethal strains.

After Telfer. Morbidity Mortality Weekly Report. (1995) Imported Dengue – US 1993–4. *MMWR* **44**(18): 353–356.

After Telfer. Morbidity Mortality Weekly Report. (1996) Ebola-Reston Virus infection among quarantined non-human primates — Texas 1996. *MMWR* **45** (15): 314–315.
Update of episodes of viral hemorrhagic fever among non-human primates imported from the Philippines to the USA. Disease control requirements facilitated early diagnosis and limited exposure of associated personnel.

After Telfer. Morbidity Mortality Weekly Report. (1998) Rift Valley Fever—East. *MMWR* **47**(13): 261–264.
Description of Rift Valley Fever in the North Eastern Province of Kenya and southern Somalia with 478 unexplained deaths from widespread hemorrhage in man and also domestic animals.

Muerhoff, A.S., Leary, T.P., Simons, J.N., *et al.* (1995) Genomic organization of GB viruses A and B: two new members of the flaviviridae associated with GB agent hepatitis. *J. Virol.* **69**: 5621–5630.
Discovery of GBV-A and GBV-B.

Mushahwar, I.K., Erker, J.C., Muerhoff, A.S., *et al.*, (1999) Molecular and biophysical characterization of TT virus: Evidence for a new virus family infection humans. *Proc. Natl. Acad. Sci. USA* **96**: 3177–3182.
Completion of TT virus genomic sequence and recognition of relatedness to circoviruses.

Nishizawa, T., Okamoto, H., Konishi, K., Yoshizawa, H., Miyakawa, Y., Mayumi. M. (1997) A novel DNA virus (TTV) associated with elevated transaminase levels in post-transfusion hepatitis of unknown etiology. *Biochem. Biophys. Res. Commun.* **241**(1): 92–97.
Discovery of TTV.

Ockenga, J., Tillmann, H.L., Trautwein, C. *et al.* (1997) Hepatitis B and C in HIV-infected patients. *J. Hepatol.* **27**: 18–24.
Comprehensive study of epidemiology and impact of HBV and HCV infection on survival with HIV infection and AIDS. Reduced actuarial survival in HBV infected or HCV infected patients with AIDS but not HIV seropositivity per se.

Pol, S., Lamorthe, B., Thi, N.T., *et al.* (1998) Retrospective analysis of the impact of HIV infection and alcohol use in chronic hepatitis C in a large cohort of drug users. *J. Hepatol.* **28**: 45–50.
Retrospective histopathological analysis of 210 consecutive HCV infected patients with HIV (60 patients) and alcohol excess (> 80 g/day) showing progressive risk of cirrhosis over around 12 years following viral infection. The risk of cirrhosis increases with HIV/HCV/alcohol (30%), compared with HIV/HCV/no excess alcohol (24.5%) and HCV alone (11.3% cirrhosis).

Porterfield, J.S. (1995) *Exotic Viral Infections. Handbook of Infectious Diseases.* Chapman & Hall Medical, London, pp. 1–384.
Detailed authoritative review, including immunology and treatment strategies.

Roger, P.M., Carles, M., Saint-Paul, M.C., *et al.* (1996) Comparative profitability of hepatic biopsy and microbiological tests in patients with HIV infection (in French). *Presse Medicale* **25**: 1147–1151.
Limited usefulness in detecting microbial infections via liver biopsy in addition to conventional microbiological approaches.

Rutschmann, O.T., Negro, B., Hirschel, B., *et al.* (1998) Impact of treatment with human immunodeficiency virus (HIV) protease inhibitors on hepatitis C viremia in patients coinfected with HIV. *J. Infect. Dis.* **177**: 783–785.
Highly active antiretroviral therapies can worsen liver disease in HCV infected/HIV infected patients with low (<100 μl) CD4 cell counts. HCV RNA levels increased temporarily and ALT levels increased in 19 patients treated with combinations of ritinovir, saquinavir, indinavir and reverse transcriptase inhibitors.

Simons, J.N., Leary, T.P., Dawson, G.J., *et al.* (1995) Isolation of novel virus-like sequences associated with human hepatitis. *Nature Med.* **1**: 564–569.
Discovery of GB virus C.

Soto, B., Sanchez-Quijano, A., Rodrigo, L., *et al.* (1997) Human immunodeficiency virus infection modifies the natural history of chronic parenterally-acquired hepatitis C with an unusually rapid progression to cirrhosis. *J. Hepatol.* **26**: 1–5.
Multicenter study of 547 adults with parenterally acquired HCV infection enrolled 1989–1994 with persistent ALT elevations and excluded excess alcohol intake. Although the prevalence of cirrhosis and severe liver disease were similar for anti-HIV seropositive and negative groups, the duration of infection was shorter in the HIV seropositive group indicating more rapid progression of liver disease with HIV infection.

Telfer, P., Sabin, C., Devereaux, H., *et al.* (1994) The progression of HCV-associated liver disease in a cohort of haemophiliac patients. *Br. J. Haematol.* **87**: 555–561.
Retrospective comparison (1979–1993) of 438 HCV seropositive hemophiliacs; 183 with known date of exposure (median 15.1 years) to large donor pool concentrates. The relative hazard of

developing liver failure after HIV infection was 21.4; hepatic decompensation increased after becoming positive for p24 antigen, with falling CD4 cell counts, greater use of concentrates and with increasing age.

Van Velden, D.J.J., Meyer, J.D., Olivier, J., Gear, J.H.S., McIntosh, B. (1997) Rift Valley Fever affecting humans in South Africa. A clinicopathological study. *South African Med. J.* **51**: 867–871.
First report of RVF virus causing severe hepatitis, more than 40 years after identification of the virus.

Vento, S., di Perri, G., Luzzati, R., *et al.* (1989) Clinical reactivation of hepatitis B in anti-HBs-positive patients with AIDS. *Lancet* **1**: 332–333.

Wagner, N., Rotthauwe, H.W., Becker, M., Dienes, H.P., Mertens, T., Fodisch, H.J., Brackmann, H.H. (1992) Correlation of hepatitis B virus, hepatitis D virus and human immunodeficiency virus type I infection markers in hepatitis B surface antigen positive haemophiliacs and patients without haemophilia with clinical and histopathological outcome of hepatitis. *Eur. J. Pediatr.* **151**, 90–94.

Wright, T., Hollander, H., Pu, X., (1994) Hepatitis C in HIV-infected patients with and without AIDS. *Hepatology* **20**: 1152–1155.
No significant excess mortality from hepatitis C per se with concurrent HIV infection and AIDS.

Glossary

Common terms used in the United States (USA) and United Kingdom (UK)

Acetaminophen: paracetamol.

Acinus (liver): liver acinus (syn: acinus of Rappaport) is the smallest functional unit of the liver parenchyma supplied by the portal vein and hepatic artery.

Acute-on-chronic hepatitis: acute infection of the liver on a background of chronic hepatitis; typical presentation of an acute exacerbation of chronic hepatitis C and hepatitis D superinfection in chronic hepatitis B.

Acute hepatic/liver failure: ALF is defined in Chapter 5.1.

Alpha-fetoprotein: a serum marker of primary liver cancer; positive in 60–80% of tumors associated with HCV and HBV. Used as part of serial screening program for patients with cirrhosis.

Alpha (α) interferons: type 1 interferons (cytokines), the most common interferon used in treating hepatitis B and C.

Amino and carboxyl termini: designations, on the basis of free groups on the terminal amino acids, of the ends of a polypeptide chain. Proteins are translated in the amino to carboxyl direction. Membrane-spanning proteins are oriented with an external amino terminus.

Amplicon: the product of a PCR reaction.

Amplimers: the oligonucleotide primers used in a PCR reaction.

Ampulla of Vater: entrance (os) to common bile duct; can be cannulated via endoscopic retrograde cholecyst pancreatography (ERCP).

Anicteric: without the development of jaundice.

Antigen: a molecule that stimulates an antibody response.

Antigenemia: the presence of circulating antigen (e.g. HBsAg, HBeAg).

Antimitochondrial antibodies: autoantibodies against components of the mitrochondrion and diagnostic of primary biliary cirrhosis.

Antinuclear antibodies: syn: antinuclear factor; autoantibodies seen in type I AIH (typically high titer: >1 : 160) as well as in HCV (typically in low titer, <1 : 160). See also autoantibodies, below.

Apoptosis: degeneration (in the liver) of hepatocytes by programmed cell death; seen in acute and acute-on-chronic viral hepatitis.

Ascites: accumulation of fluid in the peritoneum; a complication of cirrhosis and prone to infection (see spontaneous bacterial peritonitis).

Asterixis: recurrent flapping tremor of the limbs, a sign of hepatic encephalopathy.

Autoantibodies (in liver disease): antibodies against antigenic components of the cell characteristically found in autoimmune hepatitis (e.g. antinuclear antibodies: ANA) and primary biliary cirrhosis (anti-mitochondrial antibodies: AMA). Auto antibodies [e.g. ANA, rheumatoid factor (RhF) and anti-thyroid] may be detectable (in low titer, <1 : 160) in chronic viral hepatitis, typically hepatitis C. (See Type I and Type II autoimmune hepatitis, below.) Their pathogenic significance remains unknown.

Autoimmune hepatitis (AIH): see Type I and Type II.

Base (pair): nucleotide, or building block, of nucleic acid. In double-stranded molecules A (adenine) pairs with T (thymidine in DNA) or U (uracil in RNA) and C (cytosine) with G (guanine), also known as Watson–Crick pairing.

Beta (β) interferon: a type I interferon, cytokine, preliminary studies show potential in treatment of acute hepatitis C.

Biochemical response: reduction in transaminase levels (e.g. ALT in serum) in relation to antiviral therapy (see also virological response, below).

Budd–Chiari syndrome: a disorder resulting from obstruction of hepatic venous outflow from the liver.

Buffy coat: visible white layer of cells containing PBMCs and PMLs when whole blood is separated by gravity.

Burnt-out cirrhosis: a histological term describing a paucity of inflammatory cell activity typically seen in end-stage cirrhosis, making diagnosis of the etiology of cirrhosis difficult especially in autoimmune diseases.

Cap: post-transcriptional modification of the 5′ ends of cellular mRNAs, including methylation and 5′ to 5′ linkage of the terminal guanine. The structure, which is important for recognition for translation by the ribosome, also may be present at the 5′ end of HEV RNA.

Carboxyl termini: see amino termini.

cDNA: complementary (copy) DNA synthesized from an RNA template (such as viral RNA) using reverse transcriptase. cDNA libraries contain clones representing potentially all RNA species in a source such as liver or plasma.

Cerebral edema: swelling of the brain; a complication of intracranial hypertension in ALF. The exact cause is unknown; responds to hyperosmolar diuresis with mannitol.

Chain terminator: dideoxynucleotides lack a 3'-hydroxyl group and their incorporation into DNA precludes further chain extension. Used in the most common method of DNA sequencing (Sanger sequencing) and for antiviral therapy (ddI, ddC, 3TC).

Cholangitis: inflammation and/or scarring process involving bile ducts.

Cholestasis (Cholestatic phase of hepatitis): impaired flow of bile and bile acids from the liver, often associated with jaundice and pruritus (itching of the skin); typically occurs in the later phases of acute and chronic hepatitis.

Chronic active, lobular, and persistent hepatitis: CAH, CLH and CPH are defined in section 1.9.1.

Cirrhosis: irreversible damage to the entire liver includes fibrosis and the formation of regeneration nodules (see section 1.8.5) (*Kirrhos*: Greek for tawny). Note: fibrosis is not synonymous with cirrhosis (See fibrosis, below).

Cis-acting: molecules with enzyme activities which self-modify, for example, ribozyme cleavage of HDV RNA and viral proteinases which cleave themselves from the polyprotein.

Clustering: epidemiologically linked.

Coagulative necrosis: a histological term relating to degeneration (in the liver) of hepatocytes typically during severe acute/fulminant hepatitis, characteristically seen in acetaminophen (paracetamol) hepatotoxicity.

Codon: a triplet of nucleotides specifying an amino acid or termination signal.

Coinfection: simultaneous infection with two viruses, see section 1.4.

Complete blood count: hemoglobin, white cells and differential (proportion or % of neutrophils, lymphocytes, eosinophils, basophils) and platelet counts (CBC = WBC: whole blood count).

Conformational: see epitope, below.

Consensus interferon: a synthetic recombinant type I interferon made by assigning the most common amino acid at each position of several alpha interferons (non-allelic subtypes) to generate a consensus sequence. Doses are prescribed in micrograms (typically $3\,\mu g$ and $9\,\mu g$).

Councilman's bodies: apoptotic degeneration of hepatocytes seen in viral hepatitis, typically in acute phases. Prominent in yellow fever.

Cryoglobulinemia: detectable cryoproteins, typically in the patient with hepatitis C, and may be associated with complications outside of the liver (extrahepatic) such as glomerulonephritis and joint symptoms (joint pains).

Cryoprecipitates: serum proteins that precipitate reversibly under different temperatures.

Cryptogenic: of unknown etiology, typically applied to chronic hepatitis and cirrhosis (idiopathic, indeterminate).

Decompensated liver disease: a clinical term describing complications of cirrhosis such as development of one or more of the following: ascites, spontaneous bacterial peritonitis, hepatic encephalopathy, bleeding esophageal varices. Decompensation may be precipitated by infection, bleeding, tumor and antiviral therapy.

Disseminated intravascular coagulation: a state of impaired coagulation associated with elevated fibrin degradation products and d-dimer, characteristically seen with septic complications, multiorgan failure and renal dysfunction.

Domain: region of a protein with a particular attribute or function.

Doppler imaging: used in conjunction with ultrasound (USSD) to measure direction of flow, velocity and amplitude in blood vessels e.g. hepatic artery, portal vein.

Dupuytren's contracture: contraction of the palmar fascia leading to finger deformities; one of the cutaneous stigmata of chronic liver disease/cirrhosis. May arise spontaneously (idiopathic) or in other diseases.

EGD: esophago-gastroduodenoscopy: visualization of the esophagus, stomach, duodenum via a flexible fiberoptic endoscope passed via the mouth (syn: OGD).

Encephalopathy (hepatic): altered mental state/confusion in a patient with acute liver failure or cirrhosis. The hepatic encephalopathy in ALF is of unknown cause but is associated with cerebral edema that can be treated by hyperosmolar diuresis (e.g. mannitol). The hepatic encephalopathy complicating cirrhosis responds best to reduction of nitrogenous load from the intestine (e.g. with lactulose).

Endoscopic retrograde cholecystpancreatography (ERCP): visualization of the esophagus, stomach, ampulla of Vater and biliary tree and pancreas and pancreatic duct with cannulation via a fiberoptic endoscope introduced from the mouth.

Enhancer: *cis*-acting element with DNA which upregulates the activity of a promoter.

Envelope/envelope glycoprotein: the virions of certain viruses (including HBV, HCV, HDV and GBV-C/HGV) are composed of nucleocapsids surrounded by membranes derived from the host cell. Glycosylated viral proteins (envelope glycoproteins) which are embedded in the membrane may attach specifically to cellular receptors.

Epitope: the region of an antigen that reacts with an antibody. Linear epitopes comprise co-linear sequences of amino acids and may readily be mimicked by synthetic peptides or recombinant proteins. Secondary structure is an essential attribute of conformational epitopes, which are not necessarily composed of co-linear amino acid sequences. Conformational epitopes may not necessarily be mimicked by synthetic peptides or represented in recombinant proteins.

Esophageal varices: varicose veins at the lower end of the esophagus (consequence of portal hypertension).

Fibrosis (hepatic): excess deposition of collagen in the liver (a scarring process), a prerequisite for development of cirrhosis. Part of the histological staging of the liver (typically stage 1–4). Note, fibrosis can spare certain areas of the liver (as in congenital hepatic fibrosis) and early stages (1–2) can be reversed. Fibrosis is not synonymous with cirrhosis (stage 4: see above).

5′ and 3′ ends: designations, on the basis of the free groups on the sugar/phosphate backbone, which indicate the orientation of a strand of nucleic acid. The genetic code is read in the 5' to 3' direction. In double-stranded nucleic acid, the strands are antiparallel—they run in opposite directions.

Fulminant hepatic failure: see acute liver failure, defined in Chapter 5.1.

Genome equivalent: the mass of nucleic acid equivalent to one viral genome—an attempt to quantify viremia by estimating the viral nucleic acid content of a serum sample (expressed as copies or genome equivalents per ml, geq/ml).

Granuloma (of the liver, plural: granulomata): a cluster of epithelioid cells and histiocytes typically seen in tuberculosis, sarcoidosis and Hodgkin's disease. When involving the liver, these can resemble a lymphoid follicle, as seen in the spleen and lymph nodes. Lymphoid-like follicles resembling granulomata can be seen in chronic hepatitis C; their histopathological significance is unknown.

Ground glass cells: describing the characteristic histological appearance of hepatocytes containing HBsAg in HBV infection.

Helicase: helicases unwind the strands of duplex nucleic acid. HAV, HCV, HEV and GBV-C/HGV encode helicase activities which are believed to be involved in genome (RNA) replication.

HELLP syndrome. Hemolysis, Elevated Liver enzymes and Low Platelet count: a syndrome first described in 1982 and associated with acute fatty liver of pregnancy and pre-eclampsia; can occur in many disorders during pregnancy, especially in association with renal impairment and disseminated intravascular coagulation.

Hemochromatosis: iron overload with deposition in the liver in genetically predisposed individuals. The genetic defects in most familial forms have been identified.

Hepatitis B Immune globulin (HBIG): contains high-titer anti-HBs (and anti-HBc) antibodies used as passive immunoprophylaxis against hepatitis B. Originally obtained from harvesting plasma from patients who had seroconverted from HBsAg to anti-HBs following HBV infection. Since the 1990s, mostly harvested by plasmapheresis of volunteers previously immunized with hepatitis B vaccines.

Hepatocellular carcinoma: tumor arising from the hepatocyte (primary liver cancer, PLC). Development of HCC frequently is associated with cirrhosis associated with chronic hepatitis B and C. Tumors associated with hepatitis B tend to be monoclonal and arise from cells containing integrated viral DNA. HCC arises uncommonly without cirrhosis in chronic hepatitis B and very rarely without cirrhosis in chronic hepatitis C.

Hepatoma: a casual term for hepatocellular carcinoma.

Hepatotropic: having the propensity to infect the liver.

Horizontal transmission: mode of transmission implicated in all hepatotropic viruses: refers to person-to-person spread occurring after birth via percutaneous, parenteral, sexual or inapparent (sporadic) transfer.

Hybridization: the pairing of complementary strands of nucleic acid by the formation of hydrogen bonds between G and C and between A and T (or U).

Hypersplenism: enlarged spleen associated with leukopenia and thrombocytopenia.

ID$_{50}$: the dose required to infect 50% of a group of experimental animals (c.f. LD$_{50}$). For example, CID$_{50}$ in studies involving chimpanzees.

Idiopathic: of unknown etiology, typically applied to chronic hepatitis/cirrhosis (syn: indeterminate, cryptogenic).

Immune electron microscopy: visualization (of virus particles) in the electron microscope following agglutination by antibody.

Immune globulin (IG or NIG): contains anti-HAV and anti-HBs antibodies: used as passive immunoprophylaxis in hepatitis A.

Initiation codon: the sole codon for methionine, AUG, signals the ribosome to start translation, when in the appropriate context (optimal context, CCACCAUGG, the –3 (A) and +4 (G) positions are especially critical).

Integration (viral): covalent insertion of viral nucleic acid into chromosomal DNA.

International Normalized Ratio (INR): this is defined in section 1.6.2.4 (and is not the same as prothrombin ratio).

Jaundice (icterus): yellow discoloration of skin and conjunctivae associated with accumulation of bilirubin (>50 μmol/l) in the serum.

Kat/l: moles of substrate converted per second. To convert IU/l to μKat/l, multiply by 0.01667.

Late onset hepatic failure (LOHF): this is defined in section 5.1.

Lichen planus: a skin condition characterized by itching, scaly papules on the limbs, mucous membranes (mouth) and genital region. Patients should be tested for hepatitis C virus by enzyme immunoassay.

Lobular hepatitis: inflammatory infiltrate in the parenchyma of the liver, typically seen in acute viral hepatitis and acute-on-chronic viral hepatitis.

Mallory's hyaline: describing the histological appearance of deposits in hepatocytes characteristically seen in alcoholic liver disease and Wilson's disease.

MegaUnits (MU): dose of type 1 interferons (α-2b and β). Note the dosing of consensus interferon and long-acting interferons (e.g. PEG-IFN) is given in micrograms. These are not equivalent to MU and biological efficacies cannot be extrapolated between IFNs.

Meta-analysis: a process which provides structured and quantitative methods for the review, evaluation and synthesis of information from different studies, such as clinical trials.

Metastasis: spread of (neoplastic) disease.

Naïve patient: never previously treated with antiviral therapy for viral hepatitis.

Nested PCR: two sequential PCR reactions, the second targets the amplicons from the first, using internal (nested) primers. Allows detection down to the theoretical limit of a single target molecule.

Neutrophil: polymorphonuclear leukocyte.

Non-A, non-B hepatitis: a diagnosis of exclusion that preceded the discovery of HCV and HEV—presumed viral hepatitis not attributable to HAV, HBV or rarer causes such as herpesviruses. The term non-A–E hepatitis has been used since the description of HCV and HEV.

Non-responder: the patient whose level of viremia does not change significantly during therapy (see responder, below).

Northern blotting: transfer of RNA from agarose gel to membrane for hybridization (named as the 'opposite' of Southern blotting).

Nosocomial: spread within an institution, typically a hospital.

Nucleocapsid: the internal core of an enveloped virus, typically an icosahedral structure comprising the genome surrounded by a viral protein (the core or nucleocapsid protein).

Occult hepatitis B: HBV infection with HBsAg undetectable in serum by conventional enzyme immunoassay (see section 3.6.3).

OGD: esophago gastroduodenoscopy: visualization of the esophagus, stomach, duodenum via a flexible fiberoptic endoscope passed via the mouth (syn: EGD).

Oncogene: a gene whose product has the ability to transform eukaryotic cells. The normal homologs of oncogenes encoded in the eukaryotic genome are known as proto-oncogenes.

Open reading frame (ORF): stretch of nucleotide sequence without termination codons (UAG, UGA, UAA) with potential to be translated to protein. Typically, viral ORFs commence with the initiation (methionine) codon AUG.

Palmar erythema: red discoloration of the palms: one of the cutaneous stigmata of chronic liver disease (see spider nevus).

Pan-lobular hepatitis: hepatitis affecting all three histological zones (of Rappaport) of the liver.

Paracetamol (UK): acetaminophen (USA).

Per meter squared (m⁻²): relating to body surface area, rather than body weight (e.g. dose of drug such as interferon).

p53 gene: a tumor suppressor gene, the 'guardian of the genome'. The gene most frequently found mutated in human cancers.

Piecemeal necrosis: extension of inflammatory infiltrate beyond portal tracts into parenchyma with necrosis of hepatocytes. Histological term defining chronic active hepatitis. The presence of interface hepatitis indicates potential for progression to cirrhosis (syn. interface necrosis, interface hepatitis).

Polyadenylation: the addition of polyadenylic acid (polyA) to the 3' ends of RNA molecules.

Polymerase: enzyme activity which synthesizes nucleic acid by copying a template strand, maintaining the complementary sequence. DNA polymerases synthesize DNA and RNA polymerases, RNA. The HBV DNA polymerase can use either DNA or RNA as template (reverse transcriptase activity).

Polyprotein: many viral genomes contain long ORFs which encode polyproteins which are proteolytically cleaved (post-translation) to smaller polypeptides with various functions.

Portal hypertension: increased pressure in the portal vein, a complication of cirrhosis in chronic viral hepatitis (B and C). As a consequence, portal venous blood is diverted away from the liver to the splanchnic (gastrointestinal tract) circulation. Esophageal veins become engorged (esophageal varices) and can rupture causing bleeding and death.

Primary liver cancer: see hepatocellular carcinoma.

Primer: DNA polymerases require a primer to commence synthesis. The primer may be a short stretch of nucleic acid (as in the PCR) or a nucleotide covalently bound to a protein (HAV, HBV minus strand).

Promoter: motif within DNA which acts as a signal for the start of transcription.

Proteinase (protease): enzyme activity which breaks the bond between adjacent amino acids in a protein. HAV, HCV, HEV and GBV-C/HGV encode polyproteins which include proteinase activities which process the precursor molecule to several products.

Pruritus (in liver disease): itching of the skin, typically with cholestatic phases of liver disease and attributed to deposition of bile salts.

Purpura: hemorrhage into the skin.

Relapser: the patient who responds initially to antiviral therapy (see responder, below) but whose viremia becomes detectable after cessation of treatment.

Replicase: alternative name for the RNA-dependent RNA polymerases encoded by viruses such as HAV, HCV, HEV and GBV-C/HGV.

Representational difference analysis (RDA): a PCR-based technique that enables identification of nucleic acid sequences in the test sample but not closely related (control) sample. Used in the discovery of the GB viruses and TTV.

Responder: response to antiviral therapy is best evaluated through viremia: responders show a marked reduction in viremia on antiviral therapy, in sustained responders the reduced viremea is maintained following cessation of therapy. Virological response may be mirrored by normalization of transaminases, improvement in liver histology, etc.

Reticulocytes: immature red blood cells (erythrocytes); percentages are elevated in hemolysis typically consequent to ribavirin.

Reverse transcription: synthesis of DNA on an RNA template.

Reye's syndrome: hepatitis and encephalopathy frequently associated with viral infections in childhood and abnormal fat metabolism in the liver.

Ribozyme: RNA with complex secondary structure which can catalyze its own cleavage. The HDV genome possesses a ribozyme activity.

Sequence independent single primer amplification (SISPA): PCR-based method used to amplify cDNA derived from plasma in the discovery of HEV.

Sicca syndrome: dry eyes, dry mouth due to an inflammatory process of the lacrimal glands and ducts. Characteristically seen in autoimmune diseases.

Southern blotting: transfer of denatured DNA fragments, separated on an agarose gel, to a filter for hybridization; described by Professor Southern.

Spider nevus: capillary blemish in the skin fed from an arteriole, associated with cirrhosis, typically abundant in alcoholic cirrhosis. One of the recognized cutaneous stigmata of chronic liver disease (plural, nevi/ae).

Spontaneous bacterial peritonitis (SBP): presence of bacterial infection in ascites, a common complication of cirrhosis. Diagnosed by counting neutrophils in excess of 250/ml in a diagnostic sample of ascitic fluid; detection of bacteria is unreliable, although these may grow subsequently on culture. SBP requires antibiotic therapy and carries a high mortality rate if left untreated. SBP is prone to recurrence.

Sporadic: epidemiologically unlinked (no clustering).

Spotty necrosis (in the liver): necrotic degeneration of scattered hepatocytes, typically by apoptosis and seen during acute hepatitis and acute-on-chronic viral hepatitis.

Steatosis (of the liver): abnormal accumulation of fat in the liver. Microvesicular and macrovesicular forms relate to vesicle size. Steatosis forms part of the histopathological triad in chronic hepatitis C but usually is mild. Steatosis can be prominent in alcoholic liver disease, non-alcoholic steatohepatitis (NASH syndrome), obesity and diabetes mellitus.

Subviral particles: hepatocytes infected with HBV secrete, in addition to virions, excess HBsAg embedded in cellular membranes as 22 nm spheres and rods.

Superinfection: defined in section 1.4.

Systemic lupus erythematosus (SLE): a multisystem autoimmune disease associated with autoantibodies and cryoglobulinemia that can be confused serologically with hepatitis C.

Termination codon: three codons, UAG (amber), UAA (ochre) and UGA cause termination of protein synthesis (also called nonsense codons).

Trans-acting: acting on another molecule (c.f. *cis*-acting).

Transactivator (of transcription): protein that upregulates the activity of a promoter by binding directly to the DNA or via interaction with other proteins.

Transjugular intrahepatic portosystemic stent/shunt (TIPSS): a radiological technique used to reduce the portal pressure by placing a stent (hollow wire tube) connecting intrahepatic vein/s and hepatic artery branches. TIPPS is used as an alternative to obliteration of esophageal varices by sclerotherapy/banding to prevent future bleeding. Hepatic encephalopathy can worsen as a consequence of by-passing the normal clearance mechanisms of the liver.

TTV: TT virus, a DNA virus with no hepatotropism and no known pathogenic role. T. T. are the initials of the patient from whom the virus was isolated, not an acronym for 'transfusion transmitted'.

Tumor suppressor genes: normal expression of these genes inhibits inappropriate cell division.

Type I autoimmune hepatitis (AIH): this is characterized by high (>1 : 160) titers of anti-nuclear antibodies (ANA) and smooth muscle antibodies (SMA). Diagnosis can be problematic with additional serological features suggestive of autoimmune diseases such as cryoglobulins and elevated levels of globulins.

Type II autoimmune hepatitis (AIH): this is defined by the detection of liver, kidney, microsomal type 1 (LKM-1) autoantibodies commonly associated with anti-HCV positivity. Most patients are male, whereas most with autoimmune liver diseases are female. In type II autoimmune hepatitis (AIH), antibodies to LKM-1 react against P450IID6, the cytochrome monooxygenase enzyme. Most patients with HCV RNA have undetectable anti-P450IID6 reactivity. Specificities of autoantibodies (for example anti-P450IID6, anti-cytokeratins, anti-F actins), antigens such as the human leukocyte antigens (viz. HLA-DR4) and changes in serial titers may help to distinguish autoimmune hepatitis from viral hepatitis.

Untranslated region: regions at either end of (and sometimes between) ORFs which are not translated into protein. RNA viruses have sequences at either end of the genome which are critical for replication. The 5′ UTRs of the HAV, HCV and GBV-C/HGV genomes, which are not capped, have extensive secondary structure which bind ribosomes—internal ribosome entry sites (IRES).

USSD: ultrasound imaging technique.

Vertical transmission: transfer of virus from mother to child around the time of delivery.

Viremia: the presence of circulating virus.

Virion: virus particle, the unit of infectivity.

Virological response: reduction to undetectable levels (by PCR) of circulating viral genomes (e.g. HBV DNA, HCV RNA) in relation to antiviral therapy.

Western blotting: electrophoretic transfer of polypeptides from a polyacrylamide gel to a membrane prior to detection by antibody.

Zonal necrosis: histological term referring to targeting/sparing of one or more of the three histological regions (zones of Rappaport).

Index